Alloys: Metallurgy and Engineering

Alloys: Metallurgy and Engineering

Edited by
Nathaniel Gordon

WILLFORD PRESS

www.willfordpress.com

Published by Willford Press,
118-35 Queens Blvd., Suite 400,
Forest Hills, NY 11375, USA

ISBN: 978-1-68285-778-6

Cataloging-in-Publication Data

Alloys : metallurgy and engineering / edited by Nathaniel Gordon.
    p. cm.
Includes bibliographical references and index.
ISBN 978-1-68285-778-6
1. Alloys. 2. Metallurgy. 3. Metallic composites. 4. Metals. 5. Engineering. I. Gordon, Nathaniel.
TA483 .A45 2020
669--dc23

For information on all Willford Press publications
visit our website at www.willfordpress.com

WILLFORD PRESS

Contents

Preface

This book has been an outcome of determined endeavour from a group of educationists in the field. The primary objective was to involve a broad spectrum of professionals from diverse cultural background involved in the field for developing new researches. The book not only targets students but also scholars pursuing higher research for further enhancement of the theoretical and practical applications of the subject.

A combination of metals or a combination of one or more metals with non-metallic elements which retains the characteristics of a metal is known as an alloy. It is defined by a metallic bonding character. An alloy can be a mixture of metallic phases or a solid solution of metal elements. Alloys with a defined crystal structure and stoichiometry are intermetallic compounds. Alloys are used in a variety of applications. Steel alloys are used in automobiles and construction of buildings, aerospace industry uses titanium alloys and non-sparking tools are produced using beryllium-copper alloys. Sometimes, while preserving important properties, a combination of metals may reduce the overall cost of the material. In other cases, the combination of metals imparts synergistic properties to the constituent metal elements like mechanical strength or corrosion resistance. Steel, brass, solder, bronze, and amalgams are a few examples of alloy. This book contains some path-breaking studies on metallurgy and engineering related to alloys. It outlines the processes and applications of alloys in detail. This book includes contributions of experts and scientists which will provide innovative insights into the topic.

It was an honour to edit such a profound book and also a challenging task to compile and examine all the relevant data for accuracy and originality. I wish to acknowledge the efforts of the contributors for submitting such brilliant and diverse chapters in the field and for endlessly working for the completion of the book. Last, but not the least; I thank my family for being a constant source of support in all my research endeavours.

Editor

Finite Element Analysis of Residual Stress in the Diffusion Zone of Mg/Al Alloys

Yunlong Ding[1] and Dongying Ju ⓘ[1,2]

[1]Department of Materials Science and Engineering, Saitama Institute of Technology, Fusaiji 1690, Fukaya, Saitama 369-0293, Japan
[2]Department of Materials Science and Engineering, University of Science and Technology Liaoning, Anshan 114051, China

Correspondence should be addressed to Dongying Ju; dyju@sit.ac.jp

Academic Editor: Yuanshi Li

In this study, the finite element method was applied for analyzing the effect of annealing temperatures on residual stress in the diffusion zone of AZ31 Mg and 6061 Al alloys. The microstructure and mechanical behavior of the diffusion zone were also investigated. Simulations on the annealing of the welded specimens at 200°C, 250°C, and 300°C were conducted. Moreover, experiments such as diffusion bonding and annealing, analysis of residual stress by X-ray diffraction, elemental analysis using an electron probe microanalyzer, and microstructure investigation via scanning electron microscopy were performed for further investigation of the diffusion layers. According to the results of the simulations and experiments, the diffusion layers widen with increasing annealing temperatures, and the results of the simulations are in good agreement with those of the experiments. The microstructure and elemental distribution were the most uniform and the residual stress was the least for samples annealed at 250°C. Thus, 250°C was found to be the most appropriate annealing temperature.

1. Introduction

The finite element method (FEM) has many applications in modern industry and technology because of the extensive use of computers [1–6]. This method is presently the most popular and fastest developing numerical method in aircraft, ballistic missile, automotive, shipbuilding, machine, and electro-technics industries and is used in fields such as biomechanics, medicine, mechatronics, and materials technology. Computational methods mainly help optimize design processes [7–10].

In addition, FEM is used in plastic forming and can simulate the press forming of aluminum by selecting appropriate forming parameters for the material, such as pressure force and falling speed of the punch [11].

Recently, many investigations on the welding of Mg/Al alloys have been conducted using FEM, especially on the analysis of residual stress during welding, because magnesium and aluminum alloys are widely used in aerospace, automotive, machine, electrical, and chemical industries owing to their superior properties [12–14]. Further, the combination of the superior properties of magnesium and aluminum alloys provides insight into the research of lightweight vehicles.

However, most studies have focused on the analysis of residual stress during butt welding, laser beam welding, or friction stir welding [15, 16]. In contrast, this study considers a different welding process, diffusion bonding. To decrease the residual stress during diffusion bonding, annealing experiments, simulations, and investigations of residual stress by X-ray diffraction (XRD) were performed. To the best of our knowledge, this is the first study focusing on diffusion bonding between magnesium and aluminum alloys. Based on the results of this study, more extensive applications of FEM and diffusion bonding can be determined, and the properties of composite materials containing magnesium and aluminum alloys can be investigated. The composite materials can contribute to the realization of lightweight components. In addition, the depletion of resources and energy will decrease, thus mitigating environmental pollution.

2. Materials and Methods

In this study, AZ31 magnesium alloy and 6061 aluminum alloy were used for diffusion bonding and annealing. Simulations and experiments were performed to analyze residual stress and evaluate the microstructure.

2.1. Theoretical Analysis. During diffusion bonding and annealing, microstructures of the alloys vary with temperature, and at the same time, thermal stress is induced. If the stress exceeds the elastic limit, plastic deformation occurs. A series of varieties do not emerge individually but interact with each other. The theory for analyzing the interaction is called metallo-thermomechanics, which is the foundation of thermal treatment analysis.

When the AZ31 magnesium alloy and 6061 aluminum alloy are welded by diffusion bonding, diffusion between Mg and Al should be considered. The diffusion phenomenon can be analyzed by Fick's law and can be expressed by the following equation:

$$\frac{\partial C}{\partial t} = D \frac{\partial^2 C}{\partial x^2}, \tag{1}$$

where C is the concentration of the element and D is the diffusion coefficient representing the diffusion property of the material and is a function of C. In general, if the influence of the microstructure is ignored, the diffusion equation can be expressed based on element concentration as follows:

$$\dot{C} = \text{div}\,(D\,\text{grad}\,C). \tag{2}$$

If the energy of the object is represented as $e = g + T\eta + \text{tr}\,(\sigma \varepsilon^e)$, then the first law of thermodynamics can be written as follows:

$$\rho \dot{e} - \text{tr}\,(\sigma \dot{\varepsilon}) + \text{div}\,\mathbf{h} = 0. \tag{3}$$

If the Fourier law ($\mathbf{h} = k\,\text{grad}\,T$) is used, plastic work and latent heat of transformation are not considered, and terms related to elastic strain and hardening coefficient are ignored, then (3) can be written as follows:

$$\rho c \dot{T} - k\,\text{div}\,(\text{grad}\,T) = 0, \tag{4}$$

where ρ is the density of the material, c is the specific heat, and k is the thermal conductivity. When the coefficient of heat transfer and the temperature of the fluid in contact with the object do not change, the boundary condition is expressed by the following equation:

$$-k\,\text{grad}\,T \cdot \mathbf{n} = h(T)(T - T_\omega), \tag{5}$$

where $\mathbf{n}$ is a vector whose direction is outward from the surface of the object, T_ω is the temperature of the object's environment, and $h(T)$ is the coefficient of heat transfer between the object and the environment. Generally, the coefficient of heat transfer is a function of temperature T and

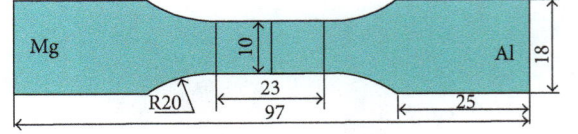

FIGURE 1: Analysis model.

FIGURE 2: Finite element mesh.

can be obtained from the experimental value of the cooling curve [17].

When plastic materials are subjected to loading, they undergo elastic or plastic deformation. Hooke's law is applicable to three-dimensional stress and strain and can be expressed as

$$\dot{\varepsilon}_{ij} = \dot{\varepsilon}_{ij}^{e} + \dot{\varepsilon}_{ij}^{p} + \dot{\varepsilon}_{ij}^{T}, \tag{6}$$

where $\dot{\varepsilon}_{ij}$ is the total strain rate, $\dot{\varepsilon}_{ij}^{e}$ is the elastic strain rate, $\dot{\varepsilon}_{ij}^{p}$ is the plastic strain rate, and $\dot{\varepsilon}_{ij}^{T}$ is the thermal strain rate. Elastic strain rate and thermal strain rate are expressed by the following equation:

$$\dot{\varepsilon}_{ij}^{e} + \dot{\varepsilon}_{ij}^{T} = E_{ijkl}^{e} \dot{\sigma}_{kl} + \beta_{ij} \dot{T},$$

$$\beta_{ij} = \frac{\partial E_{ijkl}^{e}}{\partial T} \sigma_{kl} + \alpha_{ij} \frac{\partial \alpha_{ij}}{\partial T} (T - T_0), \tag{7}$$

where α is the linear coefficient of expansion and $T - T_0$ is the temperature difference. E_{ijkl}^{e} is a coefficient that is represented by the following equation:

$$E_{ijkl}^{e} = \frac{1}{2G} \left\{ \left(\delta_{ik} \delta_{jl} + \delta_{il} \delta_{jk} \right) - \frac{\nu}{1+\nu} \delta_{ij} \delta_{kl} \right\}, \tag{8}$$

where ν and G are Poisson's ratio and the shear modulus, respectively. The plastic strain rate is expressed as follows:

$$\dot{\varepsilon}_{ij}^{p} = \frac{1}{\hat{G}} \left(\frac{\partial F}{\partial \sigma_{mn}} \dot{\sigma}_{mn} + \frac{\partial F}{\partial T} \dot{T} \right) \frac{\partial F}{\partial \sigma_{ij}},$$

$$\frac{1}{\hat{G}} = -\frac{1}{(\partial F/\partial \kappa)(\partial F/\partial \sigma_{kl})\sigma_{kl}}, \quad F = F(\sigma, T), \tag{9}$$

where F is the Mises yield function.

First, the definition of the mixture and the mixing rule are explained. Intermetallic compounds, such as Al_3Mg_2 and $Mg_{17}Al_{12}$, are formed during diffusion bonding between magnesium and aluminum alloys. It is assumed that many microstructures are present in the mixture. Further, the mixture contains N compositions whose volume fractions are ξ_I ($I = 1, 2, \ldots, N$); the sum of the volume fractions of the compositions is 1:

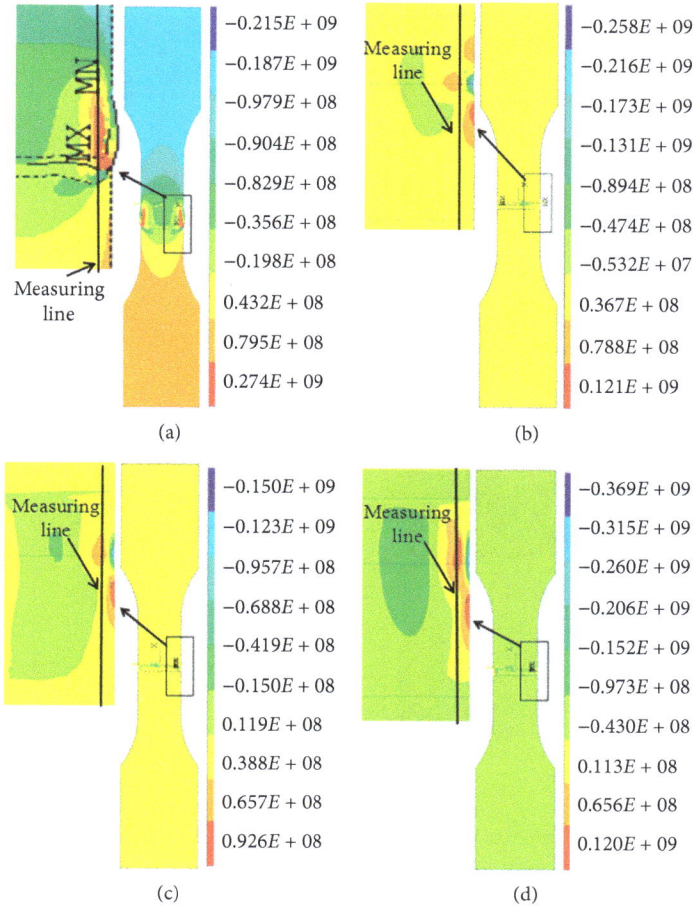

FIGURE 3: Results of simulations: (a) without annealing and annealing at 200°C (b), 250°C (c), and 300°C (d), previously presented in [20].

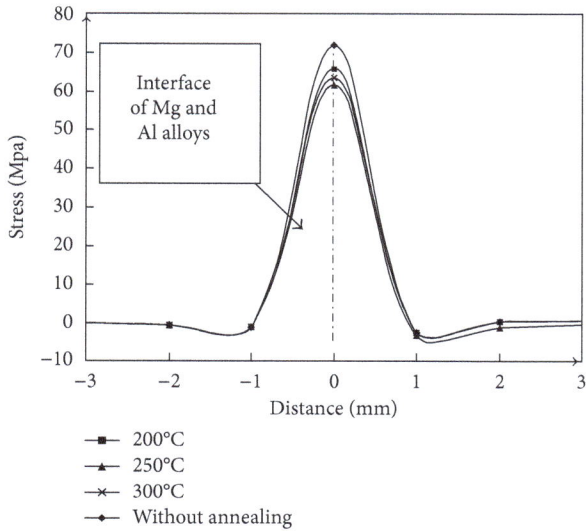

FIGURE 4: Distribution of residual stress obtained by simulation, previously presented in [20].

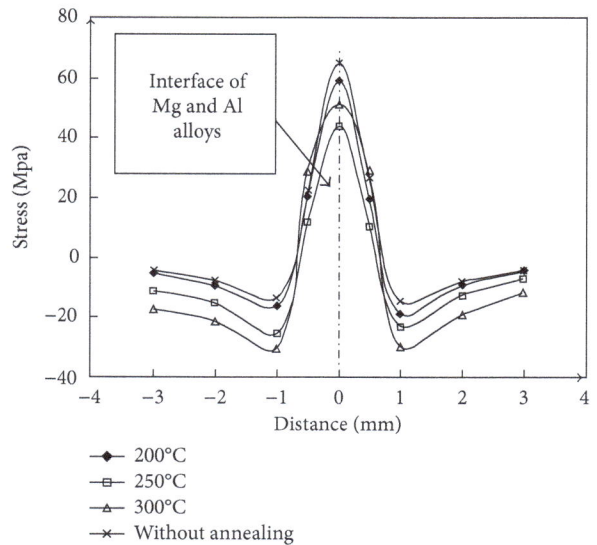

FIGURE 5: Residual stress of specimens annealed at different temperatures.

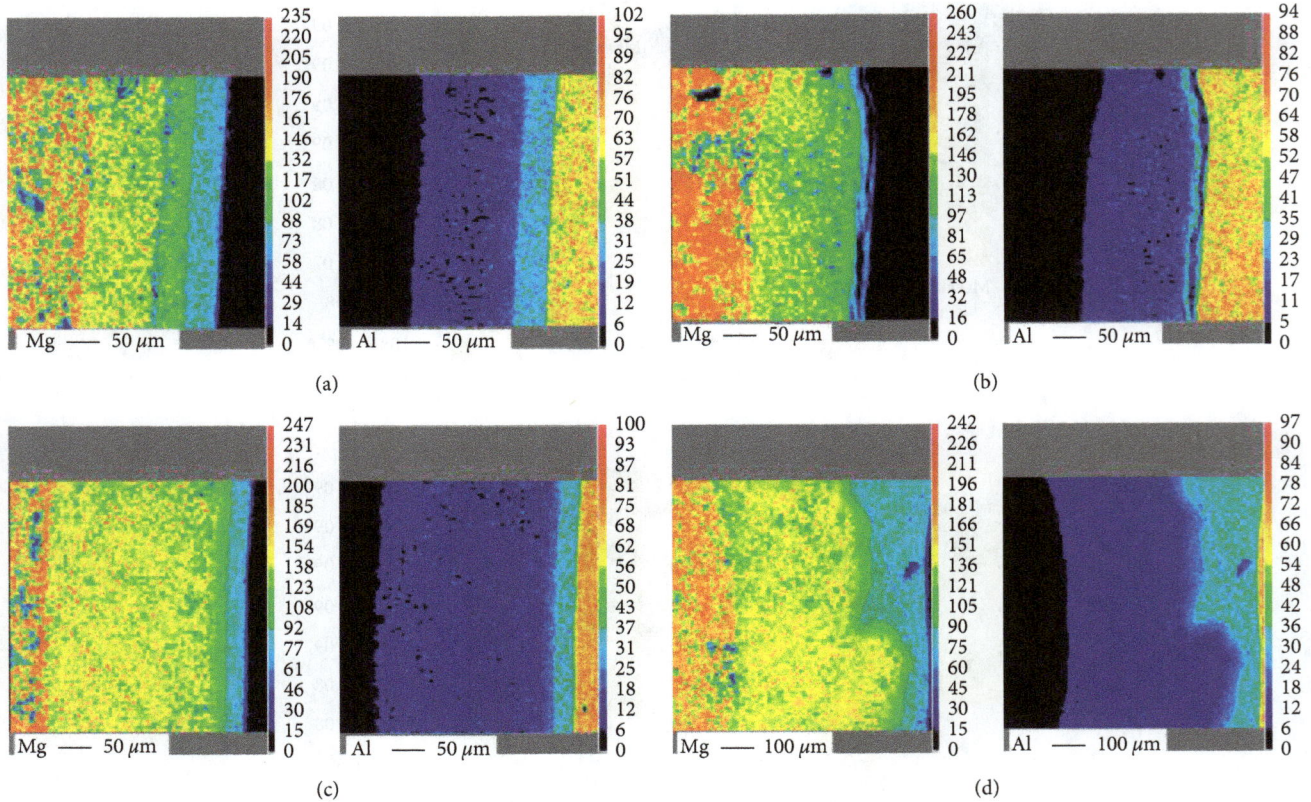

FIGURE 6: Results of elemental analysis, surface scanning: (a) without annealing and annealing at (b) 200°C, (c) 250°C, and (d) 300°C.

$$\sum_{I=1}^{N} \xi_I = 1. \tag{10}$$

If the properties of composition I are represented by χ_I, then χ denotes all the properties of N compositions. Therefore, the property χ can be expressed as (11), which is called the mixing rule [18]:

$$\chi = \sum_{I=1}^{N} \chi_I \xi_I, \quad \xi_I \ (I = 1, 2, \ldots, N). \tag{11}$$

In the intermetallic compounds of Mg and Al alloys, the ratio of atomic quantity can be expressed as the following equation:

$$\frac{M_1}{M_2} = \frac{C_1 \cdot \mathrm{Ar}_2}{C_2 \cdot \mathrm{Ar}_1}, \tag{12}$$

where M_1 and M_2 are the quantity of Mg and Al atoms. C_1 and C_2 are the atomic mass of Mg and Al. Ar_1 and Ar_2 are the relative molecular masses.

The molar concentration of Mg is expressed as follows:

$$C_{B1} = \frac{M_1}{V_0 \cdot N_A}, \tag{13}$$

where V_0 is the volume of the intermetallic compounds and N_A is the Avogadro constant. The amount of substance for Mg is represented as follows:

$$n_1 = \frac{G_1}{\mathrm{Ar}_1 \cdot g}, \tag{14}$$

where G_1 is the weight of Mg and g is the gravitational acceleration.

So the volume ratio of Mg to Al can be expressed as follows:

$$\frac{V_1}{V_2} = \frac{G_1 \cdot C_2}{G_2 \cdot C_1}, \tag{15}$$

V_1 and V_2 are the volume of Mg and Al and G_2 is the weight of Al.

According to the mixing rule and the volume ratio of Mg to Al, the coefficients of simulations can be calculated. Thermal-stress coupling field of ANSYS was applied to simulate stress at different temperatures during the diffusion process and annealing process. The element type was "Coupled Field, Vector Quad 13," and the material model was "Structural" and "Thermal." Temperature conditions were 200°C, 250°C, 300°C, and room temperature (20°C). It was assumed that the interface could not move during the diffusion process. Therefore, a fixed boundary condition was applied to the contact surface, and then the simulations were performed. The model is shown in Figure 1.

The finite element mesh is shown in Figure 2. The element size was 1 mm, and the total number of elements was 14,005.

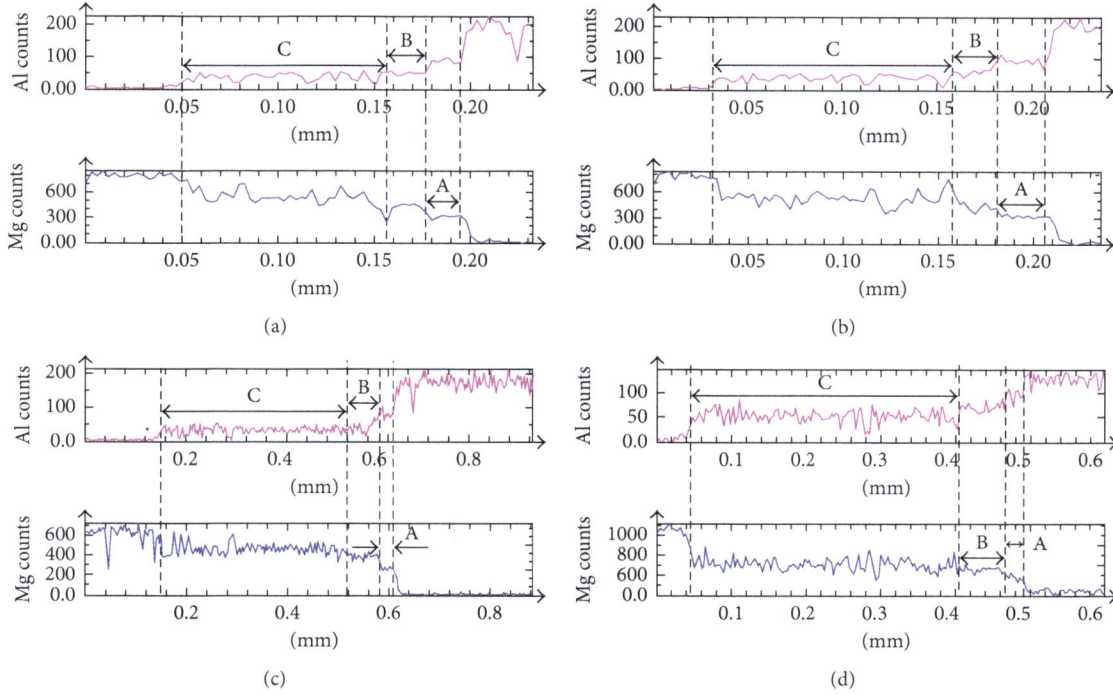

FIGURE 7: Results of elemental analysis, line scanning: (a) without annealing and annealing at (b) 200°C, (c) 250°C, and (d) 300°C.

2.2. Experimental. The AZ31 magnesium alloy sheets and 6061 aluminum alloy sheets were successfully welded using vacuum diffusion bonding. The joining temperature was 440°C. After vacuum diffusion bonding, the samples were annealed at 200, 250, and 300°C. After heat treatment, the samples were cooled to room temperature in an electric furnace.

XRD was used to investigate the residual stress distribution of the specimens annealed at different temperatures. Based on the testing principle of residual stress, the X-ray was adjusted. Initially, the specimen was radiated, and the corresponding diffraction angle 2θ was obtained, which was later used to calculate the slope M of 2θ–$\sin^2\psi$ (ψ was set as 0°, 15°, 30°, and 45°). In addition, the relationship between 2θ and $\sin^2\psi$ was obtained, and the residual stress σ was calculated according to the following equation:

$$\sigma = \frac{K \cdot \Delta 2\theta}{\Delta \sin^2\psi} = K \cdot M, \quad (16)$$

where K is the stress constant of the XRD analysis and can be expressed as follows:

$$K = -\frac{1}{2} \cdot \cot \theta_0 \cdot \frac{E}{1+v} \cdot \frac{\pi}{180}, \quad (17)$$

where E is the elastic modulus of the material, θ_0 is the diffraction angle without stress, and v is Poisson's ratio [19].

For the 6061 Al alloy, 2θ was set as 140°, the stress constant was −163.32 MPa/°, tube type was Cr, wavelength was kα, and the size of the collimator was $\varphi = 0.5$ mm. For the AZ31 Mg alloy, 2θ was set as 155° and stress constant was −79.14 MPa/°; tube type, wavelength, and the size of the

collimator were the same as that for the 6061 Al alloy. In addition, the 2θ of $Mg_{17}Al_{12}$ and Al_3Mg_2 were set as 150° and 145° and their stress constants were −98.97 MPa/° and −126.22 MPa/°, respectively. Based on these values, the values of residual stress were calculated.

Furthermore, the microstructure and elemental distribution of the diffusion zone were investigated by SEM and EPMA.

3. Results and Discussion

The Mises stress data were obtained after completion of the simulation. The simulation results of stress distribution are shown in Figures 3(a)–3(d).

Figure 3 shows that the stress values change with annealing temperature. Because stress is mainly concentrated at the edge of the interface, the values along the interface are the largest, which in turn causes premature failure [20].

In this paper, stress distribution along the line crossing the edge of the interface was investigated (the measuring line is shown in Figure 3). Based on the stress values at the nodes and the distance between the nodes, Figure 4 is obtained; this figure shows the distribution of residual stresses calculated from the simulations.

Residual stress is a vector, and in this study, its direction was along the axial direction of the specimens. According to material mechanics, tensile stress is positive, while compressive stress is negative. Therefore, it could be concluded from Figure 4 that the value of stress near the interface was positive, that is, tensile stress, and the largest. The stress near the interface of the specimens annealed at temperatures of

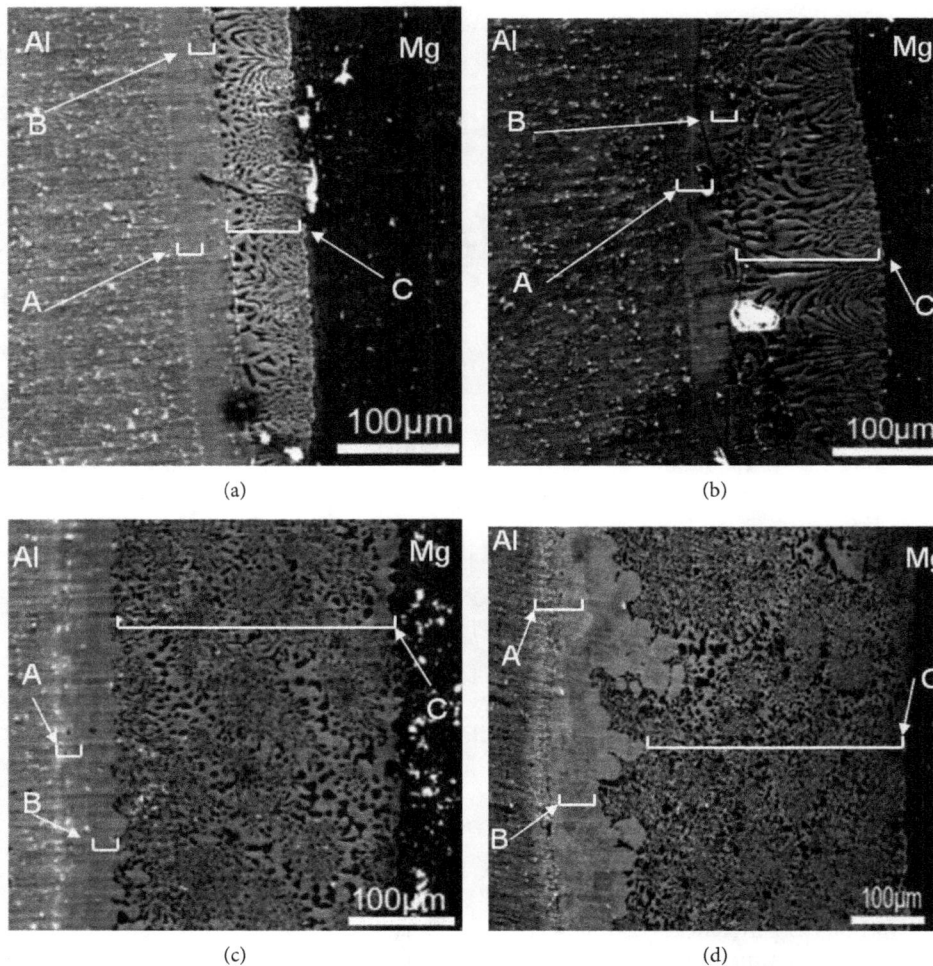

FIGURE 8: SEM micrographs of joints: (a) without annealing and annealing at (b) 200°C, (c) 250°C, and (d) 300°C; layer A: Al$_3$Mg$_2$, layer B: Mg$_{17}$Al$_{12}$, and layer C: Mg-based solid solution.

200, 250, and 300°C was 66, 61, and 63 MPa, respectively. The untreated specimens exhibited a stress value of 71 MPa. However, as the distance from the interface increased, stress decreased and even became negative, that is, compressive stress. In addition, the residual stress was almost 0 MPa far away from the interface.

The residual stress measured by XRD is shown in Figure 5. The stress of the untreated specimens was about 65 MPa, while that of the specimens annealed at 300°C and 200°C was 51 MPa and 59 MPa, respectively. However, when the specimens were annealed at 250°C, the stress was approximately 44 MPa. During the experiments, the diffraction peak for the (311) plane of Al appeared at a 2θ of about 139°, while that for the (104) plane of Mg appeared at a 2θ of 152°. Furthermore, the 2θ value of the intermetallic compounds near Al and Mg was 139.4° and 143°, respectively [21]. Thus, it could be concluded that the distribution of residual stress obtained from the experiments is nearly the same as that obtained from simulations.

The above results showed that 250°C is the most appropriate annealing temperature. This could be further confirmed by microstructure investigation. The results of the

elemental analysis are shown in Figures 6 and 7. In Figure 6, the colors represent the different amounts of elements detected, and the y-axis numbers represent the levels that the different colors stand for.

As shown in Figures 6 and 7, the diffusion layers widened with increasing annealing temperatures. As the temperature increased, the magnesium and aluminum contents varied in regions A, B, and C. Specifically, the untreated specimens exhibited a magnesium content ranging from 800 to 300 counts, while the aluminum content was 50–200 counts. For the specimens annealed at 200°C, the Mg content ranged from 600 to 300 counts in the direction from Mg to Al, while the Al content was 50–100 counts in the same direction. However, for the specimens annealed at 300°C, the Mg content varied from 800 to 650 counts, while the Al content varied within the range of 50–70 counts. The result for the specimens annealed at 250°C is shown in Figure 7(c). For this specimen, the magnesium content ranged from 500 to 300 counts, while the Al content was 50–70 counts. The reason for the variations in the diffusion layer width and element content was the difference in the annealing temperature. As the annealing temperature increased, the diffusion rate

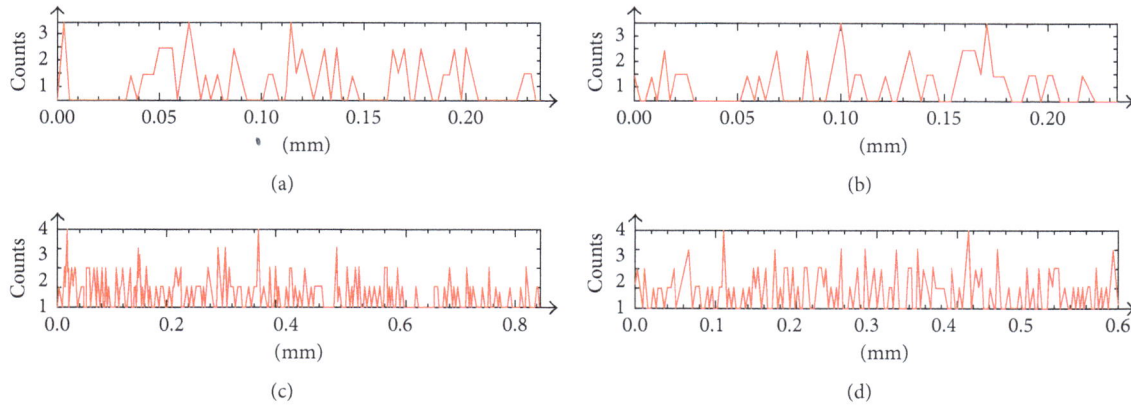

FIGURE 9: Elemental distribution of Zn, line scanning: (a) without annealing and annealing at (b) 200°C, (c) 250°C, and (d) 300°C.

increased. Therefore, the diffusion layers were the widest for the specimens annealed at 300°C. However, as shown in Figure 7, the element content for the specimens annealed at 250°C was steady. This indicates that the microstructure is more uniform at 250°C.

To verify the above results, the microstructures of the specimens annealed at different temperatures were investigated. The microstructures of the joints are shown in Figure 8.

The diffusion layers, including layers A, B, and C, which were investigated and confirmed to be Al_3Mg_2, $Mg_{17}Al_{12}$, and Mg-based solid solutions, respectively, could be clearly observed. Moreover, the width of the diffusion layers increased with increasing annealing temperature. However, the microstructure of the specimens annealed at 250°C was more uniform than that of those annealed at other conditions, thus confirming the results of the elemental distribution analysis.

As shown in Figure 9, for specimens annealed at 250°C, the Zn content was relatively steady and was greater than that for specimens annealed at other temperatures. Although Zn is a kind of rare earth element and its content is low, it greatly affects the diffusion zone of Mg and Al alloys. Zn could retard the formation of intermetallic compounds of Mg and Al. Though intermetallic compounds were formed, they precipitated and were dispersed, thus generating precipitation strength [22]. In other words, the amount of the hard and brittle phase at the interface was small. Thus, the properties of the specimens annealed at 250°C were better than those of the specimens annealed at other temperatures.

4. Conclusions

The following results were obtained from the simulations and experiments:

(1) Annealing temperatures have a great effect on the microstructure and elemental distribution. The most appropriate annealing temperature for the diffusion-bonded Mg/Al alloy is 250°C.

(2) It is difficult to obtain a sufficiently high-quality diffusion zone by diffusion bonding magnesium and aluminum alloy sheets because of the formation of intermetallic compound layers. However, this study used annealing to improve the microstructure, thus achieving such a diffusion zone.

(3) The outcomes obtained by FEM are in good agreement with those of the experiments; thus, the accuracy of FEM for analyzing residual stress during annealing is reliable.

Acknowledgments

This research work has been partially supported by the grant subsidy of the "Nano Project" for Private Universities: 2011–2014 from MEXT, Japan. This study was also supported by the "Advanced Science Research Laboratory" in Saitama Institute of Technology.

References

[1] W. Walke, Z. Paszenda, and W. Jurkiewicz, "Numerical analysis of three-layer vessel stent made from Cr-Ni-Mo steel and tantalum international," *Journal of Computational Materials Science and Surface Engineering*, vol. 1, no. 1, pp. 129–139, 2007.

[2] T. Da Silva Botelho, E. Bayraktar, and G. Inglebert, "Experimental and finite element analysis of spring back in sheet metal forming," *International Journal of Computational Materials Science and Surface Engineering*, vol. 1, no. 2, pp. 197–213, 2007.

[3] A. V. Benin, A. S. Semenov, and S. G. Semenov, "Modelling of fracture process in concrete reinforced structures under steel corrosion," *Journal of Achievements in Materials and Manufacturing Engineering*, vol. 39, no. 2, pp. 168–175, 2010.

[4] S. Thipprakmas, M. Jin, K. Tomokazu, Y. Katsuhiro, and M. Murakawa, "Prediction of Fineblanked surface characteristics using the finite element method (FEM)," *Journal of Materials Processing Technology*, vol. 198, no. 1, pp. 391–398, 2008.

[5] Y. Kim, S. Yaang, D. Shan, S. Choi, S. Lee, and B. You, "Three-dimensional rigid-plastic FEM simulation of metal forming processes," *Journal of Materials Engineering and Performance*, vol. 15, no. 3, pp. 275–279, 2006.

[6] K. Lenik and D. Wójcicka-Migasiuk, "FEM applications to the analysis of passive solar wall elements," *Journal of Achievements in Materials and Manufacturing Engineering*, vol. 43, no. 1, pp. 333–340, 2010.

[7] J. Okrajni and W. Essler, "Computer models of steam pipeline components in the evaluation of their local strength," *Journal of Achievements in Materials and Manufacturing Engineering*, vol. 39, no. 1, pp. 71–78, 2010.

[8] J. Bouzakis, G. Maliaris, and A. Tsouknidas, "FEM supported semi-solid high pressure die casting process optimization based on rheological properties by isothermal compression tests at thixo temperatures extracted," *Computational Materials Science*, vol. 59, pp. 133–139, 2012.

[9] B. Regener, C. Krempaszky, E. Werner, and M. Stockinger, "Modelling the micromorphology of heat treated Ti6Al4V forgings by means of spatial tessellations feasible for FEM analyses of microscale residual stresses," *Computational Materials Science*, vol. 52, no. 1, pp. 77–81, 2012.

[10] Z. Shiping, O. El Kerdi, K. A. Khurram, and G. Habashi, "FEM analysis of in-flight ice break-up," *Finite Elements in Analysis and Design*, vol. 57, pp. 55–66, 2012.

[11] U. E. Ozturk and G. Anlas, "Finite element analysis of expanded polystyrene foam under multiple compressive loading and unloading," *Materials and Design*, vol. 32, no. 2, pp. 773–780, 2011.

[12] J. Shang, K.-H. Wang, Q. Zhou, D.-K. Zhang, J. Huang, and J. Q. Ge, "Effect of joining temperature on microstructure and properties of diffusion bonded Mg/Al joints," *Transactions of Nonferrous Metals Society of China*, vol. 22, no. 8, pp. 1961–1966, 2012.

[13] X. Li, W. Liang, X. Zhao, Y. Zhang, X. Fu, and F. Liu, "Bonding of Mg and Al with Mg-Al eutectic alloy and its application in aluminum coating on magnesium," *Journal of Alloys and Compounds*, vol. 471, no. 1-2, pp. 408–411, 2009.

[14] F. Liu, D. Ren, and L. Liu, "Effect of Al foils interlayer on microstructures and mechanical properties of Mg-Al butt joints welded by gas tungsten arc welding filling with Zn filler metal," *Materials and Design*, vol. 46, pp. 419–425, 2013.

[15] J. R. Cho, B. Y. Lee, Y. H. Moon, and C. J. Van Tyne, "Investigation of residual stress and post weld heat treatment of multi-pass welds by finite element method and experiments," *Journal of Materials Processing Technology*, vol. 155-156, pp. 1690–1695, 2004.

[16] P.-H. Chang and T.-L. Teng, "Numerical and experimental investigations on the residual stresses of the butt-welded joints," *Computational Materials Science*, vol. 29, no. 4, pp. 511–522, 2004.

[17] D. Y. Ju, "Actuality and scope on simulation of heat treatment III: simulation of carburized and nitrided quenching process," *Journal of the Society of Materials Science*, vol. 55, no. 7, pp. 712–717, 2006.

[18] T. Inoue, "Inelastic constitutive models under plasticity-creep interaction condition—theories and evaluations," *JSME International Journal. Series 1, Solid Mechanics, Strength of Materials*, vol. 31, no. 4, pp. 653–663, 1988.

[19] J. G. Wang and D. Y. Ju, *Study on Evolution Technology of Anisotropic Mechanical Properties and Microstructure Evolution of Metal Thin Plate under Complex Stress Condition*, Saitama Institute of Technology, Saitama, Japan, 2011.

[20] Y. L. Ding, J. Wang, and D. Ju, "Simulations about the effect of heat treatment temperatures on the properties of diffusion bonded Mg/Al joints," *MATEC Web of Conferences*, vol. 130, p. 06004, 2017.

[21] D. Zhang and J. He, *Residual Stress Analysis by X-Ray Diffraction and it's Functions*, Xi'an Jiaotong University, Xi'an, China, 1st edition, 1999.

[22] L. M. Zhao and Z. D. Zhang, "Effect of Zn alloy interlayer on interface micro-structures and strength of diffusion-bonded Mg-Al joints," *Scripta Materialia*, vol. 58, no. 4, pp. 283–286, 2008.

Effect of Solidification Rate and Rare Earth Metal Addition on the Microstructural Characteristics and Porosity Formation in A356 Alloy

M. G. Mahmoud,[1] A. M. Samuel,[1] H. W. Doty,[2] S. Valtierra,[3] and F. H. Samuel[1]

[1]*Département des Sciences Appliquées, Université du Québec à Chicoutimi, Chicoutimi, QC, Canada*
[2]*General Motors Materials Engineering, 823 Joslyn Ave., Pontiac, MI 48340, USA*
[3]*Nemak, S.A., P.O. Box 100, 66221 Garza Garcia, NL, Mexico*

Correspondence should be addressed to F. H. Samuel; fhsamuel@uqac.ca

Academic Editor: Francesco Delogu

The present study was performed on A356 alloy with the main aim of investigating the effects of La and Ce additions to 356 alloys (with and without 100 ppm Sr) on the microstructure and porosity formation in these alloys. Measured amounts of La, Ce, and Sr were added to the molten alloy. The results showed that, in the absence of Sr, addition of La and Ce leads to an increase in the nucleation temperature of the α-Al dendritic network with a decrease in the temperature of the eutectic Si precipitation, resulting in increasing the freezing range. Addition of 100 ppm Sr results in neutralizing these effects. The presence of La or Ce in the casting has a minor effect on eutectic Si modification, in spite of the observed depression in the eutectic temperature. It should be noted that Ce is more effective than La as an alternate modifying agent. According to the atomic radius ratio, r_{La}/r_{Si} is 1.604 and r_{Ce}/r_{Si} is 1.559, theoretically, which shows that Ce is relatively more effective than La. The present findings confirm that Sr is the most dominating modification agent. Interaction between rare earth (RE) metals and Sr would reduce the effectiveness of Sr. Although modification with Sr causes the formation of shrinkage porosity, it also reacts with RE-rich intermetallics, resulting in their fragmentation.

1. Introduction

The main role of rare earth (RE) metals as modifiers is to change the mode of growth of the eutectic silicon. Lanthanum is the most powerful of its kind. Microadditions of RE should be enough to modify the eutectic Si particles provided that a critical cooling rate is reached. RE-treated alloys can maintain the modified structure much longer than Na-treated alloys. A reliable and persistent eutectic modification effect can be obtained with rare earth element addition [1–4]. However, the minimum amount of rare earth elements necessary to obtain proper modification is exceptionally large. Kim and Heine [5] showed that both as-cast grain size and secondary dendrite arm spacing were decreased by adding cerium (Ce) and lanthanum (La) metals.

Aguirre-De la Torre et al. [6] investigated the mechanical properties of A356 aluminum alloy modified with La/Ce.

It was suggested that an increase in the mechanical performance of the alloy could be achieved by a homogeneous dispersion of fine particles containing La/Ce phase. The work of Nogita et al. [7] on eutectic modification of Al-Si alloys with rare earth metals reveals that the impurity induced twinning model of modification, based on atomic radius alone, is inadequate and other mechanisms are essential for the modification process. Furthermore, modification and the eutectic nucleation and growth modes are controlled independently of each other.

Kinetic nucleation of primary α-Al dendrites in Al-7% Si-Mg cast alloys with Ce and Sr additions was analyzed by Chen et al. [8]. The results show that the values of activation energy and nucleation are decreased and the nucleation frequency is increased with the addition of Ce and Sr to the alloys. The nucleation temperatures of primary α-Al dendrites are decreased with the additions of Ce and Sr.

TABLE 1: Chemical composition of the base A356 alloy.

Alloy	Elements (wt.%)					
	Si	Cu	Mg	Fe	Zn	Al
A356	7.2	<0.20	0.35	<0.20	<0.10	Bal

An elaborate review was carried out by Alkahtani et al. [9] on the modification mechanism and microstructural characteristics of eutectic Si in Al-Si casting alloys. Although all rare earth elements have some effect on the eutectic silicon as they are within the atomic radii range predicted by models to be effective in producing growth twins in an Al-Si alloy ($r/r_{Si} = 1:65$), however, they only result in a minor refinement of the plate-like silicon morphology.

El Sebaie et al. [10, 11] studied the effects of mischmetal (MM), cooling rate, and heat treatment on the eutectic Si particle characteristics of A319.1, A356.2, and A413.1 Al-Si casting alloys. Measurements of the eutectic Si particles revealed that addition of mischmetal led to partial modification, while full modification was achieved with the addition of Sr in the as-cast condition, at both high and low cooling rates. The interaction between Sr and mischmetal weakened the effectiveness of Sr as a Si particle-modifying agent. This effect was particularly evident at the low cooling rate.

The present study was undertaken to investigate the combined effect of solidification rate and addition of rare earth metals with or without Sr on the characteristics of the eutectic Si structure as well as on porosity formation in cast A356 alloy.

2. Experimental Procedure

Table 1 lists the chemical composition of the base A356 alloy used in the present study. The as-received ingots were melted using an electrical resistance furnace at 750°C. The molten metal was degassed using pure, dry argon, injected into the melt by means of a graphite rotary impeller (at a speed of 130 rpm). Prior to degassing, measured amounts of Sr, La, and Ce were added. The three elements were introduced into the molten alloy in the form of Al-10% Sr, Al-20% La, and Al-20% Ce master alloys. At the end of the degassing period, the molten alloy was poured into three different molds which provided different solidification rates:

(1) A graphite mold preheated at 600°C used for carrying out thermal analysis for obtaining the solidification curve (see Figure 1(a)) [12]

(2) A variable angle metallic mold (0°, 5°, and 15°) heated at 350°C (see Figure 1(b))

(3) A step-like metallic mold heated at either 200°C or 400°C (see Figure 1(c))

(4) Setup for measuring the solidification curves in the variable angle mold (see Figure 1(d))

For each pouring/casting, samplings for chemical analysis were also taken, to determine the exact composition of the melt. Chemical analyses were carried out at General Motors facilities in Warren, MI, and the results are listed in Table 2. Samples for metallography were also sectioned from all castings. These samples were individually mounted in bakelite using a Struers LaboPress-3 machine, subjected to grinding and polishing procedures using a TegraForce-5 machine, and subsequently polished to a fine finish using 1 μm diamond suspension. The polished samples were examined using an Olympus PMG3 optical microscope-Clemex Vision PE image analysis system. The secondary dendrite arm spacing (SDAS) values were measured using the line intercept method shown in Figure 1(e). Table 3 summarizes the measured SDAS values obtained from the examined samples. Phase identification was carried out using an electron probe microanalyzer (EPMA) in conjunction with energy dispersive X-ray (EDX/EDS) analysis and wavelength dispersive spectroscopic (WDS) analysis where required, integrating a combined JEOL JXA-8900l WD/ED microanalyzer operating at 20 KV and 30 nA, where the size of the spot examined was ~2 μm.

3. Results and Discussion

3.1. Thermal Analysis. Solidification kinetics of an unmodified and Sr-modified near-eutectic Al-Si alloy were analyzed by Aparicio et al. [13] who found that there are changes in the solidification rate during eutectic nucleation followed by similar solidification rate evolutions during growth, suggesting that this parameter is governed principally by the heat extraction conditions. The work of Hengcheng et al. [14] on the effects of Sr and solidification rate on eutectic grain structure in an Al-13 wt.% Si alloy revealed that the characteristic temperature of eutectic nucleation (T_N), the minimum temperature prior to recalescence (T_M), and the growth temperature (T_G) during cooling as determined by quantitative thermal analysis are continuously decreased with increasing Sr content. As mentioned previously, Nogita et al. [15] reported that all rare earth elements had some effect on the eutectic silicon; however, europium was the only element to cause fully modified, fine fibrous silicon, whereas the other elements only produced a minor refinement of the plate-like silicon morphology.

Ferdian et al. [16] studied the effect of cooling rate on the eutectic modification in A356 alloy. Figure 2 shows the parameters taken from thermal analysis cooling curves for characterizing the (Al)-Si eutectic arrest which, according to the authors, comprises the minimum eutectic temperature ($T_{e,min}$), maximum eutectic temperature ($T_{e,max}$), and recalescence (ΔT_e). In case of no recalescence, $T_{e,max}$ was obtained as the temperature for which the absolute value of the cooling rate (time derivative of the cooling curve) was the highest. $\Delta T_d = T_R - T_{e,max}$ is the eutectic depression, where T_R is the equilibrium eutectic temperature calculated using (1) which was obtained by updating the equation proposed by Mondolfo [17].

$$T_R \ (°C) = 577 - \frac{12.5}{w_{Si}} \cdot \left(4.59 \cdot w_{Mg} + 1.37 \cdot w_{Fe} \right.$$
$$\left. + 1.65 \cdot w_{Cu} + 0.35 \cdot w_{Zn} + 2.54 \cdot w_{Mn} + 3.52 \cdot w_{Ni}\right). \tag{1}$$

Applying this equation for the present alloy shows that T_R for A356.1 alloy is about 570.8°C.

TABLE 2

(a) Modifier additions and corresponding A356 alloy codes, graphite mold castings

| Alloy | Mold type | Mold temp. (°C) | Alloy code | Modifier addition (wt.%) | | | | | |
| | | | | Aimed | | | Actual | | |
				Sr	La	Ce	Sr	La	Ce
			TB	0	0	0	0	0	0
			T10	0	0.2	0	0	0.165	0
			T1	0	0.5	0	0	0.356	0
			T2	0	1	0	0	0.685	0
			T3	0	1.5	0	0	1.025	0
			T11	0	0	0.2	0	0.002	0.082
			T4	0	0	0.5	0	0.006	0.185
			T5	0	0	1	0	0.006	0.317
			T6	0	0	1.5	0	0.009	1.088
			T7	0	0.5	0.5	0	0.442	0.282
			T8	0	1	1	0	0.781	0.377
A356	Graphite	600	T9	0	1.5	1.5	0	1.073	0.531
			T10S	0.01	0.2	0	0.0079	0.165	0
			T1S	0.01	0.5	0	0.0109	0.356	0
			T2S	0.01	1	0	0.0077	0.685	0
			T3S	0.01	1.5	0	0.0077	1.025	0
			T11S	0.01	0	0.2	0.0073	0.002	0.182
			T4S	0.01	0	0.5	0.0078	0.006	0.358
			T5S	0.01	0	1	0.0081	0.006	0.817
			T6S	0.01	0	1.5	0.0081	0.009	1.288
			T7S	0.01	0.5	0.5	0.0083	0.442	0.382
			T8S	0.01	1	1	0.0083	0.781	0.877
			T9S	0.01	1.5	1.5	0.0089	1.073	1.231

(b) Actual Sr concentrations obtained for the step-like and variable angle mold castings of A356 alloy

Mold	Alloy code	Sr (ppm)[*]
Step-likemold	S12	83
	S22	95
	S32	84
	S42	105
	S52	108
	S14	79
	S24	73
	S34	77
	S44	88
	S54	89
Variable angle mold	DBLS	127
	D10LS	79
	D1LS	109
	D2LS	74
	D3LS	74
	D11LS	73
	D4LS	70
	D5LS	71
	D6LS	77
	D7LS	75
	D8LS	78
	D9LS	70

[*] Aimed amount was 100 ppm ± 20 ppm.

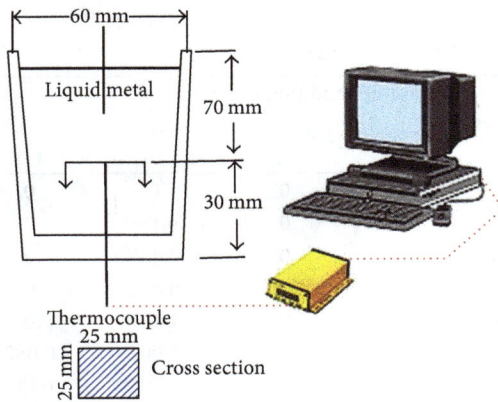

(a) Setup showing the graphite mold used for carrying out thermal analysis

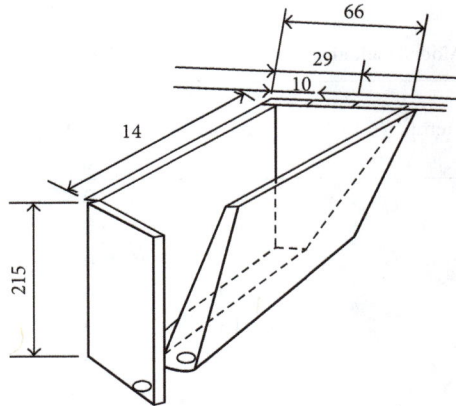

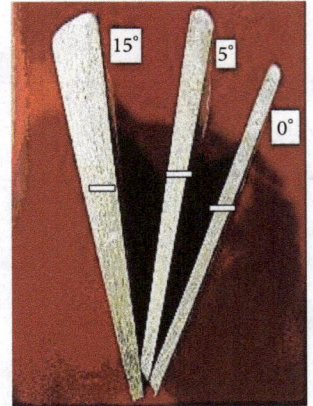

(b) (L) Variable angle metallic mold (dimensions are in mm) and (R) obtained castings; the white blocks indicate the sampling positions almost halfway along the casting height

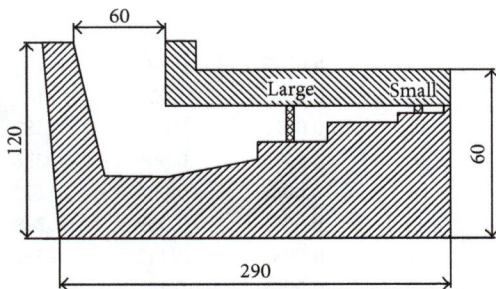

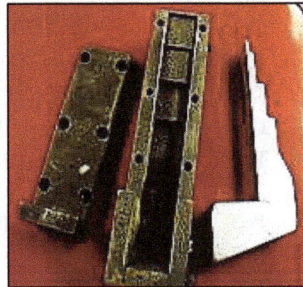

(c) (L) Step-like metallic mold (dimensions are in mm) and (R) the obtained casting

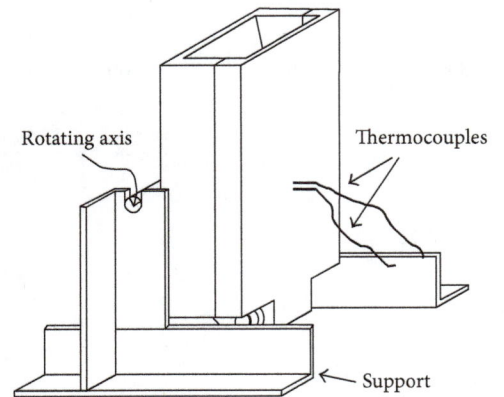

(d) Setup showing the variable angle mold used for carrying out thermal analysis

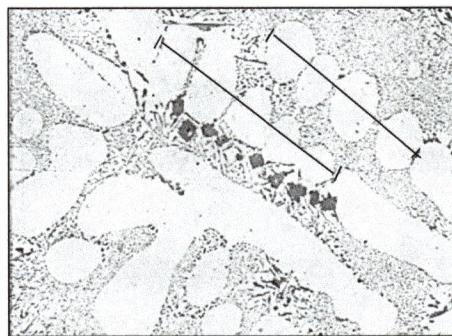

(e) Intercept method used to measure the secondary dendrite arm spacing

FIGURE 1

Figure 3(a) presents the solidification curve and its first derivative obtained from the base A356 alloy using the preheated graphite mold while Figure 3(b) shows the effect of 100 ppm Sr addition to the base alloy. It is inferred from Figure 3 that addition of Sr resulted in increasing the nucleation temperature of α-Al from about 614.4°C to 616.4°C, that is, 2°C, with a decrease in the eutectic temperature from 569.65°C to 562.23°C (approximately 7.4°C) with a total increase in the freezing zone by about 9.5°C. Modification with Sr also resulted in the merging of peaks (3) and (4) observed in Figure 3(a). The studies carried out by Samuel et al. [18, 19] revealed that Sr addition causes fragmentation of the π-iron phase during the course of solidification which explains the disappearance of peak (2) seen in Figure 3(b). Due to the directional solidification nature of castings made using the variable angle mold (Figure 3(c)), it is difficult

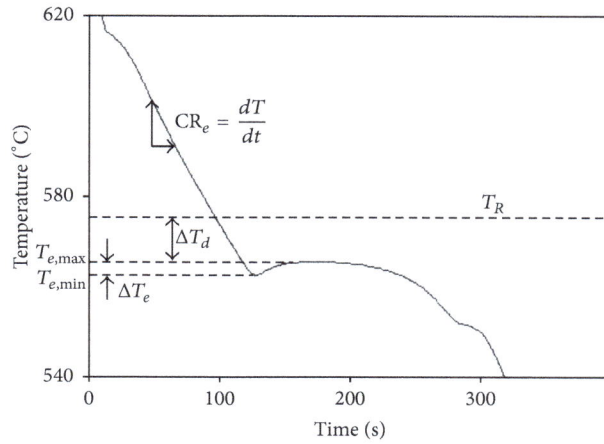

FIGURE 2: Solidification parameters of 356 alloy [16].

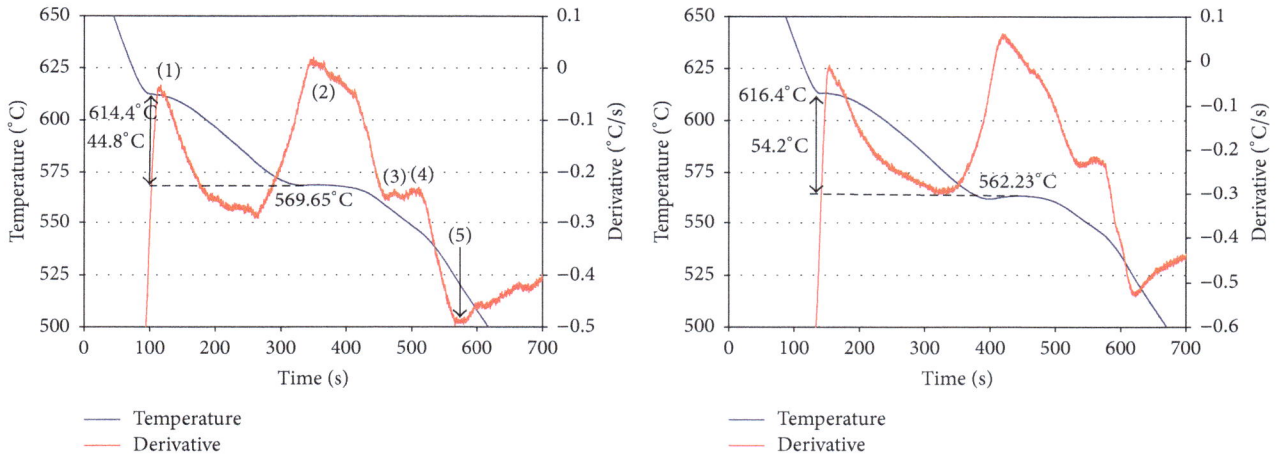

(a) Temperature-time curve and first derivative of the A356 base alloy showing (1) precipitation of α-Al dendrite network, (2) Al-Si eutectic reaction, (3) transformation of β-Fe to π-Fe phase, (4) precipitation of Mg_2Si, and (5) end of solidification [31]

(b) Temperature-time curve and its first derivative obtained from the A356 alloy modified with about 100 ppm Sr

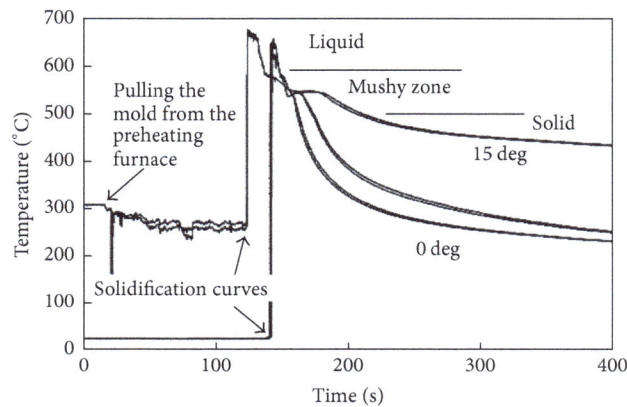

(c) Temperature-time curves obtained from the variable angle mold at different angles

FIGURE 3

TABLE 3: Average secondary dendrite arm spacing of the examined A356 alloy samples.

Mold	Mold temp. (°C)	Mold section	SDAS* (µm)	
			Average	SD
Graphite	600	Center	68.63	5.49
Step-like	200	Large	31.56	2.92
	200	Small	16.17	1.89
	400	Large	41.66	3.02
	400	Small	22.46	3.96
Variable angle	325	Large	53.62	5.7
	325	Small	25.04	3.0

*Secondary dendrite arm spacing obtained over 20 measurements per sample. SD: standard deviation.

FIGURE 4: Backscattered electron images of A356 alloy in the as-cast condition: (a) nonmodified and (b) modified with 100 ppm Sr, showing π-phase and β-phase particles.

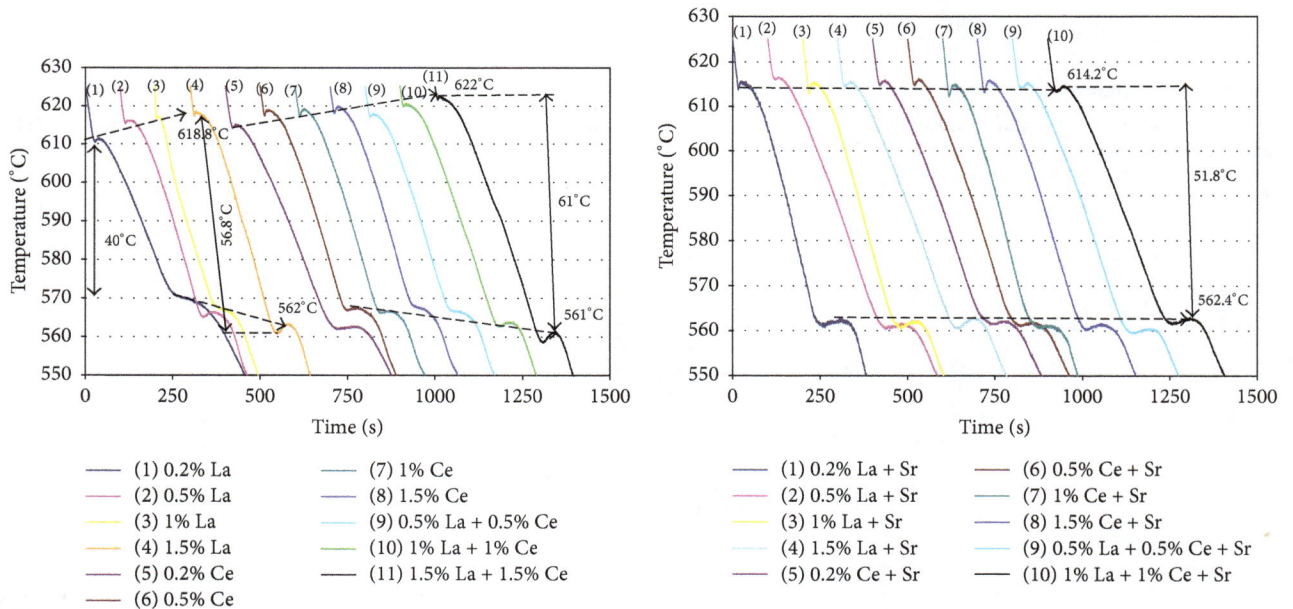

(a) Solidification curves obtained from A356 alloys containing different La and Ce additions (no Sr was added). The black arrow indicates the start of solidification [18, 19]

(1) 0.2% La
(2) 0.5% La
(3) 1% La
(4) 1.5% La
(5) 0.2% Ce
(6) 0.5% Ce
(7) 1% Ce
(8) 1.5% Ce
(9) 0.5% La + 0.5% Ce
(10) 1% La + 1% Ce
(11) 1.5% La + 1.5% Ce

(b) Solidification curves obtained from A356 alloys containing different La and Ce additions (~100 ppm Sr was added)

(1) 0.2% La + Sr
(2) 0.5% La + Sr
(3) 1% La + Sr
(4) 1.5% La + Sr
(5) 0.2% Ce + Sr
(6) 0.5% Ce + Sr
(7) 1% Ce + Sr
(8) 1.5% Ce + Sr
(9) 0.5% La + 0.5% Ce + Sr
(10) 1% La + 1% Ce + Sr

FIGURE 5

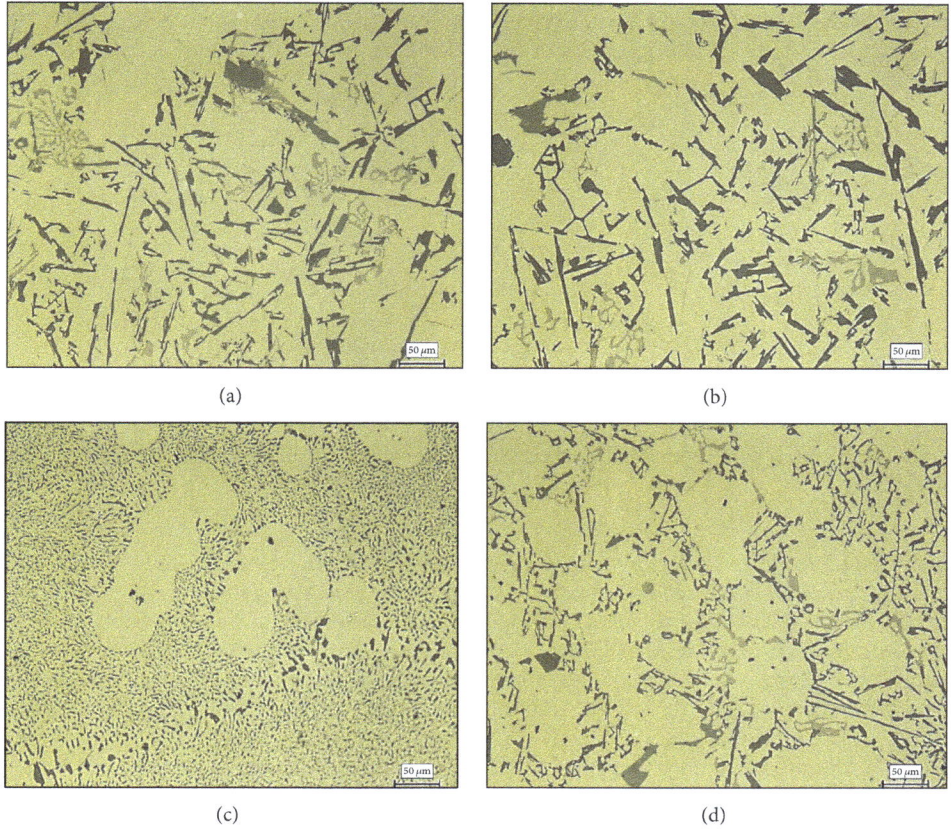

FIGURE 6: Variation in the eutectic Si morphology as a function of melt treatment: (a) no addition, (b) addition of 3% RE (alloy #11 in Figure 4(a)), (c) modification with 100 ppm Sr, and (d) addition of 2% RE to Sr-modified alloy (alloy #10 in Figure 4(b)).

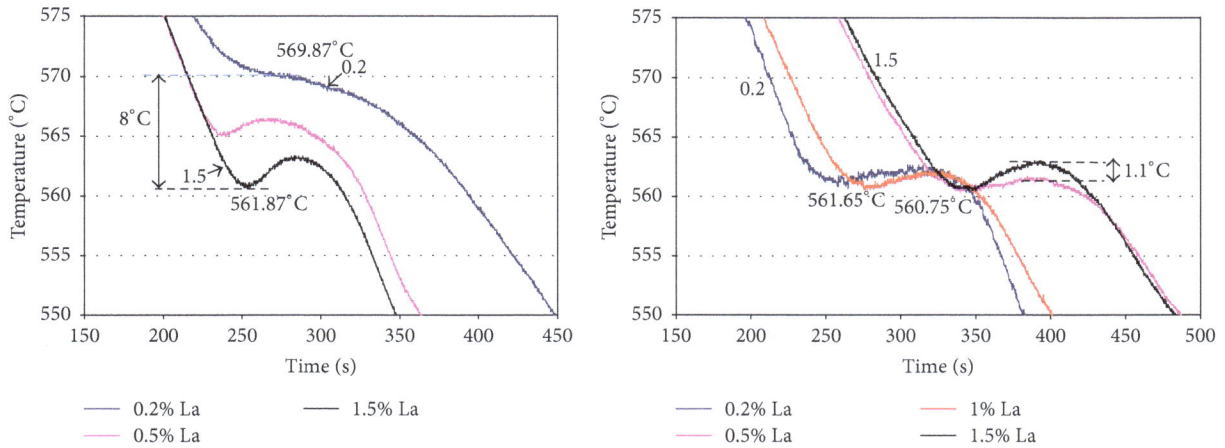

(a) Effect of La addition on the depression in the eutectic temperature in non-Sr-modified A356 alloys

(b) Effect of La addition on the depression in the eutectic temperature in non-Sr-modified A356 alloys

FIGURE 7

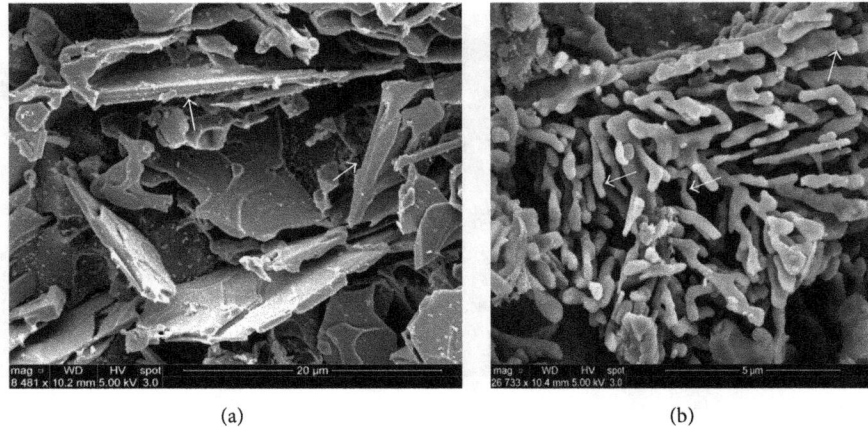

(a)   (b)

FIGURE 8: Eutectic Si morphology revealed in deep etched samples of (a) base alloy without Sr and (b) base alloy + Sr.

to identify the sequence of reactions taking place during solidification. Thus, in this case, the values of SDAS are based on the polished samples. Figure 4 illustrates the effect of the Sr modification on the microstructure of the A356 alloy.

Figure 5(a) illustrates the changes in the characteristics of the temperature-time curves as a function of the added amount of La and Ce. In Figure 5(a), addition of 1% La resulted in increasing the freezing zone from 44.8°C (Figure 3(a)) to 56.8°C, with a further increase to 61°C and 623°C with the use of 1.5% La + 1.5% Ce (see Table 2(a) for actual concentrations). This behavior depends on the added element, that is, La or Ce. The reported increase in the freezing zone is brought about by the increase in the nucleation temperature of α-Al coupled with the depression in the eutectic temperature. The increase in the freezing zone is expected to result in poor feedability of the alloy and hence more porosity [20, 21]. It should be noted that La is more effective than Ce in increasing the freezing zone. It should be emphasized here that the addition of the RE elements changes the nucleation process of the molten alloy and, hence, the undercooling of the melt will be influenced. Modifying the alloys with about 100 ppm Sr resulted in neutralizing the effect of RE on the change in both α-Al and eutectic precipitation temperatures, as shown in Figure 5(b), which is very close to that presented in Figure 3(b). In this case, the solidification temperature range is about 53°C. Reis et al. [20] presented a model for prediction of shrinkage defects in long and short freezing range materials. Their results showed that internal and external shrinkage defects depend on the freezing range of the metal. A short freezing range results mainly in internal shrinkage whereas the long freezing range shows more external shrinkage. Figure 6 presents the microstructure of the eutectic Si when the alloy was subjected to different melt treatments. It is evident from Figure 6(b) that addition of 3% RE to the base alloy has no significant modification effect in spite of the observed decrease in the eutectic temperature. Thus, in this case, the observed depression in the eutectic temperature should not be used as an indicator of modification [11]. When the base alloy is treated with 100 ppm Sr (Figure 6(c)), the microstructure is

fully modified. Due to RE-Sr interaction, the Si particles tend to lose their fibrous morphology; that is, they are partially modified (Figure 6(d)). These observations will be discussed in more detail in the next section.

The depression of the eutectic temperature is the feature that is used most often in thermal analysis as an indicator of modification [22]. As eutectic temperature is easy to measure, it is generally employed to assess whether or not a melt is properly modified. However, if temperature alone is used as the criterion for proper modification, it is difficult to detect overmodified structures, because the greatest change of temperature occurs in the unmodified-to-modified transition [23]. Figure 7(a) exhibits the depression in the eutectic temperature with the addition of La in the absence of Sr. In this case, addition of 1.5% La resulted in decreasing the eutectic temperature by about 8°C. Modification of the same alloys with 100 ppm Sr neutralized this effect, as seen from Figure 7(b). Thus, the depression in the eutectic temperature is mainly due to Sr modification (6°C) with undercooling of about 2°C. It should be mentioned here that the addition of 0.2% RE has no noticeable effect on the eutectic temperature.

3.2. Eutectic Si Particle Characteristics

3.2.1. Variable Angle Mold.
Figure 8(a) shows the morphology of the eutectic Si in deep etched samples of A356 alloy. In the case of the nonmodified alloy sample, the Si appears in the form of platelets with sharp edges (arrow). Addition of 80 ppm Sr (Figure 8(b)) resulted in changing the Si morphology into an interconnected "branched tree" type, with the particles exhibiting necking (arrow). Table 4 lists the eutectic Si particle characteristics measured from the variable angle mold at two tilting angles, that is, 0 and 15 degrees. At zero tilting angle (SDAS ~ 25 μm), the Si particles were refined due to the high solidification rate, as documented in Table 4(a). It is evident that the addition of RE up to 3% has no significant refining effect. Also, the standard deviation is noticeably large, indicating a marked variation in the Si particle size. Increasing the mold tilting angle to 15 degrees, thereby reducing the solidification rate (SDAS ~ 52 μm), the

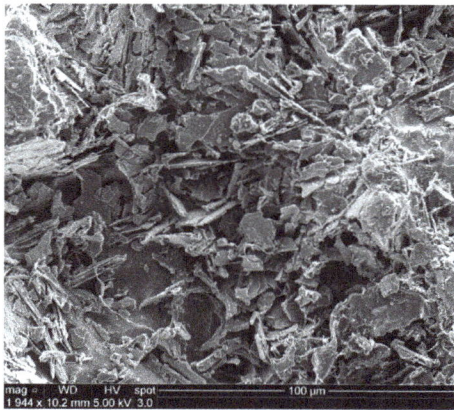

(a) Nonmodified alloy, DBL alloy

(b) After Sr addition, DBLS alloy

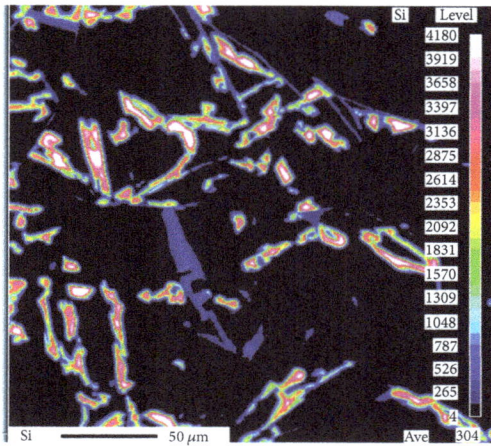

(c) D7L alloy

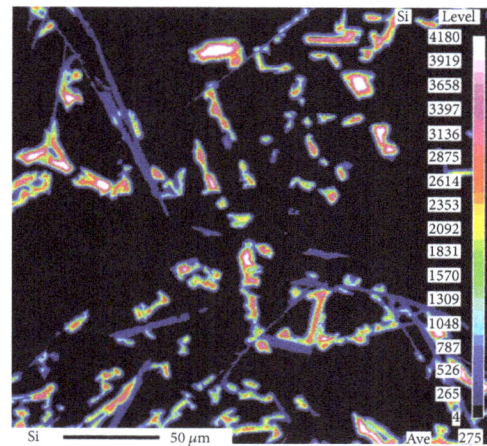

(d) D9L alloy

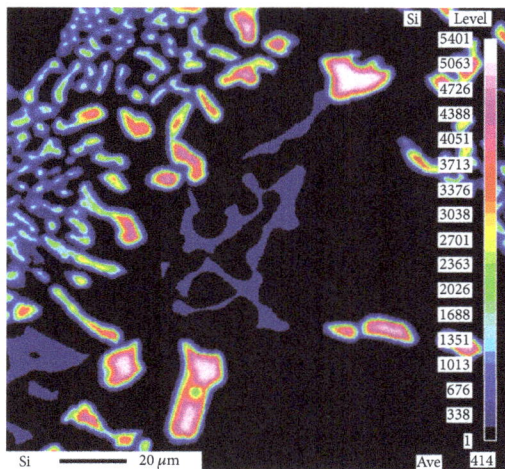

(e) D7LS alloy

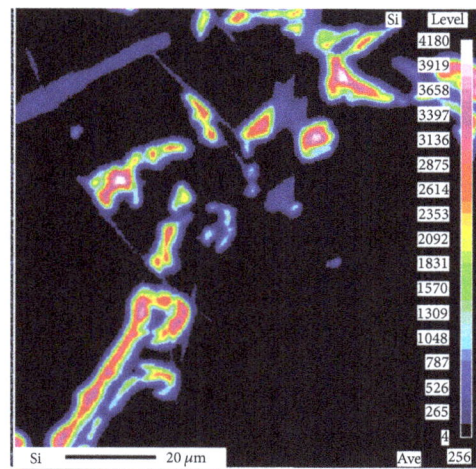

(f) D9LS alloy

FIGURE 9: Examples of the eutectic Si particle size and distribution after different melt treatments/additions in A356 alloy samples obtained from the variable angle mold (large) castings.

TABLE 4

(a) Eutectic Si particle characteristics observed in variable angle mold (small*) samples of A356 alloys containing La and Ce additions

| Alloy code | RE content | | Silicon particle characteristics | | | | | |
| | Actual | | Area (μm^2) | | Length (μm) | | Aspect ratio | Density (number/mm^2) |
	La (wt.%)	Ce (wt.%)	Average	SD	Average	SD		
DBS	0	0	5.990	7.032	5.957	7.421	2.671	8240
D10S	0.165	0	8.805	10.020	8.536	10.30	2.468	6037
D1S	0.356	0	6.492	6.958	6.459	11.70	2.282	8174
D2S	0.685	0	7.08	8.56	5.409	6.78	2.288	4201
D3S	1.025	0	3.211	5.164	3.101	5.82	2.221	10149
D11S	0.032	0.082	7.154	7.915	6.870	9.75	2.553	6307
D4S	0.136	0.185	6.413	7.472	6.307	8.53	2.533	7330
D5S	0.046	0.317	5.683	6.376	4.967	6.37	2.451	7902
D6S	0.089	1.088	8.105	9.673	6.094	9.01	2.367	6795
D7S	0.44	0.282	5.058	5.868	6.820	10.80	2.219	8331
D8S	0.781	0.877	5.917	6.754	7.555	9.96	2.355	6253
D9S	1.073	0.831	4.426	4.914	5.254	7.60	2.124	16277

Note. Mold temperature: 325°C; *small: angle of variable angle mold is 0°.

(b) Eutectic Si particle characteristics observed in variable angle mold (large*) samples of A356 alloys containing La and Ce additions

| Alloy code | RE addition | | Silicon particle characteristics | | | | | |
| | Actual | | Area (μm^2) | | Length (μm) | | Aspect ratio | Density (number/mm^2) |
	La (wt.%)	Ce (wt.%)	Average	SD	Average	SD		
DBL	0	0	21.45	29.53	9.467	9.397	2.586	35521
D10L	0.165	0	28.48	36.16	19.38	10.230	2.275	2349
D1L	0.356	0	18.24	28.87	11.42	9.498	2.237	5521
D2L	0.685	0	24.43	37.46	15.3	10.95	2.287	3754
D3L	1.025	0	19.59	13.96	9.374	5.804	2.195	9446
D11L	0.032	0.082	16.37	23.23	12.42	8.741	2.600	7352
D4L	0.136	0.185	16.19	25.41	10.23	8.960	2.499	5924
D5L	0.046	0.317	16.09	28.03	8.752	9.570	2.367	5668
D6L	0.089	1.088	18.66	31.05	9.310	10.410	2.305	5236
D7L	0.44	0.282	11.93	18.95	12.01	7.605	2.343	7515
D8L	0.781	0.877	15.95	24.51	13.200	8.162	2.332	4122
D9L	1.073	0.831	8.20	12.82	5.880	5.416	2.185	12862

Note. Mold temperature: 325°C; *large: angle of variable angle mold is 15°.

(c) Characteristics of eutectic Si particles in variable angle mold (large*) samples of Sr-modified A356 alloys containing La and Ce additions

| Alloy code | Addition | | | Silicon particle characteristics | | | | | |
| | | Actual | | Area (μm^2) | | Length (μm) | | Aspect ratio | Density (number/mm^2) |
	Sr (ppm)	La (wt.%)	Ce (wt.%)	Average	SD	Average	SD		
DBLS	127	0	0	1.765	3.906	2.161	2.431	2.061	64199
D10LS	69	0.165	0	2.138	3.930	2.345	2.378	1.952	58454
D1LS	109	0.356	0	2.385	5.157	2.494	2.958	1.974	42526
D2LS	74	0.685	0	4.240	2.470	2.625	8.348	2.370	5374
D3LS	74	1.025	0	5.775	2.633	3.815	4.688	2.262	18163
D11LS	73	0.033	0.082	3.549	6.440	3.207	3.862	2.114	38504
D4LS	70	0.136	0.185	1.876	3.542	2.165	2.053	1.937	55283
D5LS	71	0.046	0.317	2.524	4.593	2.565	2.854	1.906	49097
D6LS	77	0.089	1.088	1.885	4.197	2.155	2.435	1.927	61875
D7LS	75	0.44	0.282	3.119	5.461	2.931	3.235	1.927	36082
D8LS	78	0.781	0.877	7.421	12.000	4.937	5.439	2.184	17488
D9LS	72	1.073	0.831	4.622	7.183	3.840	4.001	2.151	26061

Note. Mold temperature: 325°C; *large: angle of variable angle mold is 15°.

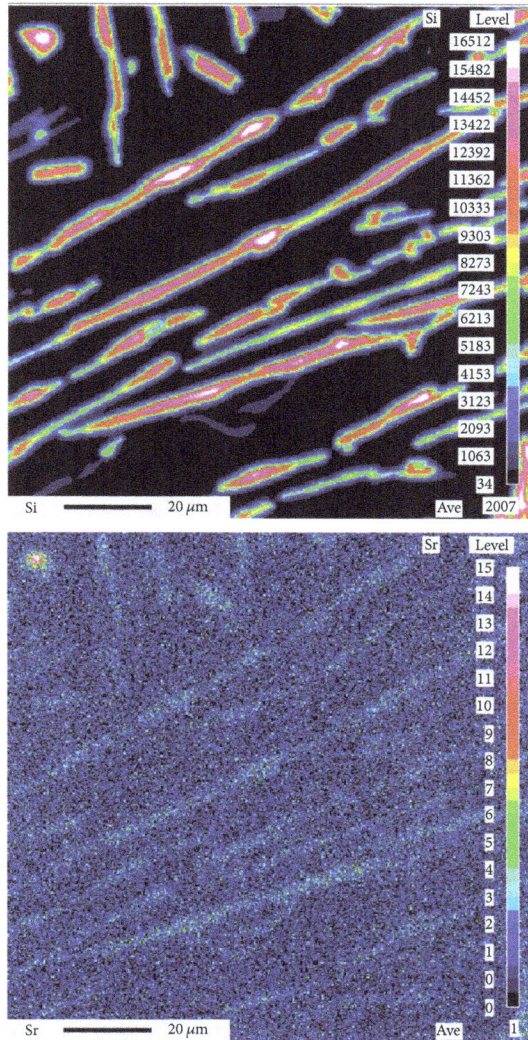

FIGURE 10: Distribution of Sr in the modified A356 alloy.

addition of RE exhibited a tendency for Si modification as may be seen from Table 4(b). As in the previous case (zero tilting angle), the standard deviation is also high in this case. It should be noted that Ce is relatively more effective than La as an alternate modification agent. Figures 9(a)–9(f) show examples of the eutectic Si morphology in A356 alloy samples treated with different amounts of RE, without (a, c, d) and with (b, e, f) Sr addition. It should be mentioned here that, according to the atomic radius ratio, r_{La}/r_{Si} is 1.604 and r_{Ce}/r_{Si} is 1.559, which theoretically confirms the abovementioned observation [23].

Based on the data presented in Tables 4(a) and 4(b), the observed depression in the eutectic temperature is not necessarily related to modification of the eutectic Si particles. In confirmation of this conclusion, Table 4(c) lists the variation in the eutectic Si particles when the alloys were treated with 100 ppm Sr. It is obvious that Sr is the most effective agent for modification compared to the other RE elements analyzed in the present study from both size and distribution aspects, as represented by the standard deviation. Figure 10 shows the distribution of Sr within the eutectic Si particles in the DBLS alloy. Figures 11 and 12 reveal the distribution of La, Ce, and Sr in RE-rich platelets, which explains the partial modification of the surrounding Si particles as less Sr is available for modification of the eutectic Si. It may be noted from Figures 9 and 10 that the affinity of La to react with Sr is relatively higher than that of Ce.

3.2.2. Step-Like Mold. Since the modification effect of La is similar to that provided by Ce, only La will be considered in this section. Table 5(a) lists the characteristics of the eutectic Si particles as a function of solidification rate and the level of La addition. Apparently, the solidification rate is the main factor affecting modification compared to the added La content. With the increase in the SDAS from 16 μm to 41 μm, the average Si particle area increases from 3.7 μm^2 to 14 μm^2 with an increase in the La concentration to about 0.7 wt.%. It should also be noted that the corresponding standard

TABLE 5

(a) Silicon particle characteristics of La-containing A356 alloy samples obtained from step-like mold castings

Alloy code	Mold temp. (°C)	Mold section	RE addition Actual La (wt.%)	Area (μm^2) Average	SD	Length (μm) Average	SD	Aspect ratio	Density (number/mm^2)
12S			0	3.663	5.680	3.444	3.567	2.272	13495
22S			0.165	1.685	3.740	2.132	2.376	2.044	27050
32S	200	Small (SDAS: 16 μm)	0.356	1.364	3.825	1.898	2.229	2.029	38893
42S			0.685	1.464	3.665	1.952	2.119	2.022	33243
52S			1.025	1.506	4.175	1.879	2.306	1.978	35101
12L			0	7.876	12.290	5.591	5.970	2.548	6471
22L			0.165	6.763	11.410	4.743	5.394	2.230	7416
32L	200	Large (SDAS: 31 μm)	0.356	5.334	7.426	4.475	3.772	2.107	11179
42L			0.685	5.194	10.910	4.318	4.639	2.162	6644
52L			1.025	2.727	6.007	2.678	2.962	2.129	11145
14S			0	5.245	7.768	4.360	4.452	2.393	9361
24S			0.165	3.371	4.856	3.148	2.908	2.147	13055
34S	400	Small (SDAS: 22 μm)	0.356	5.152	8.234	4.119	4.040	2.129	10819
44S			0.685	4.517	7.080	3.671	3.844	2.092	13439
54S			1.025	2.680	5.351	2.648	3.153	2.034	25475
14L			0	15.800	24.850	8.113	9.591	2.565	3457
24L			0.165	13.830	21.330	7.083	8.016	2.463	4007
34L	400	Large (SDAS: 41 μm)	0.356	15.160	24.430	7.668	9.044	2.385	3555
44L			0.685	14.240	23.500	7.072	8.191	2.411	3788
54L			1.025	9.514	16.030	5.746	6.027	2.363	5904

Note. Small: sample sectioned from step of 5 mm height. Large: sample sectioned from step of 20 mm height. SD: standard deviation.

(b) Silicon particle characteristics of La-containing Sr-modified A356 alloy samples obtained from step-like mold castings

Alloy code	Mold temp. (°C)	Mold section	Composition Actual Sr (ppm)	La (wt.%)	Area (μm^2) Average	SD	Length (μm) Average	SD	Aspect ratio	Density (number/mm^2)
S12S			83.3	0	0.875	3.592	1.268	1.212	1.84	85236
S22S			95.0	0.165	0.860	4.796	1.235	1.964	1.92	57253
S32S	200	Small	84.1	0.356	0.669	3.963	1.108	1.330	1.81	90306
S42S			105.8	0.685	0.689	3.756	1.205	1.473	1.84	79410
S52S			108.3	1.025	0.853	4.438	1.246	1.685	1.85	74495
S12L			83.3	0	1.653	3.738	1.948	1.661	1.89	48426
S22L			95.0	0.165	1.606	4.175	1.897	1.976	1.96	33046
S32L	200	Large	84.1	0.356	1.401	3.898	1.763	1.820	1.95	39323
S42L			105.8	0.685	1.364	4.134	1.766	2.043	1.95	46551
S52L			108.3	1.025	1.406	4.180	1.853	2.229	2.02	37318
S14S			79.1	0	0.996	3.526	1.548	1.465	1.83	74841
S24S			73.1	0.165	1.020	3.681	1.389	1.431	1.81	65911
S34S	400	Small	77.0	0.356	0.899	3.619	1.409	1.485	1.88	75695
S44S			88.0	0.685	0.768	3.504	1.288	1.282	1.90	58987
S54S			89.5	1.025	1.042	5.243	1.539	2.948	1.91	65816
S14L			79.1	0	2.532	5.228	2.640	2.254	1.90	32512
S24L			73.1	0.165	2.262	5.044	2.245	2.223	1.91	34871
S34L	400	Large	77.0	0.356	1.679	3.691	2.003	1.896	1.96	43197
S44L			88.0	0.685	2.121	5.302	2.287	2.652	1.99	35591
S54L			89.5	1.025	2.471	3.718	2.600	1.748	2.06	30309

Note. Small: sample sectioned from step of 5 mm height. Large: sample sectioned from step of 20 mm height. SD: standard deviation.

(a)

(b)

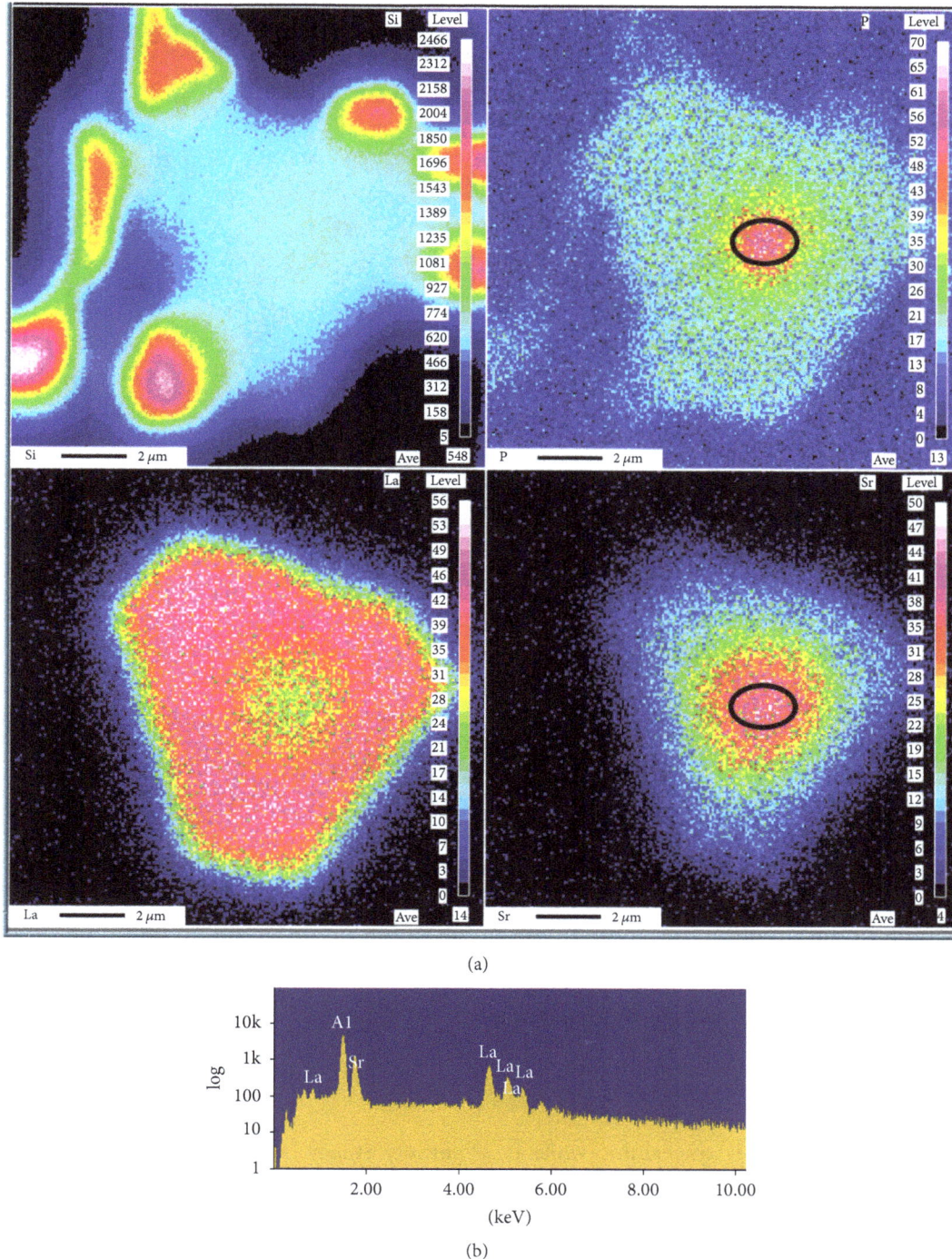

FIGURE 11: (a) Elemental distribution of Si, La, Sr, and P corresponding to a backscattered electron image showing La-rich platelet and surrounding Si particles in 100 ppm Sr-modified A356 alloy containing 1.025% wt.% La; (b) EDS spectrum of the La-rich platelet. Note the presence of P at the center of the Sr (circled areas).

deviation increased from ±5.7 to ±23.5, indicating a wide range of Si particle sizes and distributions. At ~1 wt.% La level, a tendency for modification is observed, as demonstrated by Figures 13(a) and 13(c). The variations in the Si particle characteristics in the same series of alloys with the addition of ~100 ppm Sr are shown in Table 5(b). The average Si particle area is about $1.5\,\mu m^2 \pm 3.6$ regardless of the solidification rate and La concentration as displayed in Figures 13(b) and 13(d).

3.3. Porosity. Liao et al. [24] studied the effect of Sr addition on porosity formation in directionally solidified A356 alloy. Their results showed that the growth rate of pores decreases

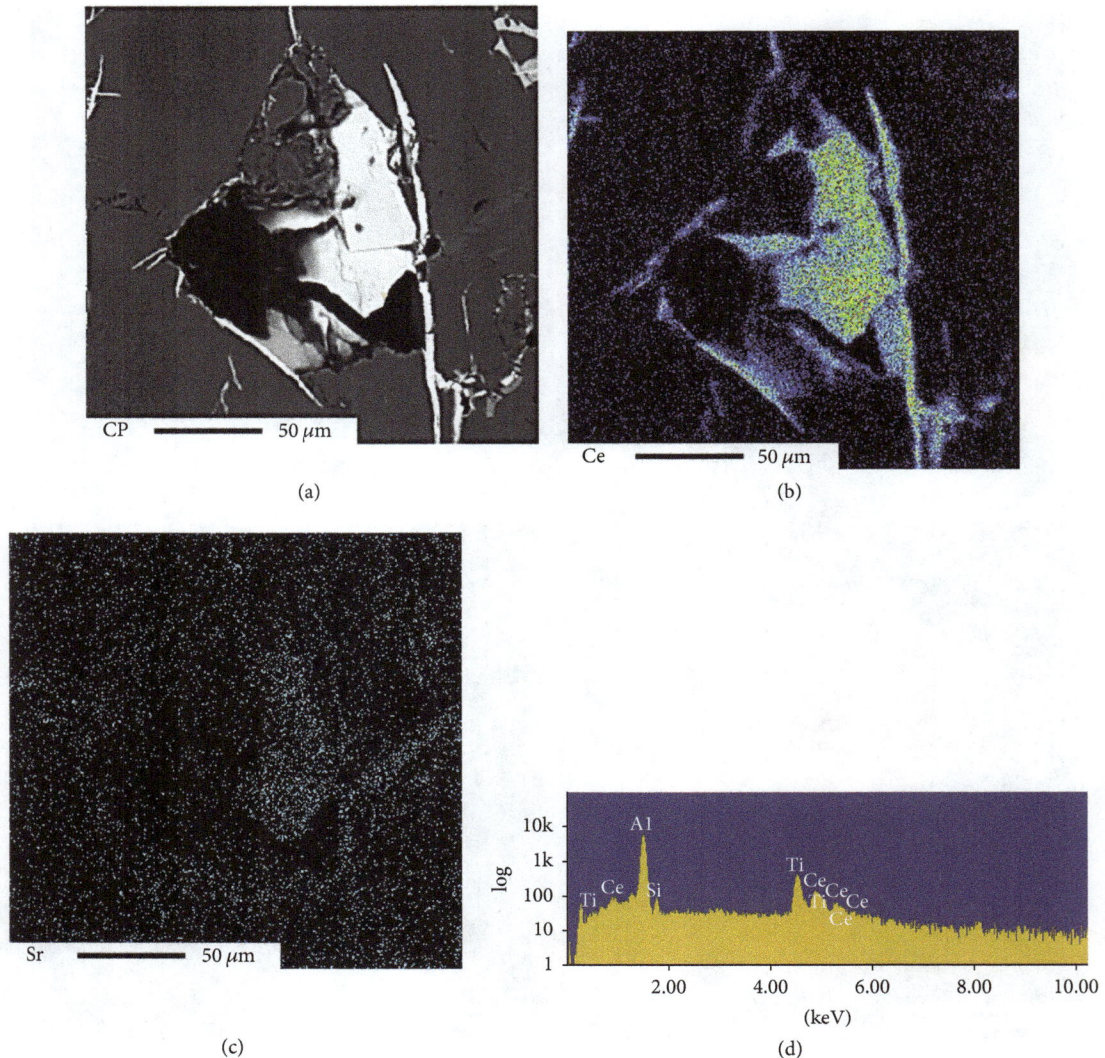

FIGURE 12: (a) Backscattered electron image showing Ce-rich platelets in Sr-modified A356 alloy containing 1.025% wt.% Ce and elemental distribution of (b) Ce, (c) Sr, and (d) EDS spectrum corresponding to (a).

with the reduction in local liquid temperature, while it fluctuates violently during directional solidification. Addition of Sr weakens this fluctuation and decreases the average growth rate of pores. Addition of Sr has a considerable influence on the size distribution of pores.

The modification of Al-Si casting alloys was analyzed by Sigworth [25]. Modification may change the relative formation of porosity and shrinkage in a casting. The modifiers strontium and sodium are poisoned by phosphorus, antimony, and bismuth. Consequently, the levels of these impurities should be monitored carefully in secondary alloys.

The effect of Sr content on porosity formation in a directionally solidified Al-12.3 wt.% Si alloy was investigated by Hengcheng et al. [26] who suggested that Sr solute in liquid Al-Si alloys can diffuse into oxide inclusions to form loose oxide aggregations which are more active nucleation sites for

porosity. Stunov [27] concluded that, in Al-Si-Mg alloys, Sr has a negative effect on the level of gas porosity and on the distribution of shrinkage porosity.

The effect of the metallurgical parameters on porosity formation examined in the present study was limited to the variable angle mold and the step-like mold. Figure 14 shows the general distribution of porosity in Sr-free degassed A356 alloy castings obtained from the variable angle mold heated at 350°C, using different angles. As can be seen, the high solidification rate associated with the thin plates of the zero-degree angle castings may lead to the formation of hot spots within the casting as denoted by the black circled areas.

Figures 15(a) and 15(b) show two examples of the porosity observed in samples of Sr-modified A356 alloys containing La obtained from the 15° variable angle mold. These pores are associated with thick oxide films (white arrows), as inferred

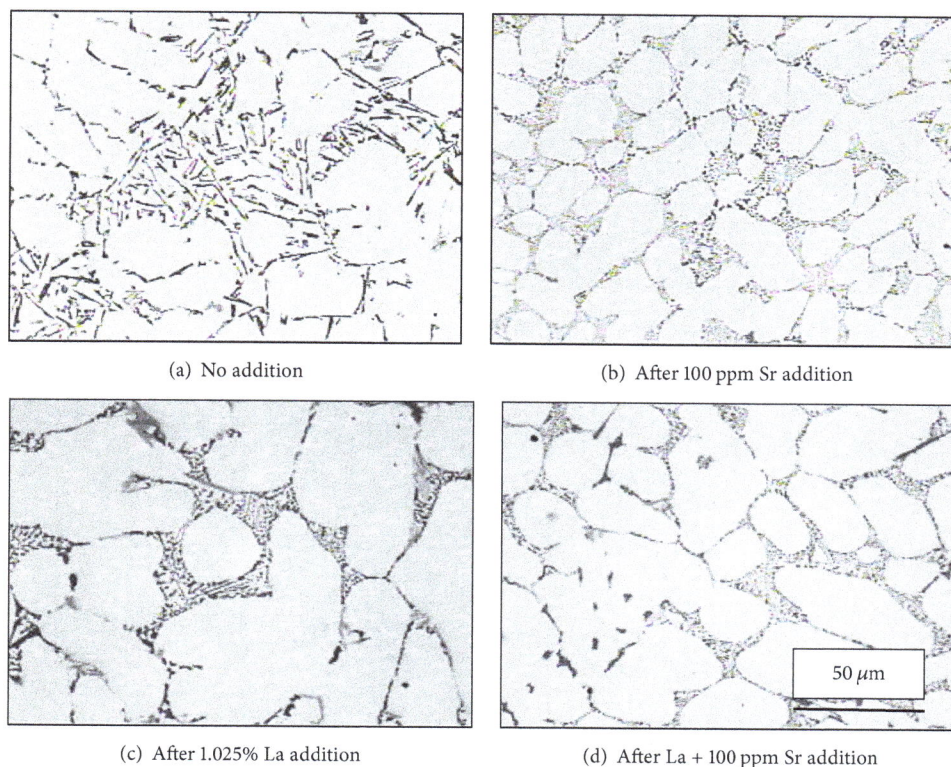

(a) No addition

(b) After 100 ppm Sr addition

(c) After 1.025% La addition

(d) After La + 100 ppm Sr addition

FIGURE 13: Examples of the eutectic Si particles observed in A356 alloy following addition of Sr, La, or Sr + La in samples obtained from the small section (200°C) of the step-like mold casting.

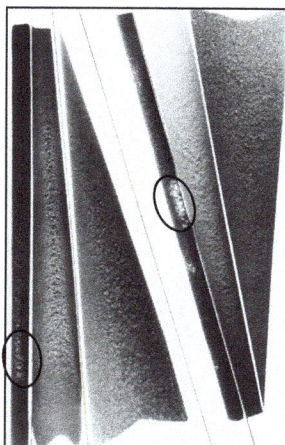

FIGURE 14: Porosity distribution in degassed A356 alloy castings obtained from the variable angle mold heated at 350°C.

from the X-ray images of oxygen and strontium distribution corresponding to Figure 15(b), as shown in Figures 15(c) and 15(d). It should be kept in mind that these X-ray images were taken from deep pores. Thus, the distribution may have been affected by the pore morphology. Another important observation noted from Figure 15 is that the fragmentation of the La-rich platelets appears to occur in a similar manner to that reported for the fragmentation of β-Al$_5$FeSi platelets [28].

Porosity measurements were carried out on nonmodified and Sr-modified La-containing A356 alloys as a function of La content, using samples obtained from the step-like mold castings. The obtained data is documented in Figure 16. Since the molten metal was properly degassed (humidity in the vicinity of the melt was about 13%), the amount of dissolved hydrogen in the molten alloy would be at its threshold value. Thus, porosity in this case would be caused mainly by

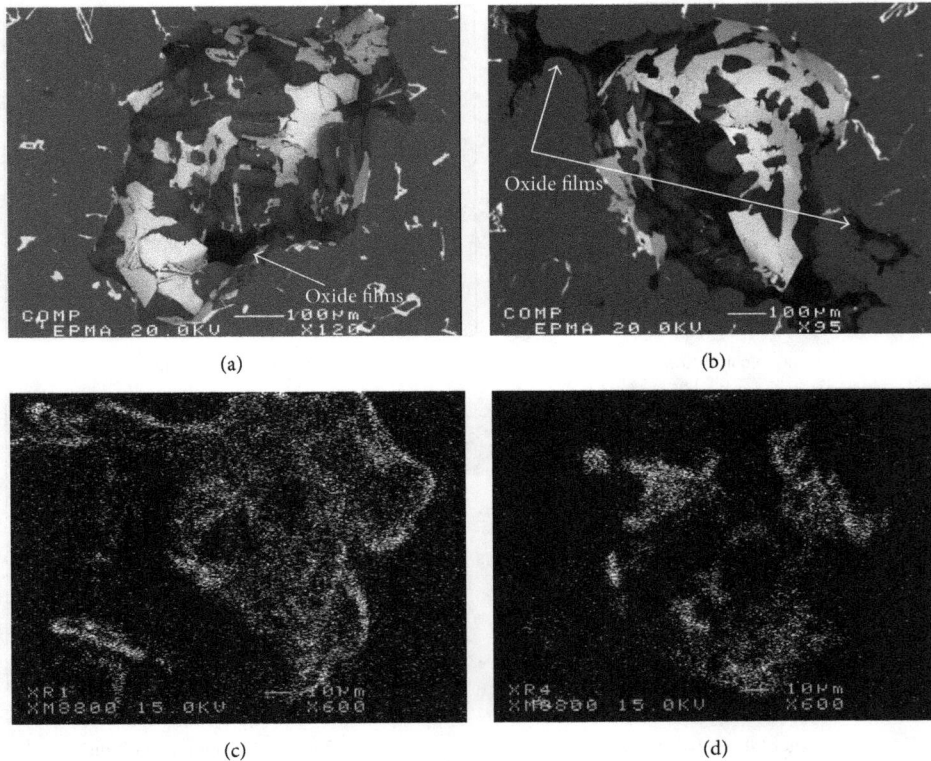

FIGURE 15: (a, b) Backscattered electron images showing porosity in Sr-modified A356 alloy samples containing (a) 0.8% La and (b) 1.025% La; note the presence of thick oxide films (white arrows) (sample SDAS ~ 42 μm). (c, d) X-ray images showing the distribution of (c) oxygen and (d) strontium in (b).

shrinkage due to volumetric change during solidification. Liu et al. [29] reported that Sr oxide is one of the main sources of porosity due to the high affinity of Sr to react with oxygen. As was observed from Figure 5, the addition of RE resulted in an increase in the freezing range so that poor feedability of the molten metal would be expected.

Figure 16(a) presents the variation in percentage porosity as a function of solidification rate and the levels of La and Sr present in the alloy, where "Small" and "Large" correspond to samples obtained from the small and large steps of the step-like mold casting (see Figure 1(c)), and the suffixes 200 and 400 indicate the two temperatures of the mold during casting. In the absence of Sr, the total percentage porosity is less than 0.2% which decreases to 0.1% with the increase in the amount of La to 1.025 wt.%.

With 100 ppm Sr addition, a noticeable increase in the percentage porosity is observed, especially at low solidification rate (Figure 16(a)). It should be noted also that increasing the La content has no specific effect on percentage porosity. Figure 16(b) illustrates the increase in pore density (measured by number of pores/mm^2) emphasizing the strong role of modification with Sr coupled with low solidification rate in the intensity of porosity in the final casting [30].

Figure 17 shows an example of the shape and size of the type of porosity observed in a 1.025 wt.% La-containing A356

alloy sample obtained from the large section of the step-like mold (mold heated at 400°C). The La-based intermetallic phase is seen to precipitate in the form of platelets, surrounded by partially modified eutectic Si. The EDS spectrum corresponding to the La-containing platelet is shown in Figure 17(b) indicating that it is composed of Al and La. The WDS analysis revealed that the composition of these platelets is Al$_3$La (77.4 at.% Al, 22.6 at.% La). Figure 18 shows the fragmentation of a La-rich platelet under the same casting conditions.

4. Conclusions

Based on an analysis of the results obtained in this study, the following conclusions may be drawn:

(1) Addition of La and Ce rare earth (RE) metals leads to an increase in the α-Al precipitation temperature, lowering at the same time the Al-Si eutectic temperature, resulting in long freezing temperature ranges and hence poor feedability.

(2) The changes in both α-Al and Si eutectic temperatures are independent of the undercooling.

(3) Addition of Sr to RE-treated alloys eliminates their effect on the microstructure during solidification.

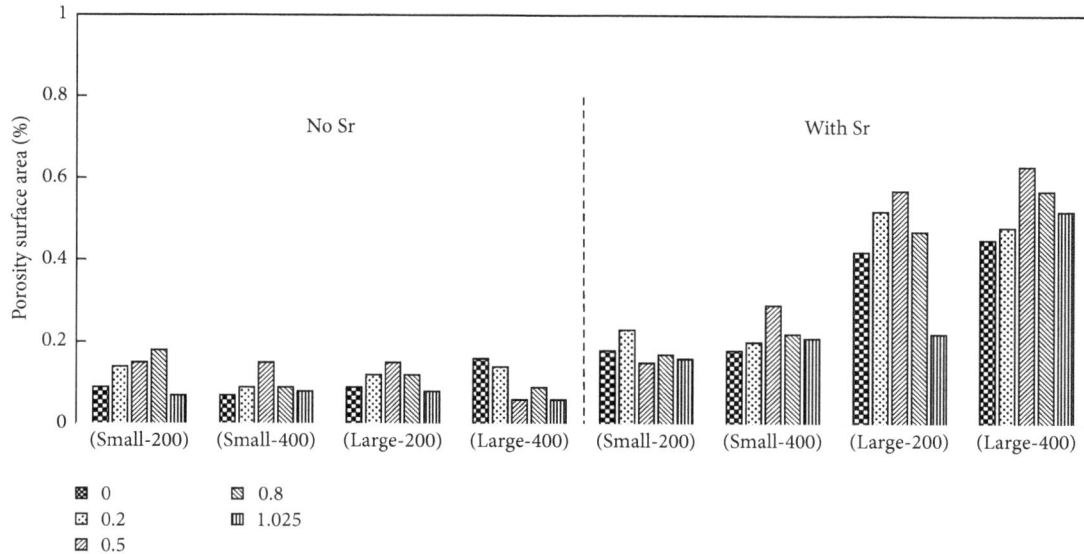

(a) Variation in surface area percentage porosity as a function of La content and casting/solidification condition of the A356 alloy samples (legend at the top center indicates the La level of the samples)

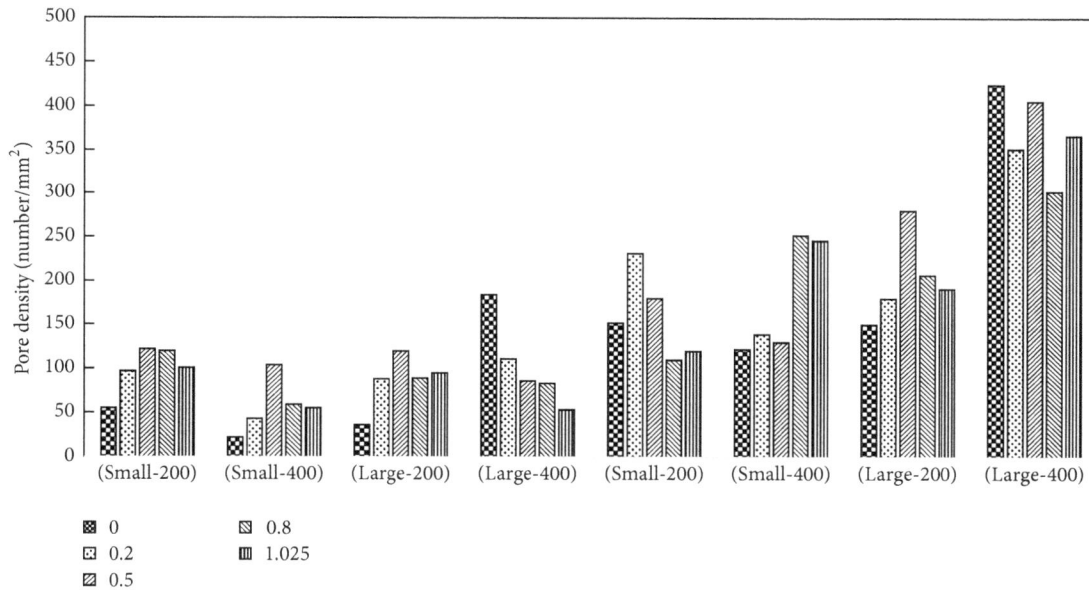

(b) Variation in pore density as a function of La content and casting/solidification condition of the A356 alloy samples (legend at the top center indicates the La level of the samples)

FIGURE 16

(4) Rare earth metals have a minor effect on the refinement of the eutectic silicon particles whereas Sr is the dominating modifying agent.

(5) According to the atomic radius ratio, r_{La}/r_{Si} is 1.604 and r_{Ce}/r_{Si} is 1.559, theoretically, which shows Ce to be a more effective modifier than La.

(6) Strontium has a strong affinity for reacting with rare earth metals, which results in reducing its effectiveness as a eutectic Si modifier.

(7) The presence of traces of P observed in the A356 alloy samples leads to a reaction between P and Sr in the Sr-modified alloys, forming a complex compound which acts as a nucleation site for the precipitation of RE-rich phases.

(8) Porosity occurs mainly due to shrinkage and oxide films. In the absence of Sr, addition of rare earth metals, especially at high concentration, reduces the percentage of shrinkage porosity.

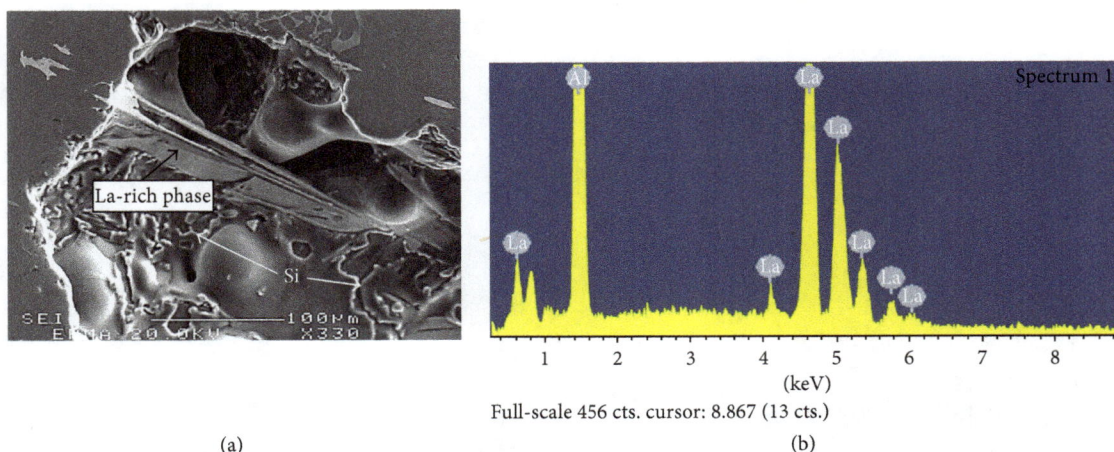

Full-scale 456 cts. cursor: 8.867 (13 cts.)

(a)   (b)

FIGURE 17: (a) Secondary electron image showing porosity in A356 alloy containing 1.025 wt.% La and no Sr (sample SDAS ~ 42 μm). (b) EDS spectrum corresponding to the La-rich phase platelet (black arrow) in (a).

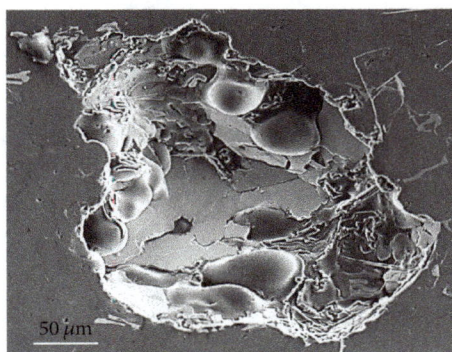

FIGURE 18: Secondary electron image of La-rich platelet within a pore in Sr-modified, high (1.025%) La-containing A356 alloy (sample SDAS ~ 42 μm).

(9) Addition of Sr causes fragmentation of the rare earth intermetallics similar to that observed for the β-Al_5FeSi iron intermetallic phase.

Acknowledgments

The authors would like to thank Amal Samuel and Hicham Farid for enhancing the quality of the images and figures presented in this article.

References

[1] R. Sharan and N. P. Saksena, "Rare earth additions as modifiers of aluminum silicon alloys," *AFS International Cast Metals Journal*, vol. 3, no. 1, pp. 29–33, 1978.

[2] Y.-C. Tsai, C.-Y. Chou, R.-R. Jeng, S.-L. Lee, and C.-K. Lin, "Effect of rare earth elements addition on microstructures and mechanical properties of A356 alloy," *International Journal of Cast Metals Research*, vol. 24, no. 2, pp. 83–87, 2011.

[3] A. M. A. A Mohamed, A. M. Samuel, F. H. Samuel, and A. M. A. Al-Ahmari, "Effects of rare earths on the As-Cast Microstructure of an Al-Si-Mg alloy," *AFS Transactions*, vol. 119, pp. 83–91, 2011.

[4] A. Saoudi, F. H. Samuel, A. M. Samuel, and H. W. Doty, "Influence of the addition of rare earth metals and of overheating on microstructure and mechanical properties of aluminium alloy A319," *Revue de Metallurgie. Cahiers D'Informations Techniques*, vol. 100, no. 12, pp. 1203–1212, 2003.

[5] C. B. Kim and R. W. Heine, "Fundamentals of modification in the aluminum-silicon system," *Journal of the Institute of Metals*, vol. 92, pp. 367–376, 1963-1964.

[6] E. Aguirre-De la Torre, R. Pérez-Bustamante, J. Camarillo-Cisneros, C. D. Gómez-Esparza, H. M. Medrano-Prieto, and R. Martínez-Sánchez, "Mechanical properties of the A356 aluminum alloy modified with La/Ce," *Journal of Rare Earths*, vol. 31, no. 8, pp. 811–816, 2013.

[7] K. Nogita, S. D. McDonald, and A. K. Dahle, "Eutectic modification of Al-Si alloys with rare earth metals," *Materials Transactions*, vol. 45, no. 2, pp. 323–326, 2004.

[8] Z.-W. Chen, X.-L. Hao, J. Zhao, and C.-Y. Ma, "Kinetic nucleation of primary α(Al) dendrites in Al–7%Si–Mg cast alloys with Ce and Sr additions," *Transactions of Nonferrous Metals Society of China*, vol. 23, no. 12, pp. 3561–3567, 2013.

[9] S. A. Alkahtani, E. M. Elgallad, M. M. Tash, A. M. Samuel, and F. H. Samuel, "Effect of rare earth metals on the microstructure of Al-Si based alloys," *Materials*, vol. 9, no. 1, article 45, 2016.

[10] O. El Sebaie, A. M. Samuel, F. H. Samuel, and H. W. Doty, "The effects of mischmetal, cooling rate and heat treatment on the eutectic Si particle characteristics of A319.1, A356.2 and A413.1 Al-Si casting alloys," *Materials Science and Engineering A*, vol. 480, no. 1-2, pp. 342–355, 2008.

[11] O. Elsebaie, F. H. Samuel, and S. Al Kahtani, "Intermetallic phases observed in nonmodified and Sr modified Al-Si cast alloys containing mischmetal," *International Journal of Cast Metals Research*, vol. 26, no. 1, pp. 1–15, 2013.

[12] G. H. Garza-Elizondo, A. M. Samuel, S. Valtierra, and F. H. Samuel, "Effect of Ni, Mn, Sc, and Zr addition on the tensile properties of 354-type alloys at ambient temperature," *International Journal of Metalcasting*, vol. 11, no. 3, pp. 396–412, 2017.

[13] R. Aparicio, G. Barrera, G. Trapaga, M. Ramirez-Argaez, and C. Gonzalez-Rivera, "Solidification kinetics of a near eutectic Al-Si alloy, unmodified and modified with Sr," *Metals and Materials International*, vol. 19, no. 4, pp. 707–715, 2013.

[14] L. Hengcheng, B. Juanjuan, Z. Min, D. Ke, J. Yunfeng, and C. Mingdong, "Effect of strontium and solidification rate on eutectic grain structure in an Al-13 wt% Si alloy," *China Foundry*, vol. 6, no. 3, pp. 226–231, 2009.

[15] K. Nogita, J. Drennan, and A. K. Dahle, "Evaluation of silicon twinning in hypo-eutectic Al-Si alloys," *Materials Transactions*, vol. 44, no. 4, pp. 625–628, 2003.

[16] D. Ferdian, J. Lacaze, I. Lizarralde, A. Niklas, and A. I. Fernández-Calvo, "Study of the effect of cooling rate on eutectic modification in A356 aluminium alloys," *Materials Science Forum*, vol. 765, pp. 130–134, 2013.

[17] L. F. Mondolfo, *Aluminum Alloys, Structure and Properties*, Butterworth, London, England, 1979.

[18] A. M. Samuel, G. H. Garza-Elizondo, H. W. Doty, and F. H. Samuel, "Role of modification and melt thermal treatment processes on the microstructure and tensile properties of Al-Si alloys," *Materials and Design*, vol. 80, pp. 99–108, 2015.

[19] A. M. Samuel, H. W. Doty, S. Valtierra, and F. H. Samuel, "Effect of Mg addition of microstructure of 319 type alloys," *International Journal of Cast Metals Research*, vol. 26, no. 6, pp. 354–363, 2013.

[20] A. Reis, Z. A. Xu, R. V. Toi et al., "Model for prediction of shrinkage defects in long and short freezing range materials," *International Journal of Cast Metals Research*, vol. 20, no. 3, pp. 171–175, 2007.

[21] S. Hegde and K. N. Prabhu, "Modification of eutectic silicon in Al-Si alloys," *Journal of Materials Science*, vol. 43, no. 9, pp. 3009–3027, 2008.

[22] E. Ogris, A. Wahlen, H. Lüchinger, and P. J. Uggowitzer, "On the silicon spheroidization in Al-Si alloys," *Journal of Light Metals*, vol. 2, no. 4, pp. 263–269, 2002.

[23] H. Qiu, H. Yan, and Z. Hu, "Modification of near-eutectic Al-Si alloys with rare earth element samarium," *Journal of Materials Research*, vol. 29, no. 11, pp. 1270–1277, 2014.

[24] H. C. Liao, W. Song, Q. G. Wang, L. Zhao, R. Fan, and F. Jia, "Effect of Sr addition on porosity formation in directionally solidified A356 alloy," *International Journal of Cast Metals Research*, vol. 26, no. 4, pp. 201–208, 2013.

[25] G. K. Sigworth, "The modification of Al-Si alloys: Important practical and theoretical aspects," *International Journal of Metalcasting*, vol. 2, no. 2, pp. 19–40, 2008.

[26] L. Hengcheng, W. Yuna, F. Ran, and W. Qigui, "Effect of Sr content on porosity formation in directionally solidified Al-12.3wt.%Si alloy," *China Foundry*, vol. 11, pp. 435–439, 2014.

[27] B. B. Stunov, "Strontium as a Structure Modifier for Non-binary AlSi Alloy," *ActaPolytechnica*, vol. 52, pp. 26–32, 2012.

[28] A. M. Samuel, H. W. Doty, S. Valtierra, and F. H. Samuel, "Metallographic Studies on the Intermetallic Phases in the Al-Si Near Eutectic and Eutectic Alloys," in *Proceedings of the 120th AFS Metalcasting Congress*, 12 pages, Minneapolis, Minn, USA, April 2016, Paper 16-095.

[29] L. Liu, A. M. Samuel, F. H. Samuel, H. W. Doty, and S. Valtierra, "Influence of oxides on porosity formation in Sr-treated Al-Si casting alloys," *Journal of Materials Science*, vol. 38, no. 6, pp. 1255–1267, 2003.

[30] Z. Ma, A. M. Samuel, F. H. Samuel, and H. W. Doty, "Effect of Fe content and cooling rate on the impact toughness of cast 319 and 356 aluminum alloys," *AFS Transactions*, vol. 111, pp. 255–265, 2003.

[31] L. Bäckerud, G. Chai, and J. Tamminen, *Solidification Characteristics of Aluminum Alloys, Vol. 2: Foundry Alloys*, vol. 2, AFS/Skanaluminium, Des Plaines, Illinois, USA, 1990.

3

Effect of Heat Treatment on the Microstructure, Phase Distribution, and Mechanical Properties of AlCoCuFeMnNi High Entropy Alloy

Gulhan Cakmak

Metallurgical and Materials Engineering, Muğla Sitki Kocman University, 48100 Muğla, Turkey

Correspondence should be addressed to Gulhan Cakmak; glhnckmk@gmail.com

Academic Editor: Patrice Berthod

The present paper reports the synthesis of AlCoCuFeMnNi high entropy alloy (HEA) with arc melting process. The as-cast alloy was heat treated at 900°C for 8 hours to investigate the effect of heat treatment on the structure and properties. Microstructural and mechanical properties of the alloy were analyzed together with the detailed phase analysis of the samples. The initially as-cast sample was composed of two separate phases with BCC and FCC structures having lattice parameters of 2.901 Å and 3.651 Å, respectively. The heat-treated alloy displays microsized rod-shaped precipitates both in the matrix and within the second phase. Rietveld refinement has shown that the structure was having three phases with lattice parameters of 2.901 Å (BCC), 3.605 Å (FCC1), and 3.667 Å (FCC2). The resulting phases and distribution of phases were also confirmed with the TEM methods. The alloys were characterized mechanically with the compression and hardness tests. The yield strength, compressive strength, and Vickers hardness of the as-cast alloy are 1317 ± 34 MPa, 1833 ± 45 MPa, and 448 ± 25 Hv, respectively. Heat treatment decreases the hardness values to 419 ± 26 Hv. The maximum compressive stress of the alloy increased to 2123 + 41 MPa while yield strength decreased to 1095 ± 45 with the treatment.

1. Introduction

Traditional alloy design approach is based on mixing of one or two primary elements with known properties and minor elements are added in order to improve their properties. According to conventional strategy for developing new alloy systems, multiprincipal elements can lead to the formation of intermetallic compounds and complex microstructures. Yeh et al. [1] and Cantor [2] have broken this thinking pattern with the development of high entropy alloys (HEA). High entropy alloys are the systems with usually more than five principle elements at levels of 5–35 at % [3–5]. Due to high configurational entropy of these systems, FCC and BCC solid solutions were generally formed rather than intermetallics and complex phases [6–8]. It has been shown that some alloy systems designed with HEA approach show excellent mechanical [8–10] and high-temperature properties [6] and wear [11] and high-temperature softening resistance [12] and

so forth and they are suitable for many applications. In many cases the best balance of properties may be achieved for a multiprincipal element alloy consisting of more than one solid solution phase [8, 13].

Within the high entropy alloy systems, AlCoCrFeNi alloy is the one which has extensively been studied [14, 15]. It is well established that the properties of the alloy system depend on the composition. In the current study, we modified this alloy system by replacing Cr with Cu and Mn was added as new alloying element. The system was synthesized with arc melting method and characterized with respect to crystal structure, phase stability, microstructure, and mechanical properties. The sample was heat treated at 900°C for 8 hours to see high-temperature properties. The resultant microstructures and phases were analyzed before and after heat treatment with the method of X-ray diffraction, scanning electron microscopy, transmission electron microscopy, compression test, and microhardness test.

2. Experimental Method

Equiatomic amount of high purity Al (99.9%), Co (99.9%), Cu (99.9%), Fe (99.9%), Ni (99.9%), and Mn (99.9%) was melted in an arc furnace under argon atmosphere. Approximately batches of 3 grams were melted in a water-cooled copper mold. The ingot was remelted three times in order to increase homogeneity. The specimen was evaluated at as-cast condition and heat treated at 900°C in an argon atmosphere for 8 hours and furnace cooled.

The phases are characterized with X-ray diffraction method using a Bruker D8 ADVANCE X-ray diffractometer (XRD) with copper target operated at 40 kv 30 mA. The lattice parameters and crystal structures as well as volume fractions were obtained from the Rietveld refinement of X-ray data with MAUD program [16]. Compositional and microstructural analysis were carried out using FEI Nova NanoSEM 430 equipped with EDS detector.

Thin-foil specimens were prepared by mechanical thinning followed by ion milling and were observed under a transmission electron microscope (200 kV TEM, JEM-2100F, JEOL, Tokyo, Japan).

Mechanical properties were evaluated by uniaxial compression tests on samples with 3 mm diameter and 4.5 mm length by using an Instron 5582 testing system with a strain rate of 10^{-3} s^{-1}. Three compression tests were performed to obtain average value. The hardness values were measured using 4.903 N load for 10 sec. The reported hardness value was an average of at least 10 measurements.

3. Results and Discussion

3.1. Phase Analysis. Figure 1(a) shows the X-ray diffraction (XRD) pattern of the as-cast sample with the aforementioned composition. Detailed Rietveld refinement of the XRD data shows that as-cast sample is composed of two solid solution phases with body centered cubic (BCC) and face centered cubic (FCC) crystal structure. The lattice parameters of the BCC solid solution and FCC solid solutions are found to be 2.901 Å and 3.651 Å, respectively. The relative fractions of these two phases are 65.66% (BCC) and 34.34% (FCC).

The alloy was heat treated at 900°C for 8 hours. After heat treatment, a second FCC phase with similar lattice parameter appeared on XRD data with BCC phase seen in as-cast sample remaining in the structure (Figure 1(b)). The lattice parameters of the two FCC phases are calculated to be 3.667 Å and 3.605 Å, while the lattice constant of BCC phase is determined to be 2.893 Å. The relative fractions of these phases are found to be 26.82%, 19.53%, and 53.65, respectively. The corresponding phases, their lattice parameters, and volume fractions of these two conditions have been listed in Table 1. Rietveld refined X-ray diffractogram of the heat-treated sample is given in Figure 2. It should be noted here that we see the transformation of some fraction of BCC and FCC phases to another FCC phase which we will call FCC-2 from now on.

3.2. Microstructure and Chemical Composition Analysis. Figure 3(a) presents the backscattered electron image of the as-cast alloy. It is seen from the image that the microstructure

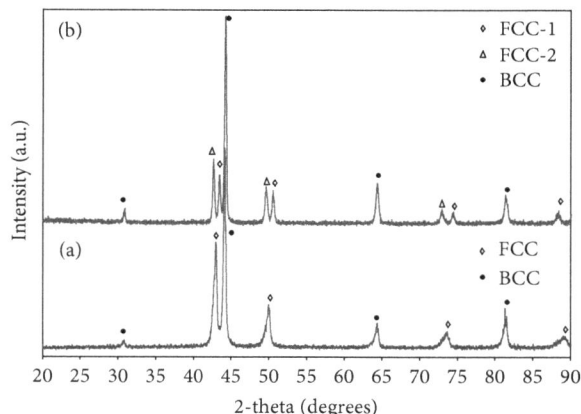

FIGURE 1: X-ray diffractograms for AlCoCuFeMnNi (a) before heat treatment and (b) after heat treatment.

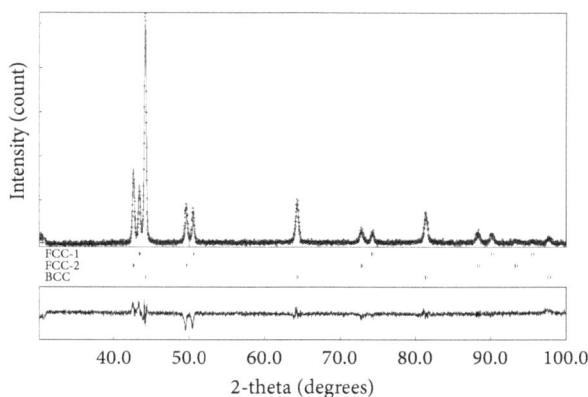

FIGURE 2: Rietveld refined X-ray diffractogram of heat-treated sample.

contains a main phase exhibiting relatively darker contrast and another phase with brighter contrast, which is consistent with XRD results. The overall atomic composition which is obtained from energy dispersive spectroscopy (EDS) analysis of the as-cast sample and point analysis of the darker and brighter phases is given in Table 2. It can be seen that the darker phase is enriched with Al while brighter regions are enriched with Fe and Cu.

Heat treatment of the sample leads to the formation of more complex microstructure, which is supported by X-ray diffraction analysis, Figure 3(b). A closer look to the microstructure reveals that rod-shaped precipitates are visible within both darker and brighter areas (refer to Figure 3(c)).

As it can be seen from the SEM image there seems to be four regions. These regions can be named as (1) the dark matrix phase (MP) and (2) bright precipitates within matrix phase (PM) and (3) bright secondary phase (SP) and (4) dark precipitates within secondary phase (PS). To understand the nature of these regions, point elemental analyses are applied to each region and these values are also included in Table 2. Although the resulting quantitative values have

TABLE 1: Rietveld Refinement results of the as-cast and heat-treated sample.

	Phase	Lattice parameter (Å)	Phase fractions (%)
As-cast	BCC	2.901	65.66
	FCC	3.651	34.34
Heat-treated	BCC	2.893	53.64
	FCC-1	3.667	26.83
	FCC-2	3.605	19.53

TABLE 2: Chemical composition of the as-cast and heat-treated alloys.

	Al	Mn	Fe	Co	Ni	Cu
As-cast						
Bright	12,04	17,28	20,32	16,40	13,94	20,04
Dark	21,49	15,36	15,12	16,07	17,35	14,61
Heat-treated						
SP	9,92	15,67	14,87	12,51	12,76	34,28
PS	19,12	15,53	17,12	16,00	16,41	15,82
PM	16,25	15,41	17,12	14,96	14,19	22,08
MP	21,78	15,28	17,07	18,03	18,25	9,60
Overall	17,92	16,70	16,19	15,87	16,27	17,07

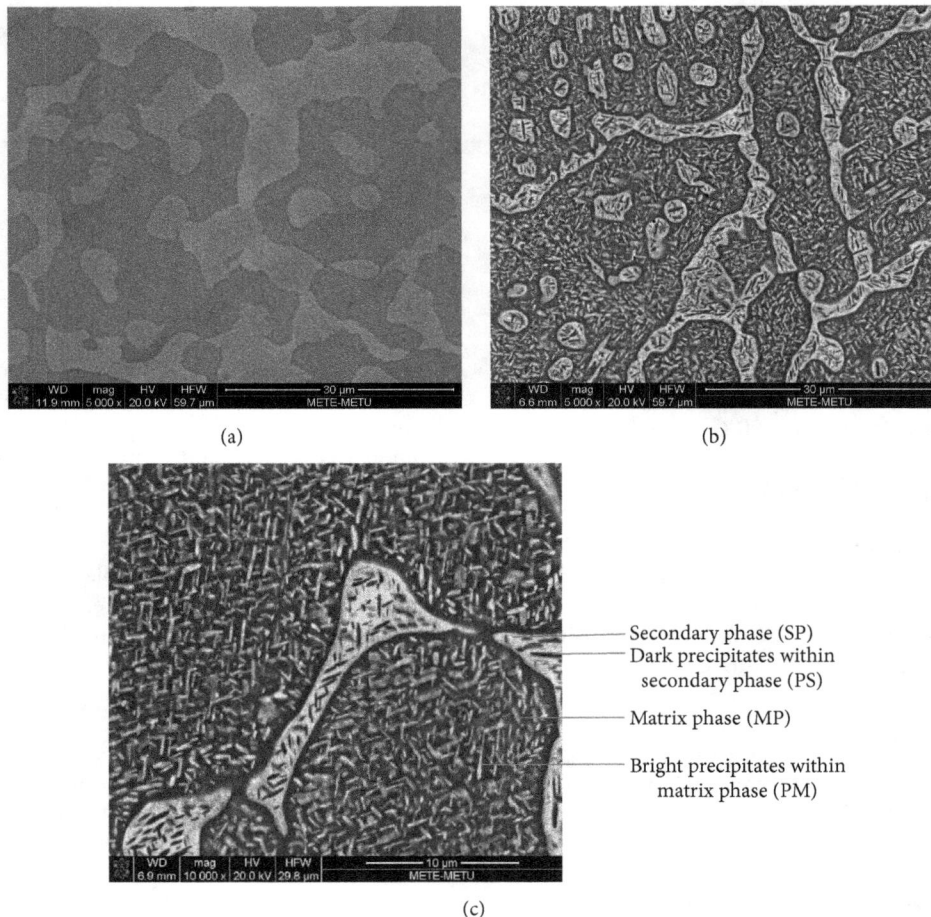

(a)

(b)

(c)

— Secondary phase (SP)
— Dark precipitates within secondary phase (PS)
— Matrix phase (MP)
— Bright precipitates within matrix phase (PM)

FIGURE 3: Backscattered SEM images of the (a) as-cast and (b) heat-treated sample. (c) A closer look to the heat-treated sample.

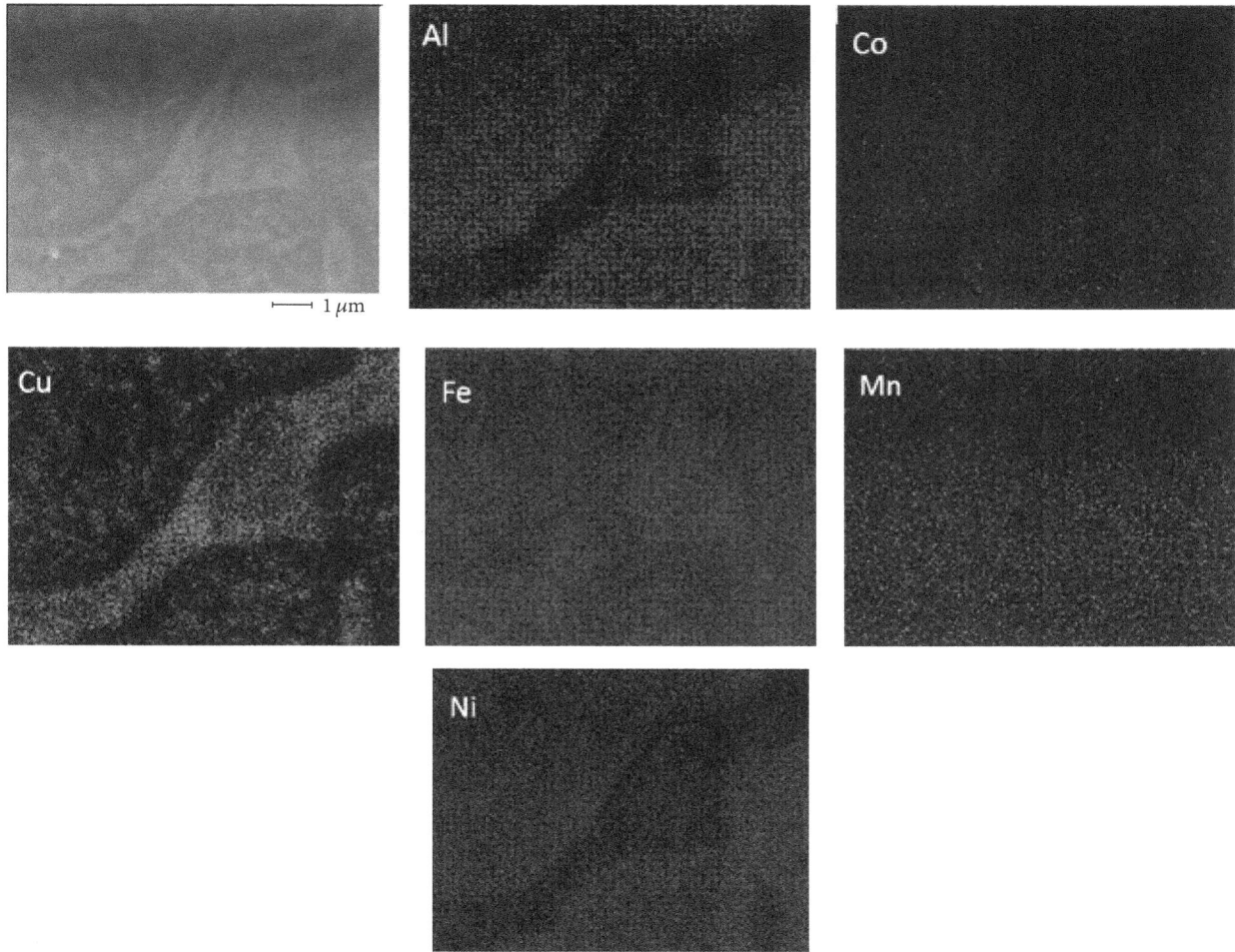

FIGURE 4: Elemental mapping of the heat-treated high entropy alloy.

restricted accuracy due to small sizes of phases, the values can give rough information related to compositional distribution among four regions. The values in Table 2 show that composition of the matrix phase (MP) is found to be rich in Al while bright secondary phase (SP) is rich in Cu.

In order to verify the compositional variation within the microstructure, the elemental mapping was carried out and the colored maps are given in Figure 4. It is seen that the matrix phase has more Al, Co, and Ni than the secondary phase which is rich in Cu. The needle-like precipitates within the matrix are enriched with Cu.

3.3. Transmission Electron Microscopy Analysis. The obtained phase distribution has further been verified with TEM analysis. The similar microstructure obtained from SEM analysis is observed in TEM. Bright field TEM image for the heat-treated sample is shown in Figure 5(a). The upper inset (b) shows diffraction pattern from matrix region (MP) and is found to have BCC structure with [102] BCC zone axis. Similarly lower inset (c) has been obtained from the precipitate in matrix (PM) and has FCC structure from [031] FCC zone axis. Crystal structure of the second phase is also determined using

selected area diffraction pattern. The diffraction patterns are obtained from the bright field image shown in Figure 5(d) and the patterns are indexed for the second-phase particle (upper inset, (e)) and for the precipitate in second phase (lower inset, (f)). The structures have been found to be FCC from [211] FCC zone axis and BCC from [311] BCC zone axis, respectively. To summarize TEM images and diffraction patterns, it can be said that the matrix phase (MP) and dark precipitates within secondary phase (PS) are found to be BCC with the same lattice parameter while secondary phase (SP) and bright precipitates within matrix phase (PM) are found to be FCC phases with two different but close lattice parameters.

3.4. Mechanical Analysis. Figure 6 shows the engineering stress-strain curve of heat treated AlCoCuFeMnNi HEA under compression at room temperature. The yield strength (σy) and compressive strength (σmax) values of AlCoCuFeMnNi HEA are 1095 ± 45 MPa and 2123 ± 41 MPa, respectively. The values before heat treatment were also shown in Figure 6 and measured to be 1317 ± 34 and 1833 ± 45 MPa, respectively. Increase in compression strength of the

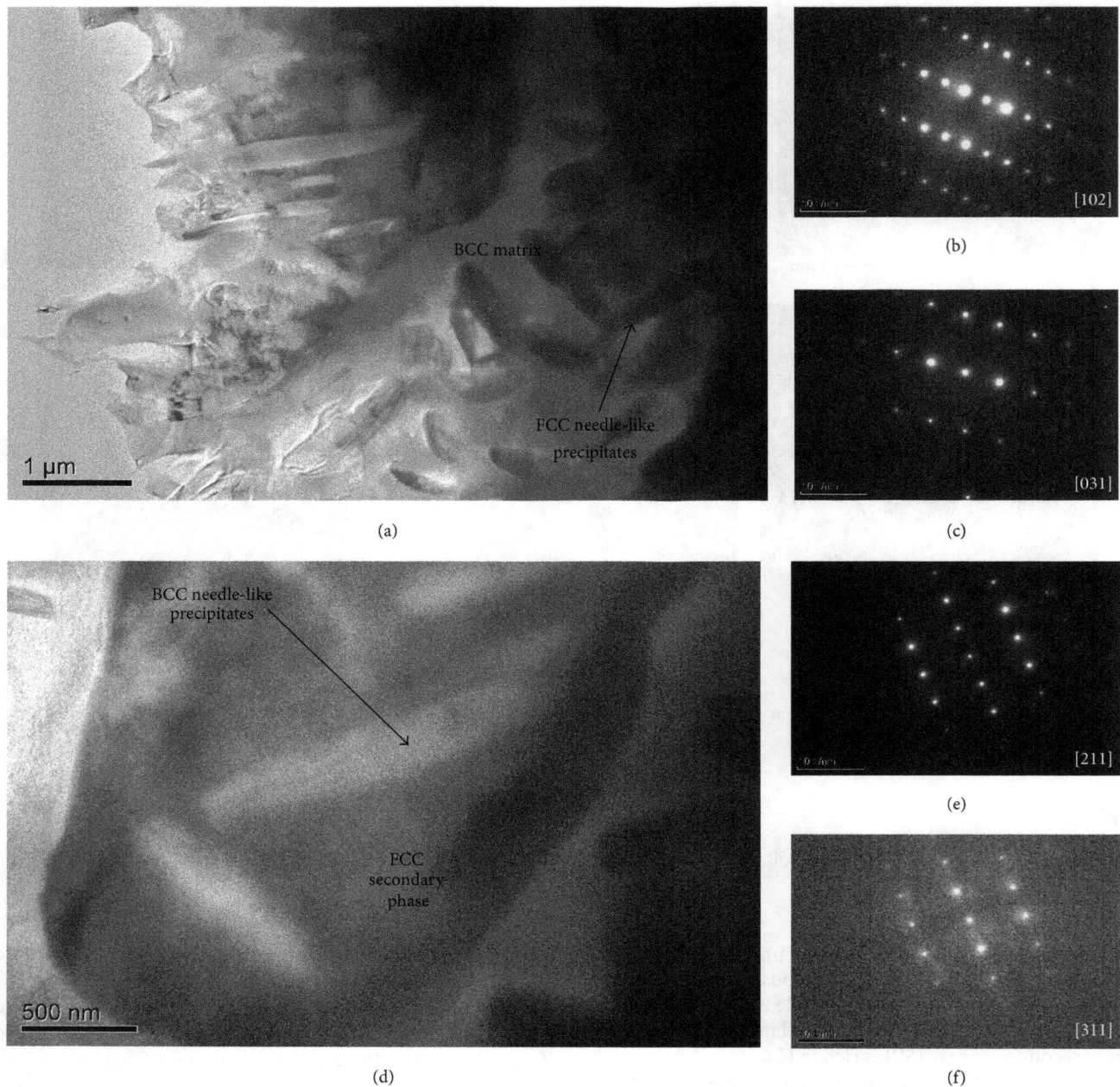

(a)

(b)

(c)

(d)

(e)

(f)

FIGURE 5: TEM image and corresponding SAED patterns of bulk AlCoCuFeMnNi HEA after heat treatment. (a) Bright field image showing the primary and secondary phases. (b) SAED pattern of the matrix from $[102]_{BCC}$ zone axis; (c) SAED pattern of FCC needle-like precipitates from $[031]_{FCC}$ zone axis; (d) bright field image showing the second phase and needle-like precipitates within; (e) SAED pattern of the second phase particle from [211] FCC zone axis; (f) SAED pattern of the needle-like precipitates within secondary phase from $[311]_{BCC}$ zone axis.

material can be related to the small sized precipitates within the heat-treated sample [17].

The microhardness value of the AlCoCuFeMnNi alloy was 448 ± 25 Hv while with heat treatment it was measured to be 419 ± 26 Hv. A decrease was observed in the microhardness values with heat treatment since there is a decrease in the volume fraction of BCC phase, which exhibits higher hardness/strength than that of the FCC phases, consistent with what has been previously reported [17–19].

4. Conclusion

In the current study, we investigate the effect of heat treatment on the microstructure and phase distribution and on the mechanical properties of the alloy. We observe that two phase structures which were initially FCC and BCC structure were converted into three phase structures with one BCC and two FCC structures with heat treatment. The SEM micrographs show that the phases of HEA exhibit rod-shaped precipitates

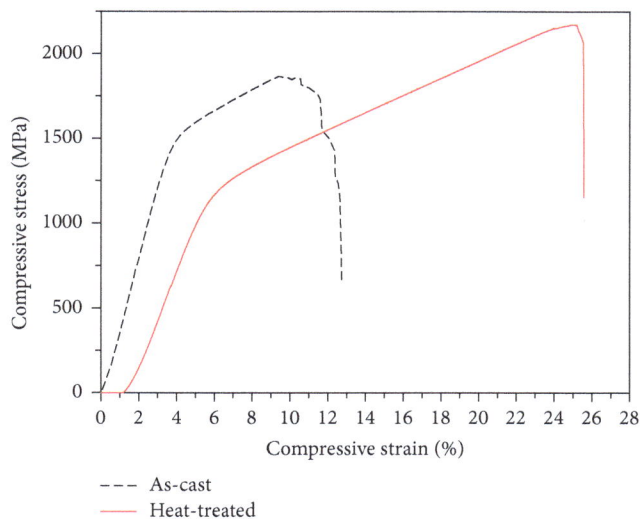

FIGURE 6: Engineering stress-strain curve of bulk AlCoCuFeMnNi HEA under compression.

within both the matrix and second phase. TEM analyses verify that these three phases were distributed as the FCC precipitates (PM) in BCC matrix (MP) and as BCC precipitates (PS) in FCC second phase (SP). These two FCC phases have close lattice parameters as 3.605 Å and 3.667 Å. This change in microstructure and phase constitution was accompanied by a corresponding progressive decrease in microhardness with decreasing BCC content, clearly indicating that the BCC microstructure is substantially harder than the FCC microstructure. The corresponding increase on mechanical strength value was thought to be related to small sized precipitates within the structure.

Conflicts of Interest

The author declares that there are no conflicts of interest regarding the publication of this paper.

References

[1] J.-W. Yeh, S.-K. Chen, S.-J. Lin et al., "Nanostructured high-entropy alloys with multiple principal elements: novel alloy design concepts and outcomes," *Advanced Engineering Materials*, vol. 6, no. 5, pp. 299–303, 2004.

[2] B. Cantor, "Multicomponent and high entropy alloys," *Entropy*, vol. 16, no. 9, pp. 4749–4768, 2014.

[3] X. F. Wang, Y. Zhang, Y. Qiao, and G. L. Chen, "Novel microstructure and properties of multicomponent CoCrCuFeNiTix alloys," *Intermetallics*, vol. 15, no. 3, pp. 357–362, 2007.

[4] J.-W. Yeh, S.-J. Lin, T.-S. Chin et al., "Formation of simple crystal structures in Cu-Co-Ni-Cr-Al-Fe-Ti-V alloys with multiprincipal metallic elements," *Metallurgical and Materials Transactions A*, vol. 35, no. 8, pp. 2533–2536, 2004.

[5] C.-W. Tsai, M.-H. Tsai, J.-W. Yeh, and C.-C. Yang, "Effect of temperature on mechanical properties of $Al_{0.5}$CoCrCuFeNi

wrought alloy," *Journal of Alloys and Compounds*, vol. 490, no. 1-2, pp. 160–165, 2010.

[6] S. Varalakshmi, M. Kamaraj, and B. S. Murty, "Processing and properties of nanocrystalline CuNiCoZnAlTi high entropy alloys by mechanical alloying," *Materials Science and Engineering A*, vol. 527, no. 4-5, pp. 1027–1030, 2010.

[7] C.-J. Tong, M.-R. Chen, J.-W. Yeh et al., "Mechanical performance of the Al_xCoCrCuFeNi high-entropy alloy system with multiprincipal elements," *Metallurgical and Materials Transactions A*, vol. 36, no. 5, pp. 1263–1271, 2005.

[8] Y. J. Zhou, Y. Zhang, Y. L. Wang, and G. L. Chen, "Solid solution alloys of AlCoCrFeNiTi$_x$ with excellent room-temperature mechanical properties," *Applied Physics Letters*, vol. 90, no. 18, p. 181904, 2007.

[9] W.-R. Wang, W.-L. Wang, S.-C. Wang, Y.-C. Tsai, C.-H. Lai, and J.-W. Yeh, "Effects of Al addition on the microstructure and mechanical property of Al xCoCrFeNi high-entropy alloys," *Intermetallics*, vol. 26, pp. 44–51, 2012.

[10] X. Ye, M. Ma, W. Liu et al., "Synthesis and Characterization of High-Entropy Alloy Al_xFeCoNiCuCr by Laser Cladding," *Advances in Materials Science and Engineering*, vol. 2011, Article ID 485942, 7 pages, 2011.

[11] P.-K. Hung, J.-W. Yeh, T.-T. Shun, and S.-K. Chen, "Multi-principal-element alloys with improved oxidation and wear resistance for thermal spray coating," *Advanced Engineering Materials*, vol. 6, no. 1-2, pp. 74–78, 2004.

[12] C.-Y. Hsu, C.-C. Juan, W.-R. Wang, T.-S. Sheu, J.-W. Yeh, and S.-K. Chen, "On the superior hot hardness and softening resistance of AlCoCrxFeMo0.5Ni high-entropy alloys," *Materials Science and Engineering A*, vol. 528, no. 10-11, pp. 3581–3588, 2011.

[13] D. B. Miracle, J. D. Miller, O. N. Senkov, C. Woodward, M. D. Uchic, and J. Tiley, "Exploration and development of high entropy alloys for structural applications," *Entropy*, vol. 16, no. 1, pp. 494–525, 2014.

[14] H.-P. Chou, Y.-S. Chang, S.-K. Chen, and J.-W. Yeh, "Microstructure, thermophysical and electrical properties in Al_xCoCrFeNi (0≤x≤2) high-entropy alloys," *Materials Science and Engineering: B*, vol. 163, no. 3, pp. 184–189, 2009.

[15] Y. P. Wang, B. S. Li, M. X. Ren, C. Yang, and H. Z. Fu, "Microstructure and compressive properties of AlCrFeCoNi high entropy alloy," *Materials Science and Engineering: A*, vol. 491, no. 1-2, pp. 154–158, 2008.

[16] L. Lutterotti, S. Matthies, and H. R. Wenk, Int Union Crystallogr Comm Powder Diffr Newslett 1999, 21, p. 14 MAUD Version 2.072, 2008 http://maud.radiographema.eu/.

[17] T. Borkar, B. Gwalani, D. Choudhuri et al., "A combinatorial assessment of AlxCrCuFeNi2 (0 < x < 1.5) complex concentrated alloys: Microstructure, microhardness, and magnetic properties," *Acta Materialia*, vol. 116, pp. 63–76, 2016.

[18] J.-M. Wua, S.-J. Lin, J.-W. Yeh, S.-K. Chen, Y.-S. Huang, and H.-C. Chen, "Adhesive wear behavior of Al_xCoCrCuFeNi high-entropy alloys as a function of aluminum content," *Wear*, vol. 261, no. 5-6, pp. 513–519, 2006.

[19] M.-H. Tsai and J.-W. Yeh, "High-entropy alloys: a critical review," *Materials Research Letters*, vol. 2, no. 3, pp. 107–123, 2014.

4

Microstructure and Mechanical Properties of Electron Beam-Welded Joints of Titanium TC4 (Ti-6Al-4V) and Kovar (Fe-29Ni-17Co) Alloys with Cu/Nb Multi-Interlayer

Yong-jian Fang,[1] Xiao-song Jiang (iD),[1] De-feng Mo,[2] Ting-feng Song,[1] Zhen-yi Shao,[1,3] De-gui Zhu,[1] Ming-hao Zhu,[1] and Zhi-ping Luo (iD)[4]

[1]School of Materials Science and Engineering, Southwest Jiaotong University, Chengdu, Sichuan 610031, China
[2]Key Laboratory of Infrared Imaging Materials and Detectors, Shanghai Institute of Technical Physics, Chinese Academy of Sciences, Shanghai 200083, China
[3]Department of Material Engineering, Chengdu Technological University, Chengdu, Sichuan 611730, China
[4]Department of Chemistry and Physics, Fayetteville State University, Fayetteville, NC 28301, USA

Correspondence should be addressed to Xiao-song Jiang; xsjiang@home.swjtu.edu.cn and Zhi-ping Luo; zluo@uncfsu.edu

Academic Editor: Donato Sorgente

Electron beam welding of a titanium alloy (Ti-6Al-4V) and a kovar alloy (Fe-29Ni-17Co) was performed by using a Cu/Nb multi-interlayer between them. Microstructure and composition of welded joints were analyzed by means of optical microscopy, scanning electron microscopy, energy dispersive spectroscopy, and X-ray diffraction. Mechanical properties of welded joints were evaluated by microhardness and tensile strength tests. Results indicated that in case of 0.22 mm thickness of Nb foil, microstructure of the titanium alloy side was mainly composed of Ti solid solution and some intermetallic compounds such as FeTi and $CuTi_2$, whereas in case of 0.40 mm thickness of Nb foil, the appearance of weld was more uniform and hardness of the weld zone decreased sharply. However, tensile strength of welded joints was increased from 88.1 MPa for the 0.22 mm Nb foil to 150 MPa for the 0.40 mm Nb foil. It was found that thicker Nb foil could inhibit diffusion of Fe atoms towards the titanium alloy side, thus promoting the formation of Ti solid solution and a small amount of $CuTi_2$ and eliminating FeTi. In addition, in both cases, $Cu_{0.5}Fe_{0.5}Ti$ was found in the fusion zone of the titanium alloy side, which had an adverse effect on mechanical properties of welded joints.

1. Introduction

Titanium alloys are widely used in nuclear, aerospace, and chemical industries because of their high corrosion resistance, light weight, high specific strength, and other high-temperature properties [1]. It is known that kovar alloys have similar coefficient of thermal expansion (CTE) with hard glass (borosilicate) and are highly applied in the field of electronic industry [2, 3]. Joining titanium and kovar alloys can achieve a combination of their advantages to improve performance of products to meet the needs for aerospace

and electronic packaging applications [4, 5]. In previous studies, friction welding [6], diffusion welding [7], and laser welding [8] were used to weld titanium alloys and iron-base alloys intimately without interlayers between them. Kundu et al. [9] showed that when titanium was diffusion bonded to a 17-4PH stainless steel, FeTi and Fe_2Ti were produced. If friction stir welding of titanium and steel was carried out in an overlapped manner, the interface exhibited a layer-shaped structure containing FeTi and Fe_2Ti, which were unfavorable to the quality of welded joints [10]. Satoh et al. [4] found that coarse intermetallic TiFe dendrites formed in the main fusion zone, which was adverse to properties of welded joints. Chen et al. [11] joined 201 stainless steel and TC4 titanium alloy by laser welding. It was found that if the laser beam was offset on the side of titanium alloy, more intermetallic compounds (IMCs) and cracks were produced. According to these research results, it appears that the challenge in connecting titanium and kovar alloys is the formation of brittle IMCs (Fe_2Ti and FeTi).

In order to avoid the production of excessive brittle IMCs, it is desirable to control the mixing of chemical components by adding a metal interlayer. Pardal et al. [12] found that the generation of IMCs could not be avoided when titanium and 316L stainless steel were joined by using cold metal transfer welding technology with CuSi-3 filler. However, IMCs were produced, which had lower hardness than Fe-Ti IMCs. Özdemir and Bilgin [13] prepared solid-state diffusion-bonded joints of a TC4 titanium alloy and a stainless steel by using copper foil as an interlayer. It was found that the increase of the strength of welded joints was attributed to the formation of Cu-Ti and Cu-Ti-Fe IMCs, which had a lower hardness than Fe-Ti IMCs. Tashi et al. [14] investigated the diffusion bonding process of a titanium alloy and a stainless steel using an Ag-based alloy interlayer. It was found that the toughness of welded joints was improved because of the production of TiAg. Similarly, TiAg was also produced at the interface during diffusion welding of a TC4 titanium alloy and a stainless steel using an Ag interlayer [15]. Yildiz et al. [16] studied the joint performance of diffusion-bonded pure titanium onto a ferritic stainless steel by inserting a Ni interlayer, and Ti-Ni IMCs were found in the joint. In addition, Oliveira et al. [17] showed that if TC4 titanium alloy was laser welded to NiTi alloy with Nb interlayer, the (Ti, Nb) solid solution was produced in the titanium alloy side and no brittle IMCs were observed, which were beneficial to the increase of the quality of welded joint. Similarly, Zhou et al. [18] applied the laser welding technique to join the TC4 titanium alloy and NiTiNb alloy by adding Nb filler wire. It was found that the addition of a Nb filler wire could inhibit the amount of IMCs, such as NiTi and Ti_2Ni, and all the welds are composed of a dendritic (Nb, Ti) solid solution and some IMCs.

Technically, it is viable to connect these materials using electron beam welding, which possesses certain advantages such as high energy density, accurate control of the heating position, and production of a narrowed heat-affected area. In order to prevent dispersion of electron beam and oxidation of materials, the welding process is carried out in a vacuum environment [19, 20]. Hence, electron beam welding was used to connect titanium and iron-base alloys [21, 22]. Tomashchuk et al. [23] attempted to weld titanium alloy to 316L stainless steel with Cu filler by electron beam welding. They only found a small amount of Fe_2Ti at the interface between the fusion zone and stainless steel. Wang et al. [24] applied the electron beam welding technique to join a titanium alloy and a stainless steel by inserting a metallic foil such as V, Ni, Cu, and Ag. It was found that filler metals could inhibit the formation of Fe-Ti IMCs. Moreover, Wang et al. [25] investigated the electron beam welding process of dissimilar Ti/steel metals using a V/Cu

TABLE 1: Chemical composition of base metals (wt.%).

Alloy	Mn	Si	C	Fe	Co	Ni	Ti	Al	V	N	H	O
TC4	—	0.15	0.1	0.3	—	—	Bal	6	4	0.05	0.015	0.2
4J29	0.4	0.2	0.02	Bal	17.3	29	—	—	—	—	—	—

multi-interlayer, and found that the formation of Ti-Cu and Ti-Fe phases was evidently prevented so as to improve the quality of welded joints.

Developments of aerospace and electronic packaging industries have demanded for multifunctional components with high mechanical and physical properties and flexibilities in design while at a low cost. In this study, the TC4 titanium alloy (Ti-6Al-4V) and 4J29 kovar alloy (Fe-29Ni-17Co) which are used in aerospace and electronic packaging industries were diffusely bonded by an electron beam welding to investigate their interfacial microstructures and intermetallic compounds. Cu/Nb multi-interlayers was used to improve the welding quality. When other welding parameters were identical, effects of the thickness of Nb foil on the quality of welded joints could be determined. The influence of the thickness of Nb foil on weld appearance, microstructure, and mechanical properties of welded joints were examined, and the fracture and forming mechanism of the welded joints were analyzed. To the best of our knowledge, electron beam welding of dissimilar TC4 and 4J29 with Cu/Nb multi-interlayer has not been reported previously. Our work demonstrated that the electron beam welding method can be used effectively to control the formation of brittle IMCs, thus improving the joint quality and mechanical properties.

2. Experimental Procedure

2.1. Materials and Preparation. Materials used in this experiment were a TC4 titanium alloy and a 4J29 kovar alloy, which were machined into 50 mm × 30 mm × 2 mm plates. Their composition and some physical properties are listed in Tables 1 and 2, respectively. The interlayer used in the welding process was a composite of bilayers composed of a pure Cu foil and a pure Nb foil. Prior to welding, metal surfaces were ground, polished, washed with acetone and alcohol, and dried in air. Two kinds of Cu/Nb multi-interlayers were prepared. One was a composite of a 0.53 mm thick pure Cu foil and a 0.22 mm thick pure Nb foil, named as EBW-1 sample; and another one was a composite of a 0.53 mm thick pure Cu foil and a 0.40 mm thick pure Nb foil, named as EBW-2 sample.

2.2. Welding Method and Processing Parameters. In this experiment, an electron beam welding machine (Germany, PTR EBW 300/5-60 PLC) was used to join materials, and a schematic diagram of the electron beam welding is given in Figure 1. Firstly, the interface between Cu foil and kovar alloy was welded, and electron beam was shifted by 0.1-0.2 mm to the kovar alloy side to achieve the better bonding. Then, the electron beam was placed in the middle of the Nb foil. The welding parameters are shown in Table 3.

TABLE 2: Physical properties of base metals.

Alloy	Density (g·cm^{-3})	Melting point (°C)	Linear CTE (10^{-6} K^{-1})	Tensile strength at room temperature (MPa)
TC4	4.55	1,725	7.14	978
4J29	8.1	1,460	4.7	548

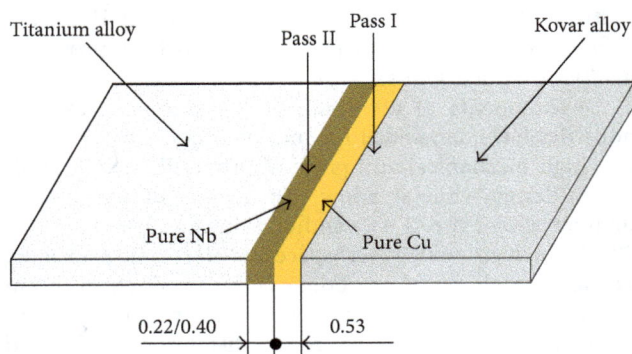

FIGURE 1: Schematic experimental conditions of dual-pass welding procedure.

2.3. Test Work. The overall fusion of welds was evaluated and analyzed. Specimens for microstructural observation were prepared and etched using two reagents. The cross sections of the titanium alloy side was corroded by a mixture acid solution (20 mL HNO$_3$ 20 mL HF, and 80 mL H$_2$O), and the kovar alloy side was corroded by the aqua regia (25 mL HNO$_3$ and 75 mL HCl). Microstructural observations of welded joints were investigated by an optical microscope (OM, Axio Cam MRc5) and a scanning electron microscope (SEM, QUANTA FEG 250). Spot and line scan analyses were used to evaluate distribution of elements by the SEM equipped with an X-ray energy-dispersive spectrometer (EDS). The micro-Vickers (HXD-100TM/LCD) hardness measurement was conducted across the weld zone under a test load of 100 gf and a dwell time of 15 s. The tensile test was performed by electronic universal testing machine (WDW-3100) at room temperature, and the displacement speed was 0.1 mm/min. At last, fracture morphology was observed and analyzed using SEM. Phase identifications were conducted at the fracture surface of tensile samples by micro X-ray diffraction (XRD, PANalytical-Empyrean). The operating current was 25 mA and the voltage was 50 kV using a Cu target. The scanning range was 30–90° (2θ) at a speed of 2°/min.

3. Results and Discussion

3.1. Weld Appearances. Figure 2 shows the surface appearance and the cross sections of welds using two Cu/Nb multi-interlayers with different Nb thickness. It is found that the thickness of Nb foil can alter the distribution and transfer of energy in the weld, affecting the welding quality. In both cases, it is found that the weld surfaces are flat and continuous without obvious defects such as pits, spatter, or cracks. However, the melting amount of metals on both sides is uneven, and the Cu zone appears as a convex shape. The energy absorbed by copper is too intense which results in

quick melting of metals. During the cooling process, the volume of solidified metal is too large to form a uniform weld. Wang et al. [25] investigated the quality of dissimilar metal joints of a titanium alloy and a stainless steel, obtained by electron beam welding. It was found that large energy input led to excessive expansion of metals, and cracks were formed during its cooling. The residual stress, generated at the interface, can result in a reduction of properties of the welded joints due to the different CTEs between copper and kovar alloy. In this work, as the melting volume of copper is too large, this effect is exacerbated. Moreover, the large melting volume may contribute to the formation of IMCs in the fusion zone to decrease properties of weld joints. Generally, the occurrence of the convex shape on copper should be avoided [26]. In case the thickness of Nb foil is 0.40 mm, the weld appearance is more uniform, and convex shape almost disappears, as shown in Figure 2(b). Therefore, a thicker Nb foil effectively improves the phenomenon of convex shape.

3.2. Microstructure of the Weld Zone of Welded Joints. Figures 3 and 4 show the cross sections of welds using two different Cu/Nb multi-interlayers. The cross sections of welds were taken from the location, which were indicated by white dashed lines, as shown in Figures 2(a) and 2(b). In both cases, melting of welded joints is relatively uniform, and there are no obvious defects. In case the thickness of Nb foil is 0.22 mm, compared with the kovar alloy side, the fusion zone of the titanium alloy side is narrower, existed with some unmelted interlayer areas, as shown in Figure 3(a). Previously, Gao et al. [10] welded titanium to a steel by using friction stir welding and found the existence of FeTi and Fe$_2$Ti. It was reported that Fe-Ti-Cu IMCs formed in the weld zone when electron beam welding of a titanium alloy and a stainless steel was carried out with a Cu interlayer [23]. Therefore, it is possible that Fe-Ti and Fe-Ti-Cu IMCs are formed in the weld zone of our sample. In case the thickness of Nb foil is 0.40 mm, a small fraction of Nb foil is melted and the melting mainly occurs on the right side of Nb, as shown in Figure 4(b). However, the fusion zone of the kovar alloy is fully penetrated by a large area of molten metals, and some unmelted Cu regions still exist due to the improper control of beam offset or the insufficient heat input, as shown in Figure 4(a). Tomashchuk et al. [23] reported a factor responsible for the melting of metals during electron beam welding. They found that the size and direction of beam offset can alter the melting of metals and the formation of IMCs. In both cases, it is found that the interlayers are insufficiently melted in the weld, especially the Nb interlayer, which will result in the low resistance and low ductility of welded joints. The unmelted interlayers will become weak places in the weld and are easily broken by the action of external forces, which impairs the

TABLE 3: Welding parameters used in dual-pass welding.

Sample	Accelerating voltage (kV)	Beam current (mA)	Welding speed (mm/s)	Interlayer thickness (mm)
EBW-1	55	6	7	Cu: 0.53
	55	5.5	7	Nb: 0.22
EBW-2	55	6	7	Cu: 0.53
	55	5.5	7	Nb: 0.40

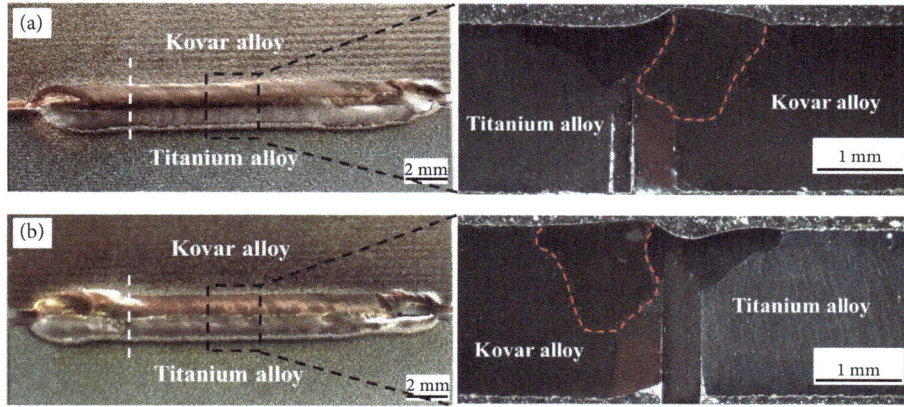

FIGURE 2: Surface appearance and the cross sections of welds: (a) EBW-1 with 0.53 mm Cu and 0.22 mm Nb; (b) EBW-2 with 0.53 mm Cu and 0.40 mm Nb.

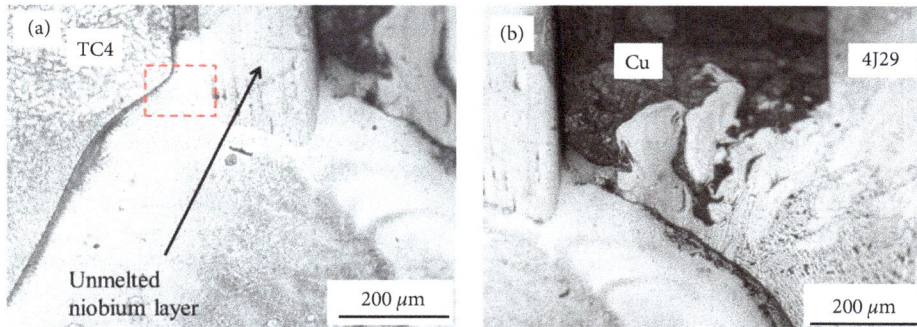

FIGURE 3: The cross section of the weld of welded joint EBW-1: (a) the titanium alloy side; (b) the kovar alloy side.

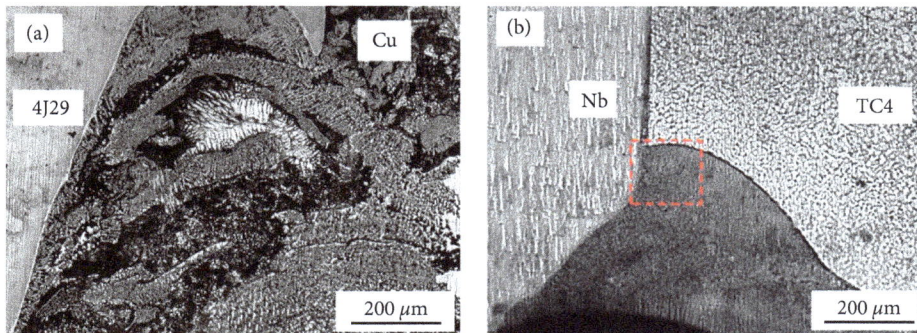

FIGURE 4: The cross section of the weld of welded joint EBW-2: (a) the kovar alloy side; (b) the titanium alloy side.

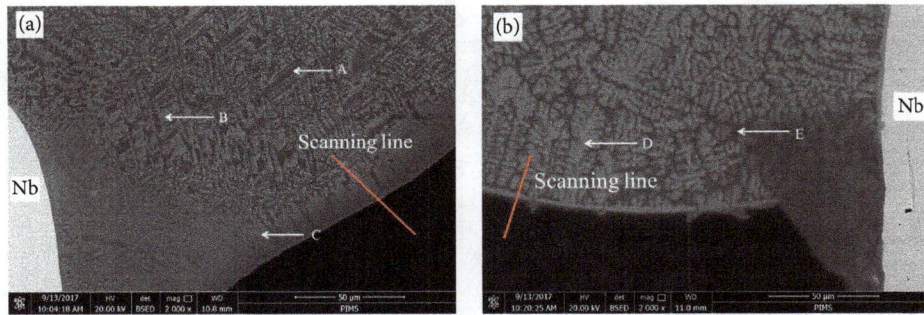

FIGURE 5: Microstructure of the fusion zone of the titanium alloy side: (a) EBW-1; (b) EBW-2.

TABLE 4: The main chemical compositions of different zones (at %) in the fusion zone.

Zone	Ti	Fe	Cu	Nb	Potential phase
A	51.5	19.5	16.0	13.0	
B	52.7	21.0	14.7	11.6	β-Ti(s,s) + FeTi
C	56.8	15.8	11.7	15.7	
D	73.6	2.2	2.0	22.1	β-Ti(s,s)
E	66.4	7.0	8.4	18.2	β-Ti(s,s) + CuTi$_2$

strength of the welded joint. However, the occurrence of this phenomenon can be improved by increasing the input of energy during the welding.

3.3. Microstructure of Titanium Alloy Side Interface.

Ti is an active element which easily forms IMCs with multiple metal elements during the welding process, as shown in the microstructure of the titanium alloy side previously [24]. For the welded joint EBW-1, Figure 5(a) shows the microstructure of the fusion zone of the titanium alloy side. It is found that the melted part is mainly concentrated on the titanium alloy side. Because of the higher thermal conductivity of Nb relative to the titanium alloy, Nb has faster energy dissipation. In addition, the specific heat of titanium alloy is higher than that of Nb, so the energy intake from the electron beam on the titanium alloy side is higher than that on the Nb side [27]. The fusion zone mainly consists of three phases according to its morphology characteristics, that is, the dark phase (A zone), gray phase (B zone), and columnar gray phase (C zone), as labelled with A, B, and C in Figure 5(a), respectively. For the welded joint EBW-2, there are just two phases in the fusion zone of the titanium alloy side. One is a dendritic phase (D zone), and the other one is a netted phase (E zone), as shown in Figure 5(b). Moreover, the dendritic phase mainly exists in the fusion zone of the titanium alloy side. The reason is that high local undercooling leads to the accelerated nucleation to refine grains and the formation of dendrites under the high cooling rate. Li et al. [28] also found that high cooling rate influenced the solidification behavior of microstructure.

To identify phase compositions of the fusion zone, EDS was performed on zones A through E in Figure 5, and the results are shown in Table 4. Referring to the compositional analysis results and relevant Cu-Fe-Ti ternary phase diagram [29] and Ti-Nb binary phase diagram [30], several phases are recognized as listed in Table 4. It is identified that the dark phase (A zone), gray phase (B zone), and columnar gray phase (C zone) are composed of β-Ti solid solution and FeTi. However, there are certain differences between these phases because of their different average atom numbers between A, B, and C zones [11]. By contrast, the dendritic phase (D zone) and netted phase (E zone) have a high content of Ti, as shown in Table 4. Similarly, it is identified that the dendritic phase is β-Ti solid solution and the netted phase is composed of β-Ti solid solution and CuTi$_2$. Therefore, it is proved that a thicker Nb foil can inhibit the diffusion of a large amount of Fe atoms towards the titanium alloy side, thus promoting the production of Ti solid solution and the deduction of IMCs. Li et al. [31] prepared laser welded joints of a TiNi shape memory alloy and a stainless steel using Ni interlayer. It was found that a thicker Ni foil could inhibit the production of brittle IMCs (TiFe$_2$ and TiCr$_2$).

Composition profiles of major elements across the titanium alloy/fusion zone are obtained from line scanning analysis in SEM, as indicated along the red lines in Figure 5, and results are presented in Figure 6. For the welded joint EBW-1, results show that the contents of Nb and Fe elements are decreased, and the content of Ti element is increased gradually from the fusion zone side to the titanium alloy side, as shown in Figure 6(a). Moreover, enrichment of Ti element is found across the interface of the titanium alloy/fusion zone, and the thickness of interface layer is about 1-2 μm. For the welded joint EBW-2, the fusion zone and interface layer are enriched in Ti and Nb elements. According to Ti-Nb binary phase diagram [30], they are mainly composed of Ti solid solution. Similar phenomenon was observed when Torkamany et al. [27] welded a pure Nb and a titanium alloy using the laser welding method. In addition, the thickness of interface layer is increased to about 2-4 μm, as shown in Figure 6(b). It is found that the thicker Nb foil suppresses energy transfer to the Cu side to increase the energy input of the Nb side. Then, the diffusion of Nb atoms is promoted, resulting in increased Nb content in interface layer, so the thickness of interface layer is increased. When Tan et al. [32] welded a magnesium alloy and

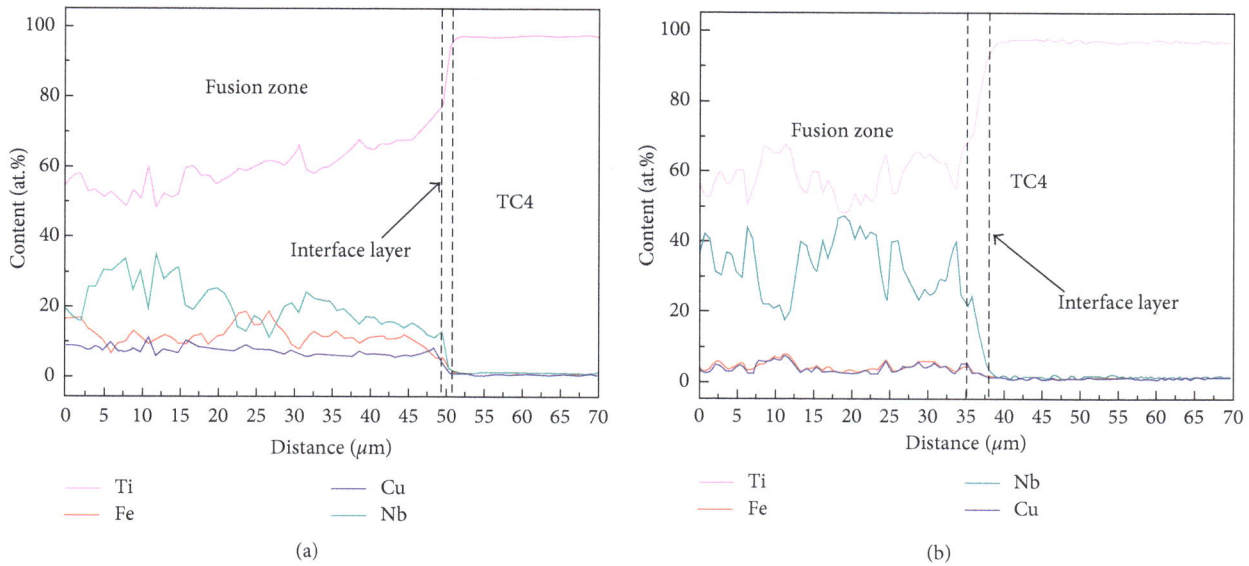

FIGURE 6: Elemental profiles of Ti, Nb, Cu, and Fe between the fusion zone (left) and the titanium alloy side (right): (a) EBW-1; (b) EBW-2.

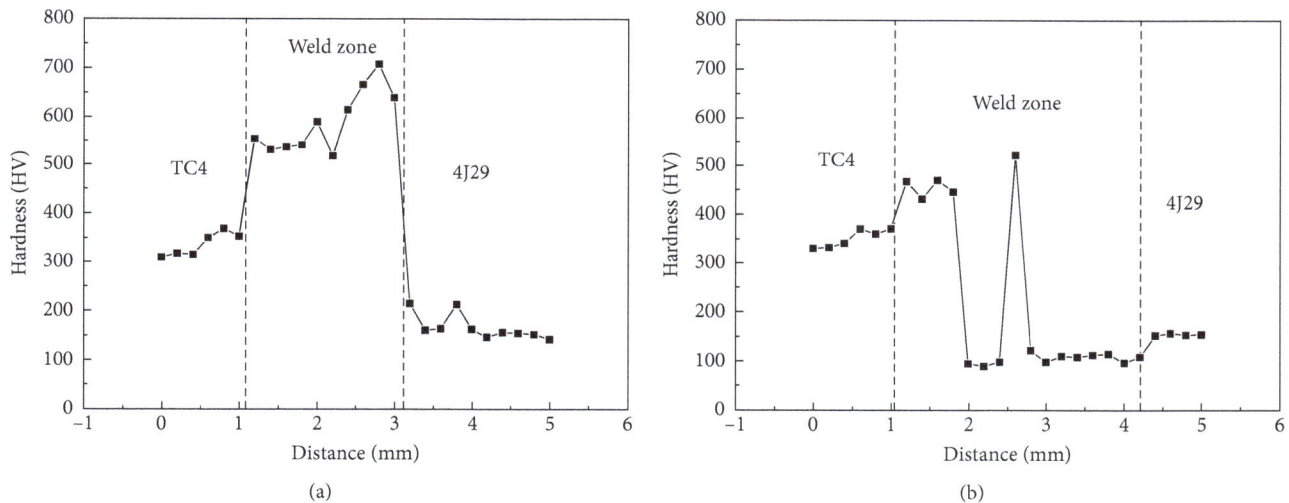

FIGURE 7: Microhardness distribution across the cross section of welded joints: (a) EBW-1; (b) EBW-2.

a titanium alloy by laser welding, similar phenomenon was found. Moreover, because the thickness of interface layer was kept far below 10 μm, metallurgical bonding can be achieved well, which is beneficial to welded joints strength. Therefore, in case the thickness of Nb foil is 0.22 mm, microstructure of the fusion zone of the titanium alloy side is mainly composed of Ti solid solution and FeTi; while in case the thickness of Nb foil is 0.40 mm, microstructure of the fusion zone of the titanium alloy side mainly consists of Ti solid solution and $CuTi_2$.

3.4. Mechanical Properties of Welded Joints

3.4.1. Microhardness Analysis of Welded Joints. Figure 7(a) shows microhardness distribution across the cross section of the welded joint EBW-1. The average hardness values of

titanium and kovar alloys are 330 HV and 160 HV, respectively. It is noticed that the maximum hardness of the fusion zone is about 710 HV, which is higher than that of the substrate. According to the previous microstructure analysis, the change in hardness value was associated with the formation of brittle FeTi, which caused the increased hardness value [12]. Figure 7(b) shows microhardness distribution across the cross section of the welded joint EBW-2. It is found that the maximum hardness of the fusion zone is about 520 HV due to the production of brittle IMCs, whereas the minimum hardness value is about 90 HV. Moreover, the hardness of the fusion zone adjacent to the kovar alloy is decreased sharply, and the hardness of the fusion zone adjacent to the titanium alloy is decreased to about 450 HV. The thicker Nb foil can inhibit the diffusion of a large amount of Fe atoms to avoid the production of the brittle IMCs, so the overall weld hardness is decreased. These

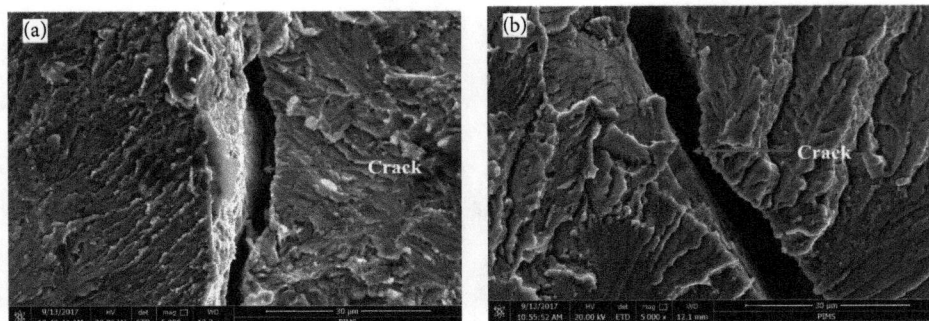

FIGURE 8: Fracture surface morphologies of welded joints: (a) EBW-1; (b) EBW-2.

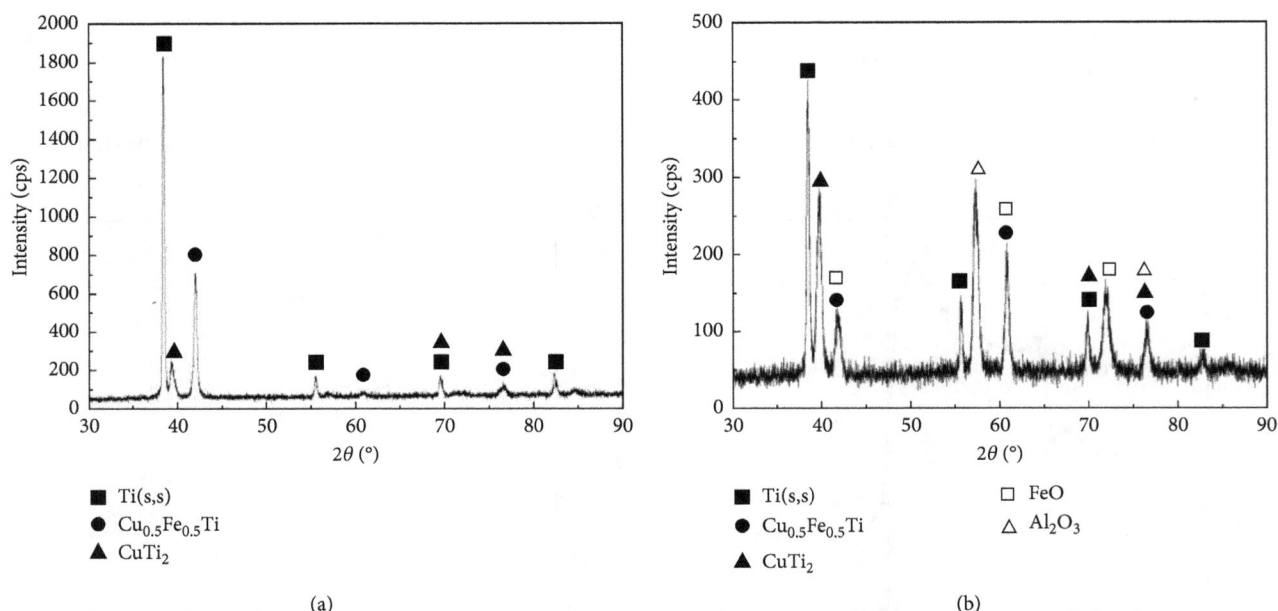

FIGURE 9: XRD patterns of fracture surface of welded joints: (a) EBW-1; (b) EBW-2.

results are consistent with the analysis in Section 3.3, and similar phenomenon occurred in a study about the influence of thickness of Cu interlayer on weld hardness [28]. Hence, it is observed that thicker Nb foil can effectively decrease the hardness of the fusion zone to enhance properties of welded joints.

3.4.2. Tensile Strength and Fracture Behavior of Welded Joints.
Figure 8 shows fracture surfaces of welded joints using two different Cu/Nb multi-interlayers. The fracture location of both of samples is located in the fusion zone of the titanium alloy side rather than the location of the maximum hardness value of welds. This fact indicates that not only IMC hardness but also IMC volume is the major factor for the failure location of welded joints. With the production of IMCs is in the fusion zone of the titanium alloy side, the failure is more likely to initiate in this zone. Pardal et al. [12] used a CuSi-3 alloy to braze titanium to

a 316L stainless steel. It was also found that the failure location was located in the zone with a high content of IMCs rather than in the zone with the maximum hardness value. Figure 8(a) exhibits the fracture surface of the welded joint EBW-1. A rough and lacerated fracture surface is found, which is the characteristic of a quasicleavage fracture. In addition, an obvious crack along the thickness direction of the weld appears at the fracture surface. In case the thickness of Nb foil is 0.40 mm, as shown in Figure 8(b), the fracture surface of welded joints still exhibits typical brittle characteristics. However, compared with the welded joint EBW-1, the fracture surface is relatively smoother.

To investigate the fracture origins of welded joints, XRD is performed on the fracture surfaces of both samples, as shown in Figure 9. The location and the intensity of some diffraction peaks are slightly different to the standards. The reasons are complex element compositions of the fusion zone and some distortion of crystal lattice [11]. The results show that the β-Ti solid solution exists on both fracture

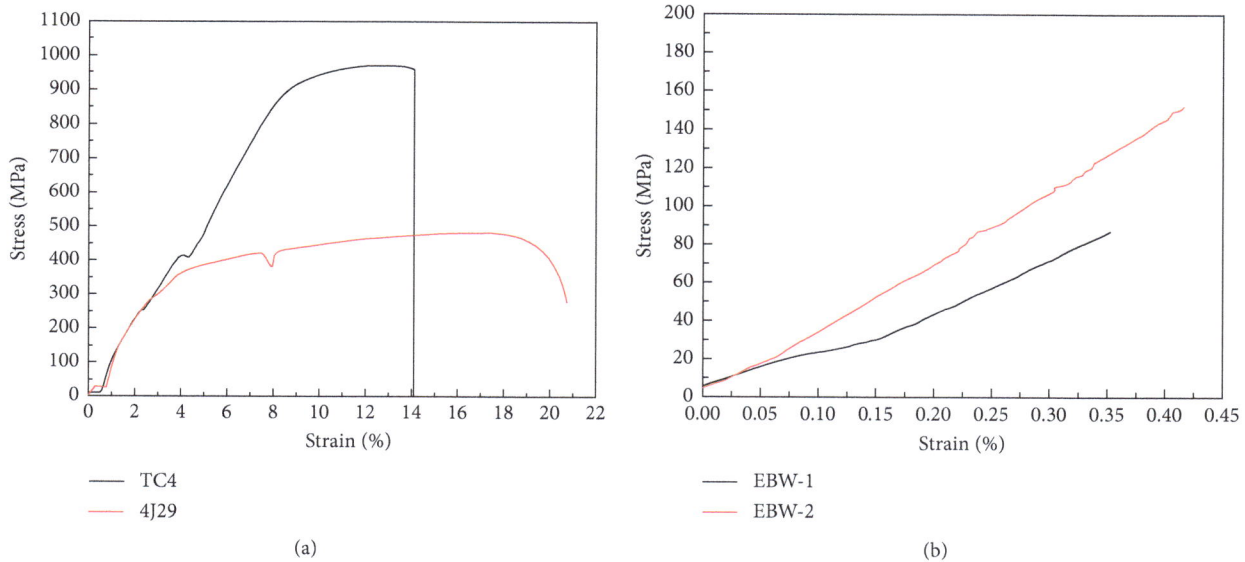

FIGURE 10: The tensile curves of the welded joint and the substrate: (a) the substrate; (b) welded joints.

surfaces, which has a higher intensity of the diffraction peaks. This confirms that the fracture location of welded joints is located in the fusion zone of the titanium alloy side.

However, for the welded joint EBW-1, it should be noted that FeTi observed in SEM is not detected by XRD, while some other IMCs ($CuTi_2$ and $Cu_{0.5}Fe_{0.5}Ti$) appear, as shown in Figure 9(a). Some of IMCs in this study are discontinuous and inhomogeneous in distribution and their contents are relatively low, and thus they may not be detectable, although some phases could be detected [11]. Besides, the fracture location is not exactly consistent with the position analyzed in SEM, and thus certain discrepancies may exist. For the welded joint EBW-2, β-Ti solid solution and $CuTi_2$ are found by XRD, while some oxides appear such as Al_2O_3 and FeO, which are produced by oxidation reaction when the sample is exposed in the air, as shown in Figure 9(b). In addition, $Cu_{0.5}Fe_{0.5}Ti$ also is found in the fracture surface which would enhance the brittleness of welded joints. The formation of Fe-Ti-Cu phase also occurs in electron beam welding of titanium alloy and stainless steel [24]. And when Tashi et al. [14] welded a titanium alloy and a stainless steel by diffusion welding, Fe-Cu-Ti IMCs were also found in the weld, which resulted in the degradation of the toughness of welded joints.

In both cases, the tensile strength of welded joints is far lower than that of the substrate, as shown in Figure 10. In case the thickness of Nb foil is 0.22 mm, tensile strength of welded joints is 88.1 MPa. With the increase of the thickness of Nb foil to 0.40 mm, tensile strength of welded joints is increased to 150 MPa. According to the above analysis, the effects of thicker Nb foil on tensile strength are mainly attributed to the production of Ti solid solution and a small amount of soft $CuTi_2$, as well as the absence of FeTi. Due to the good plasticity and toughness of Ti solid solution and lower brittleness of $CuTi_2$ than that of FeTi, welded joints' tensile strength is increased obviously. Relevant conclusion was also obtained when Li et al. [28] welded a TiNi alloy and a stainless steel using laser welding.

3.5. Bonding Mechanism. According to the above analysis, thicker Nb foil effectively inhibits the diffusion of a large amount of Fe atoms towards the titanium alloy side and restrains the formation of FeTi in the fusion zone of the titanium alloy side, and thus the tensile strength of welded joints is increased to 150 MPa. The fusion zone of the kovar alloy side is fully penetrated with large volume of melted metals, though some unmelted Cu regions exist. Combined the Cu-Fe-Ti ternary phase diagram [29] and the Ti-Nb binary phase diagram [30], the formation process of the fusion zone of the titanium alloy side mainly includes four steps as follows: firstly, partial surfaces of materials contact with each other because of the pressure of fixture, and materials have a certain deformation. Then materials begin to melt and the fusion zone is gradually formed due to energy input. The Ti, Nb, Cu, and Fe atoms diffuse and mix in the fusion zone, and the Ti element and Nb element diffuse towards the interface, as shown in Figure 11(a). Secondly, when the temperature decreases to about 1760°C, the Ti atom in the liquid at the front of liquid/solid interface saturates inducing the nucleation of β-Ti solid solution phase. Because the fusion zone has a higher temperature, the Ti, Nb, Cu, and Fe atoms are still in the state of full diffusion, as shown in Figure 11(b). Thirdly, as the temperature further decreases, from the interface to the fusion zone, dendrites begin to form due to the reduction of temperature gradient and the increase of constituent super cooling [33], as shown in Figure 11(c). Lastly, when the temperature decreases to about 850°C, remaining elements begin to precipitate along the grain boundary to form netted phases, which are composed of β-Ti solid solution and IMCs ($CuTi_2$ and $Cu_{0.5}Fe_{0.5}Ti$). Meanwhile, layer-shaped β-Ti solid solution is formed at the interface, as shown in Figure 11(d). Hence, it is concluded that connection of the titanium alloy side is achieved by physical contact, elemental diffusion, nucleation and growth of grains, and formation of β-Ti solid solution and IMCs.

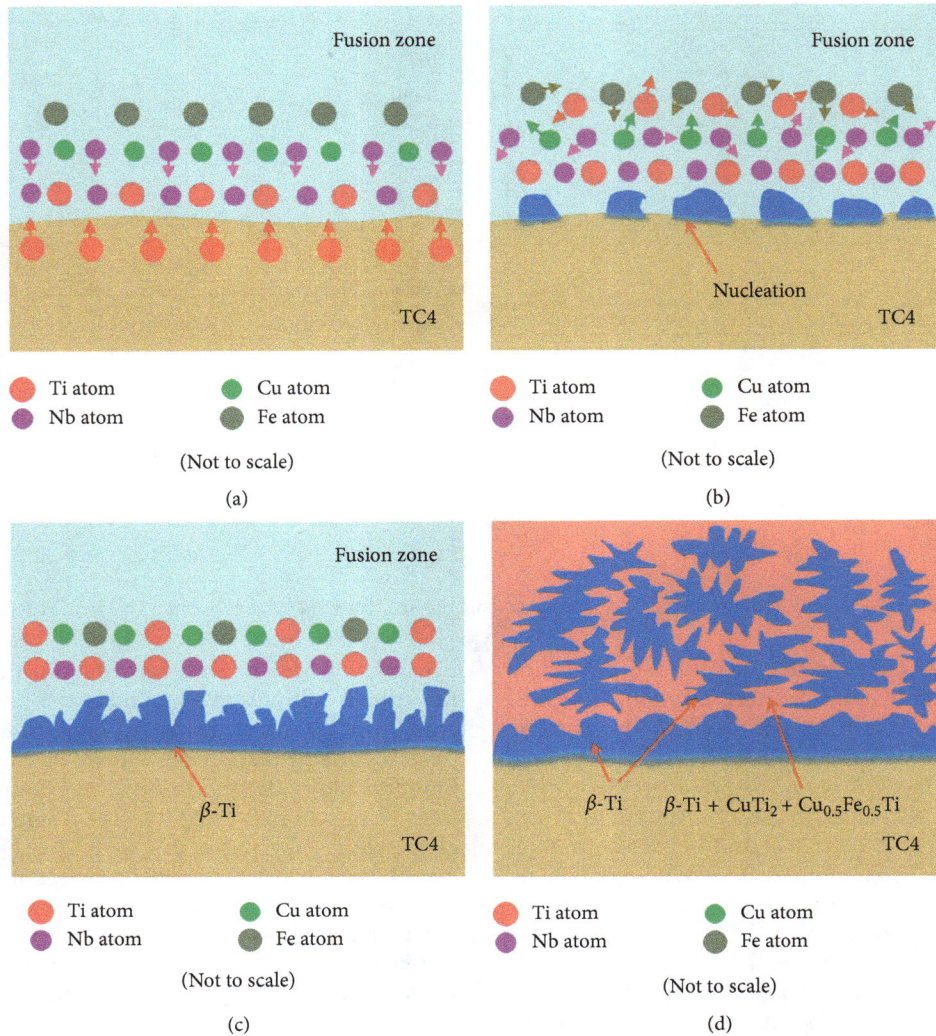

FIGURE 11: Formation process of the fusion zone of the titanium alloy side: (a) surface contact; (b) element diffusion and nucleation; (c) growth of grains; (d) formation of the solid solution and IMCs.

4. Conclusions

In this study, Cu/Nb multi-interlayers were used to conduct electron beam welding of a TC4 titanium alloy and a 4J29 kovar alloy. Microstructure and mechanical properties of welded joints were investigated, and conclusions were summarized as follows:

(a) With the addition of Cu/Nb multi-interlayers, weld surfaces were found to be flat and continuous without obvious defects. Increasing the thickness of Nb foil promoted the formation of Ti solid solution in the fusion zone, which was beneficial to the improvement of welded joints' toughness.

(b) In case the thickness of Nb foil was 0.22 mm, microstructure of the fusion zone of the titanium alloy side was mainly composed of Ti solid solution and FeTi; while in case the thickness of Nb foil was 0.40 mm, microstructure of the fusion zone of the titanium alloy side mainly consisted of Ti solid solution and $CuTi_2$. Moreover, in both cases, $Cu_{0.5}Fe_{0.5}Ti$ was formed in the fusion zone that increased the welded joint brittleness.

(c) As the thickness of Nb foil was increased from 0.22 mm to 0.40 mm, tensile strength of the welded joint was enhanced from 88.1 MPa to 150 MPa, and hardness of the weld zone decreased sharply. It was found that a thicker Nb foil could inhibit the diffusion of a large amount of Fe atoms towards the titanium alloy side, thus promoting the formation of Ti solid solution and a small amount of soft $CuTi_2$ and eliminating FeTi. Due to the good plasticity and toughness of Ti solid solution and reduced brittleness of $CuTi_2$ compared with FeTi, welded joints' tensile strength was increased obviously.

Conflicts of Interest

The authors declare that there are no conflicts of interest regarding the publication of this paper.

Acknowledgments

This work was supported by Key Laboratory of Infrared Imaging Materials and Detectors, Shanghai Institute of Technical Physics, Chinese Academy of Sciences (no. IIMDKFJJ-17-06), Sichuan Science and Technology Support Program (no. 2016FZ0079), National Natural Science Foundation of China (no. 51201143), China Postdoctoral Science Foundation (no. 2015M570794), and R&D Projects Funding from the Research Council of Norway (no. 263875/H30).

References

[1] C. H. Muralimohan, V. Muthupandi, and K. Sivaprasad, "The influence of aluminum intermediate layer in dissimilar friction welds," *International Journal of Materials Research*, vol. 105, no. 4, pp. 350–357, 2014.

[2] W. W. Zhu, J. C. Chen, and C. H. Jiang, "Effects of Ti thickness on microstructure and mechanical properties of alumina-Kovar joints brazed with Ag-Pd/Ti filler," *Ceramics International*, vol. 40, no. 4, pp. 5699–5705, 2014.

[3] Y. C. Chen, K. H. Tseng, and H. C. Wang, "Small-scale projection lap-joint welding of Kovar alloy and SPCC steel," *Journal of Chinese Institute of Engineers*, vol. 35, no. 2, pp. 211–218, 2012.

[4] G. Satoh, Y. L. Yao, and C. Qiu, "Strength and microstructure of laser fusion-welded Ti–SS dissimilar material pair," *International Journal of Advanced Manufacturing Technology*, vol. 66, no. 1–4, pp. 469–479, 2013.

[5] T. F. Song, X. S. Jiang, Z. Y. Shao et al., "Interfacial microstructure and mechanical properties of diffusion-bonded joints of titanium TC4 (Ti-6Al-4V) and Kovar (Fe-29Ni-17Co) alloys," *Journal of Iron and Steel Research, International*, vol. 24, no. 10, pp. 1023–1031, 2017.

[6] P. Li, J. L. Li, M. Salman, L. Liang, J. T. Xiong, and F. S. Zhang, "Effect of friction time on mechanical and metallurgical properties of continuous drive friction welded Ti6Al4V/SUS321 joints," *Materials and Design*, vol. 56, no. 4, pp. 649–656, 2014.

[7] C. Velmurugan, V. Senthilkumar, S. Sarala, and J. Arivarasan, "Low temperature diffusion bonding of Ti-6Al-4V and duplex stainless steel," *Journal of Materials Processing Technology*, vol. 234, pp. 272–279, 2016.

[8] R. M. Miranda, E. Assuncao, R. J. C. Silva, J. P. Oliveira, and L. Quintino, "Fiber laser welding of NiTi to Ti-6Al-4V," *International Journal of Advanced Manufacturing Technology*, vol. 81, no. 9–12, pp. 1533–1538, 2015.

[9] S. Kundu, S. M. Bhola, B. Mishra, and S. Chatterjee, "Structure and properties of solid state diffusion bonding of 17-4PH stainless steel and titanium," *Materials Science and Technology*, vol. 30, no. 2, pp. 248–256, 2014.

[10] Y. Gao, K. Nakata, K. Nagatsuka, F. C. Liu, and J. Liao, "Interface microstructural control by probe length adjustment in friction stir welding of titanium and steel lap joint," *Materials and Design*, vol. 65, pp. 17–23, 2015.

[11] S. H. Chen, M. X. Zhang, J. H. Huang, C. J. Cui, H. Zhang, and X. K. Zhao, "Microstructures and mechanical property of laser butt welding of titanium alloy to stainless steel," *Materials and Design*, vol. 53, no. 1, pp. 504–511, 2014.

[12] G. Pardal, S. Ganguly, S. Williams, and J. Vaja, "Dissimilar metal joining of stainless steel and titanium using copper as transition metal," *International Journal of Advanced Manufacturing Technology*, vol. 86, no. 5–8, pp. 1139–1150, 2016.

[13] N. Özdemir and B. Bilgin, "Interfacial properties of diffusion bonded Ti-6Al-4V to AISI 304 stainless steel by inserting a Cu interlayer," *International Journal of Advanced Manufacturing Technology*, vol. 41, no. 5-6, pp. 519–526, 2009.

[14] R. S. Tashi, S. A. A. Mousavi, and M. M. Atabaki, "Diffusion brazing of Ti-6Al-4V and austenitic stainless steel using silver-based interlayer," *Materials and Design*, vol. 54, no. 2, pp. 161–167, 2014.

[15] M. Balasubramanian, "Application of Box–Behnken design for fabrication of titanium alloy and 304 stainless steel joints with silver interlayer by diffusion bonding," *Materials and Design*, vol. 77, no. 1, pp. 161–169, 2015.

[16] A. Yildiz, Y. Kaya, and N. Kahraman, "Joint properties and microstructure of diffusion-bonded grade-2 titanium to AISI 430 ferritic stainless steel using pure Ni interlayer," *International Journal of Advanced Manufacturing Technology*, vol. 86, no. 5–8, pp. 1287–1298, 2016.

[17] J. P. Oliveira, B. Panton, Z. Zeng et al., "Laser joining of NiTi to Ti6Al4V using a Niobium interlayer," *Acta Materialia*, vol. 105, pp. 9–15, 2016.

[18] X. W. Zhou, Y. H. Chen, Y. D. Huang, Y. Q. Mao, and Y. Y. Yu, "Effects of niobium addition on the microstructure and mechanical properties of laser-welded joints of NiTiNb and Ti6Al4V alloys," *Journal of Alloys and Compounds*, vol. 735, pp. 2616–2624, 2018.

[19] C. Lusch, M. Borsch, C. Heidt, and S. Grohmann, "Qualification of electron-beam welded joints between copper and stainless steel for cryogenic application," *IOP Conference Series: Materials Science and Engineering*, vol. 102, pp. 12–17, 2015.

[20] Z. Sun and R. Karppi, "The application of electron beam welding for the joining of dissimilar metals: an overview," *Journal of Materials Processing Technology*, vol. 59, no. 3, pp. 257–267, 1996.

[21] T. Wang, B. G. Zhang, G. Q. Chen, J. C. Feng, and Q. Tang, "Electron beam welding of Ti-15-3 titanium alloy to 304 stainless steel with copper interlayer sheet," *Transactions of Nonferrous Metals Society of China*, vol. 20, no. 10, pp. 1829–1834, 2010.

[22] B. G. Zhang, T. Wang, X. H. Duan, G. Q. Chen, and J. C. Feng, "Temperature and stress fields in electron beam welded Ti-15-3 alloy to 304 stainless steel joint with copper interlayer sheet," *Transactions of Nonferrous Metals Society of China*, vol. 22, no. 2, pp. 398–403, 2012.

[23] I. Tomashchuk, P. Sallamand, N. Belyavina, and M. Pilloz, "Evolution of microstructures and mechanical properties during dissimilar electron beam welding of titanium alloy to stainless steel via copper interlayer," *Materials Science and Engineering: A*, vol. 585, no. 12, pp. 114–122, 2013.

[24] T. Wang, B. G. Zhang, H. Q. Wang, and J. C. Feng, "Microstructures and mechanical properties of electron beam-welded titanium-steel joints with vanadium, nickel, copper and silver filler metals," *Journal of Materials Engineering and Performance*, vol. 23, no. 4, pp. 1498–1504, 2014.

[25] T. Wang, B. G. Zhang, G. Q. Chen, and J. C. Feng, "High strength electron beam welded titanium stainless steel joint with V/Cu based composite filler metals," *Vacuum*, vol. 94, no. 6, pp. 41–47, 2013.

[26] S. Sam, S. Kundu, and S. Chatterjee, "Diffusion bonding of titanium alloy to micro-duplex stainless steel using a nickel alloy interlayer: Interface microstructure and strength properties," *Materials and Design*, vol. 40, no. 2, pp. 237–244, 2012.

[27] M. J. Torkamany, F. M. Ghaini, and R. Poursalehi, "Dissimilar pulsed Nd:YAG laser welding of pure niobium to Ti-6Al-4V," *Materials and Design*, vol. 53, no. 1, pp. 915–920, 2014.

[28] H. M. Li, D. Q. Sun, X. Y. Gu, P. Dong, and Z. P. Lv, "Effects of the thickness of Cu filler metal on the microstructure and properties of laser-welded TiNi alloy and stainless steel joint," *Materials and Design*, vol. 50, no. 17, pp. 342–350, 2013.

[29] J. A. Van beek, A. A. Kodentsov, and F. J. J. Van Loo, "Phase equilibria in the Cu-Fe-Ti system at 1123 K," *Journal of Alloys and Compounds*, vol. 217, no. 1, pp. 97–103, 1995.

[30] ASM International, *ASM Handbook Volume 3: Alloy Phase Diagrams*, ASM International, Cleveland, USA, 1992, ISBN: 978-0-87170-381-1.

[31] H. M. Li, D. Q. Sun, X. L. Cai, P. Dong, and W. Q. Wang, "Laser welding of TiNi shape memory alloy and stainless steel using Ni interlayer," *Materials and Design*, vol. 39, no. 1, pp. 285–293, 2012.

[32] C. W. Tan, B. Chen, S. H. Meng et al., "Microstructure and mechanical properties of laser welded-brazed Mg/Ti joints with AZ91 Mg based filler," *Materials and Design*, vol. 99, pp. 127–134, 2016.

[33] G. Li, X. F. Lu, X. L. Zhu, J. Huang, L. W. Liu, and Y. X. Wu, "The defects and microstructure in the fusion zone of multipass laser welded joints with Inconel 52M filler wire for nuclear power plants," *Optics and Laser Technology*, vol. 94, pp. 97–105, 2017.

5

Effect of Reduction in Thickness and Rolling Conditions on Mechanical Properties and Microstructure of Rolled Mg-8Al-1Zn-1Ca Alloy

Yuta Fukuda,[1] **Masafumi Noda,**[1] **Tomomi Ito,**[1] **Kazutaka Suzuki,**[2] **Naobumi Saito,**[2] **and Yasumasa Chino**[2]

[1]*Magnesium Division, Gonda Metal Industry Co. Ltd., Sagamihara, Kanagawa 252-0212, Japan*
[2]*Structural Materials Research Institute of Advanced Industrial Science and Technology (AIST), Nagoya, Aichi 463-8560, Japan*

Correspondence should be addressed to Masafumi Noda; mk-noda@s7.dion.ne.jp

Academic Editor: Jörg M. K. Wiezorek

A cast Mg-8Al-1Zn-1Ca magnesium alloy was multipass hot rolled at different sample and roll temperatures. The effect of the rolling conditions and reduction in thickness on the microstructure and mechanical properties was investigated. The optimal combination of the ultimate tensile strength, 351 MPa, yield strength, 304 MPa, and ductility, 12.2%, was obtained with the 3 mm thick Mg-8Al-1Zn-1Ca rolled sheet, which was produced with a roll temperature of 80°C and sample temperature of 430°C. This rolling process resulted in the formation of a bimodal structure in the α-Mg matrix, which consequently led to good ductility and high strength, exclusively by the hot rolling process. The 3 mm thick rolled sheet exhibited fine (mean grain size of 2.7 μm) and coarse grain regions (mean grain size of 13.6 μm) with area fractions of 29% and 71%, respectively. In summary, the balance between the strength and ductility was enhanced by the grain refinement of the α-Mg matrix and by controlling the frequency and orientation of the grains.

1. Introduction

A new approach for limiting environmental impact [1] while increasing the speed of a vehicle, by reducing its weight through the replacement of aluminum alloy with lightweight magnesium alloy, has attracted the attention of many researchers [2]. The main disadvantages of using magnesium alloys in a vehicle relate to their combustibility and unsatisfactory mechanical properties. In addition, there is a need to be able to fabricate high strength sheets with a thickness of 3 mm, which approximates the thickness of the sheets that are currently being used. Sakamoto et al. reported that, by adding calcium to a magnesium alloy, the combustion temperature can be increased by more than 250°C [3]. Improvements in the mechanical properties of the magnesium alloys have been achieved by adding different elements [4, 5], applying texture control during forging [6–8], and grain refinement [9]. In most of these studies, rolling [6–9]

was used to form the sheets, using extruded material [5, 10, 11] as the starting material.

By focusing on the rolling conditions of magnesium alloys, Sakai reported that, when an AZ31Mg alloy is processed above the recrystallization temperature, there is no tearing; the critical upset ratio is 30% per pass, and the obtained mean grain size is 6 μm after multipass rolling with reheating [12]. Using twin-rolled cast alloy, AMX1001 (mean grain size (d) = 53 μm; initial plate thickness = 3 mm), Noda et al. performed rolling at a roll temperature of 250°C and a sample temperature of 200°C and obtained an elongation of 8% at a tensile strength of 400 MPa [13]. Using extruded sheets of the AZ61 alloy (d = 19 μm) and AM60 alloy (d = 20 μm) with thicknesses of 2 mm, Huang et al. performed hot rolling to achieve a thickness of 0.8 mm and attained an elongation of 26.1% at a tensile strength of 263 MPa [6]. Kim et al. heated an extruded plate with a thickness of 2 mm to 200°C and rolled it to a thickness of 0.7 mm using

TABLE 1: Chemical composition (mass%) of the Mg-8Al-1Zn-1Ca alloy.

Al	Ca	Zn	Mn	Cu	Ni	Si	Fe	Mg
7.99	0.959	1.076	0.28	0.0014	0.0004	0.0074	0.0022	bal.

different peripheral roll speeds at a roll temperature of 200°C and obtained an elongation of 9–11% at a tensile strength of 394 MPa [7].

The hot rolling of the magnesium alloys have been performed as described above; the initial grain sizes of the samples were as small as 20–50 μm, and high strength or high ductility was achieved by thinning. In other words, there have been no studies reporting a process for fabricating a rolled sheet with a thickness of 3 mm with high strength and high ductility, starting from a cast material with a coarse structure. In this study, we rolled Mg alloys to a total reduction ratio of 75%. We used various materials and rolling temperatures and investigated the influence of the thick-plate rolling conditions on the strength, ductility, and structure.

2. Experiments

In this study, AM60B metal, Mg-30%Ca metal, pure-metal Zn (99.5%), and pure-metal Al (99.7%) were weighed and dissolved to obtain the target composition of the AZX811 alloy, Mg-8Al-1Zn-1Ca mass%. The materials were heated and melted in a steel crucible under an inert argon atmosphere. Then, 0.2 MPa Ar gas was bubbled for 20 min when the melt temperature reached 680°C. After dissolution in the Ar atmosphere and subsequent stirring, the samples were cast by antigravity suction casting [17] with a cooling rate of 12 K/s. Antigravity suction casting was conducted by sinking the down sprue 300 mm into the melt in a SS400 steel mold (95 mm in width, 15 mm in thickness, and 2 m in length), ensuring that the melt was not exposed to the atmosphere during casting. The chemical compositions of the AZX811 cast material are listed in Table 1. The compositional analysis was performed by X-ray fluorescence (JEOL JSX-1000S). To prepare samples for rolling, the cast material was machined into plates measuring 12 mm (H) × 90 mm (W) × 200 mm (L). A two-stage rolling mill was used for the rolling. The rolling samples were heated to 350°C in an electric furnace and then rolled to a thickness of 3 mm with a rolling reduction of 1 mm/pass. The samples were water-cooled after the rolling process; the heating and holding periods were 1 min in each interpass period. The roll temperature was set to 250°C, and the roll peripheral speed was set to 10 m/min.

For tensile tests, samples were cut to a gauge length of 30 mm, 5 mm in width, and 3 mm in thickness by machining, with the longitudinal direction parallel to the rolling direction. The tensile test was performed at room temperature, with an initial strain rate of 1.1×10^{-3} s^{-1}. The elongation after fracture was measured by a noncontact video extensometer (Instron, Type5565, and AVE2). Samples for microstructure observations were prepared by mechanical grinding, polishing, and subsequent etching. The samples were etched using a solution of picric acid (6 g) in ethanol

(100 mL), acetic acid (8 mL), and distilled water (10 mL). The structure was observed by optical microscopy (Keyence VHX-2000) and scanning electron microscopy (SEM, JEOL JCM-6000, and JSM-7100F) at an accelerating voltage of 15 kV. The crystallographic orientation was measured using electron backscatter diffraction (EBSD) after ion-polishing of the cross section, parallel to the rolling direction. In relation to the amount of the intermetallic compounds, the area ratio was calculated using a Sigma Scan Pro 5 image analysis software and the grain size was measured by the linear intercept method. The compound formed on the material was qualitatively analyzed by X-ray diffraction (XRD, Rigaku, Smartlab) using a sample measuring 20 mm × 20 mm.

3. Experimental Results and Discussion

3.1. Structure and Mechanical Properties of the Cast Materials.
Table 2 lists the mechanical properties of the cast materials, while Figures 1(a)–1(f) show optical and SEM images. The mean grain sizes of the gravity-cast and antigravity-suction-cast materials are 550 μm and 144 μm, respectively. With the decrease in the grain size of the magnesium, the area ratios of the intermetallic compound became 13% and 9% in the gravity-cast and antigravity-suction-cast materials, respectively, as shown in Figures 1(b) and 1(e). The ultimate tensile strength (UTS) and elongation of the antigravity-suction-cast material are 188 MPa and 2.3%, respectively. A comparison between Figures 1(c) and 1(f) showed that the intermetallic compounds were formed discontinuously along the grain boundary, in the antigravity-suction-cast material. Note that Kleiner et al. used a semisolid cast alloy of AZ origin (Al content = 7–9 wt%) to clarify that the ductility is improved by discontinuous scattering in the magnesium region or along the grain boundary, rather than because of the continuous appearance of the beta phase [16]. Yamamoto et al. reported that the strength and ductility are drastically improved when the size of the α-Mg region is reduced to less than 5 μm in the AX43 alloy fabricated by semisolid injection molding [18]. Although the elongation of the cast AZX811 alloy is as low as 2.3%, compared to the mechanical properties of the as-cast magnesium alloys [14, 17, 19] formed by other methods, as listed in Table 2, it is thought that the degree of elongation decreases and high values of the yield strength (YS) and UTS are achieved because an intermetallic compound is formed when 1 mass% of calcium is added.

3.2. Grain Refinement and Improvement of the Mechanical Properties by Multipass Rolling.
Figure 2 shows the relationship between the strength, mean grain size d, and total reduction ratio R, observed by multipass rolling to a plate thickness of 12 to 3 mm at a roll temperature (T_R) of 250°C and a sample temperature (T_S) of 350°C. Because fine grains are formed around the coarse grains at a total reduction ratio of 42% or more, the mean grain sizes for the coarse and fine grain areas are also shown in the figure. When the total reduction ratio is 42%, the YS and UTS are 274 MPa and 300 MPa, and the mean grain size decreases to 23 μm. Even if the total reduction ratio is increased to 75%, the YS and UTS are 289 MPa and 322 MPa, respectively, and the mean

TABLE 2: Mechanical properties of several Mg cast alloys.

Casting process	Materials	YS (MPa)	UTS (MPa)	Elongation (%)	Ref.
Squeeze casting	AZ91	104	183	4.5	[14]
Rheocasting	AZ91	105	171	3.4	[14]
	AZ71	98	185	4.7	[14]
Thixocasting	AZ80	102	187	3.5	[14]
	AX43	117	130	0.5	[15]
Gravity casting	AZX811	130	167	1.5	This work
Antigravity suction casting	AZX811	167	188	2.3	This work
	AMX1001	122	150	2	[16]

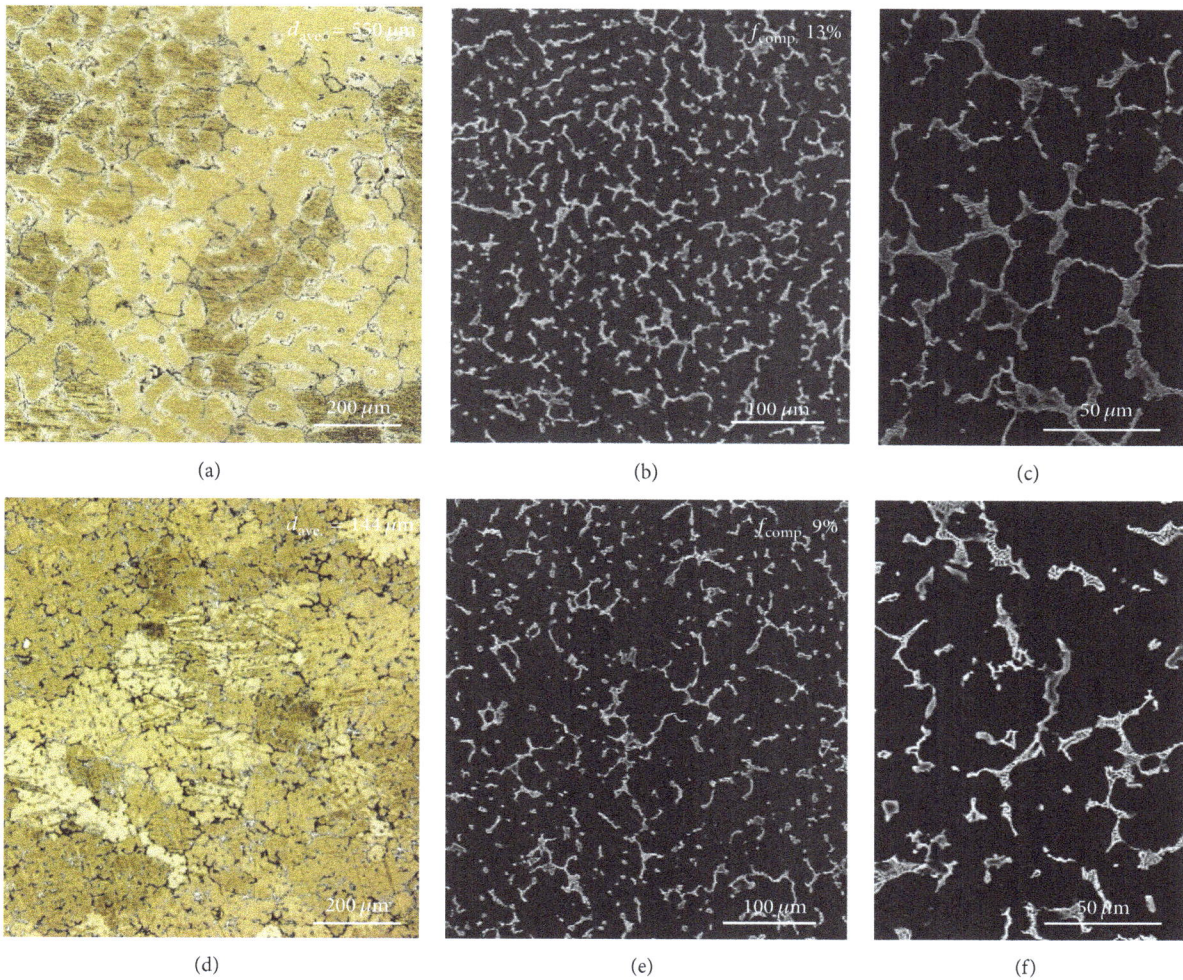

FIGURE 1: Optical and SEM micrographs of the (a)–(c) gravity-cast and (d)–(f) antigravity-suction-cast alloys. The high-magnification SEM micrographs are shown in (c) and (f).

grain size is 7.6 μm. Figure 3 shows the relationship between YS and elongation for $d^{-1/2}$, as the degree of improvement in strength and the degree of grain refinement are reduced at a total reduction ratio of 42%. The relationship between YS and $d^{-1/2}$ was divided into two linear-gradient (k value) regions. The point of contact between the two straight lines corresponds to $R = 42\%$. There have been many reports addressing the Hall-Petch equation for magnesium alloys, where the value of k is known to depend on the processing temperature [7], grain size [7, 14, 15], sampling direction [8, 15], twin formation [8, 20], and rolling texture [8, 14]. Figure 4 shows the texture variation resulting from multipass rolling. Both the OM structure and the inverse pole figure (IPF) map indicate that, in the multipass rolling of cast materials,

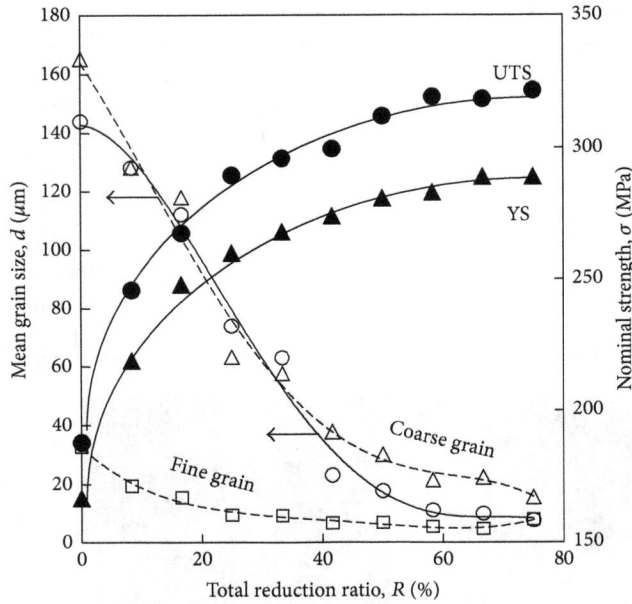

FIGURE 2: Effect of the total reduction ratio on the strength (yield strength (YS) and ultimate tensile strength (UTS)) and mean grain size of the test specimen. The mean grain sizes of the fine and coarse grain regions are shown.

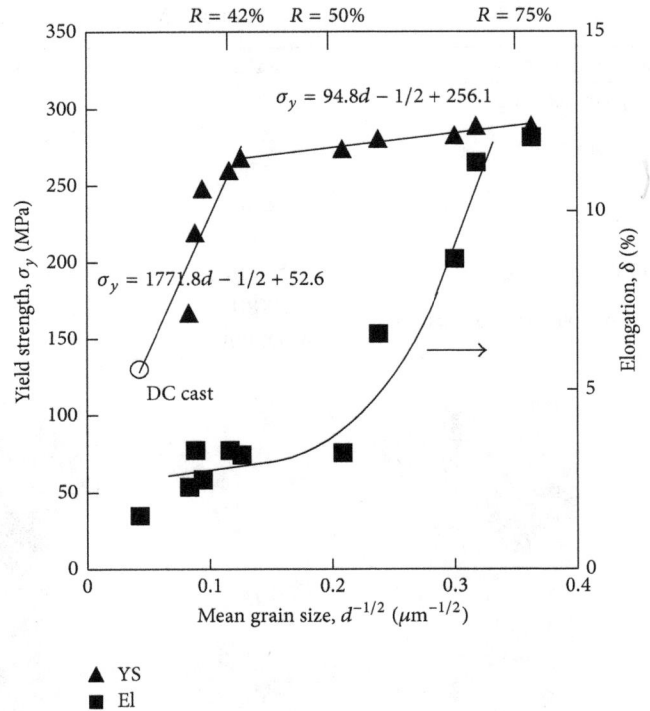

FIGURE 3: Hall-Petch plots and changes in the elongation as a function of (mean grain size)$^{-1/2}$ under various reduction ratios of the hot-rolled AZX811 alloy. The yield stress of the DC alloy and the total reduction ratio (R) are indicated by the open circle.

twinning deformation occurs inside the coarse grain if the total reduction ratio is less than 42%, thus increasing the difference between the grain boundary directions. The size reduction in the magnesium region because of dynamic recrystallization (DRX) is the dominant factor affecting the strength improvement. On the other hand, if the total reduction ratio exceeds 42%, in general, coarse grains are refined and an intermetallic compound is formed along the boundary of the grains of magnesium, acting as a stress concentration source. Fine grains of less than 7 μm are formed around the grains because of DRX. The IPF map shows that the mean grain size reaches 6.6 μm (area ratio = 44%) at $R = 42\%$ in the fine grain area and 7.9 μm ($f_{\text{fine}} = 75\%$) at $R = 75\%$. On the other hand, the grain size reached 37.7 μm ($f_{\text{coarse}} = 56\%$) in the coarse grain area and 15.6 μm ($f_{\text{coarse}} = 25\%$) at $R = 75\%$. Even in a magnesium alloy with added calcium, although the mechanism of the grain refinement acts in the same way as in a AZ series magnesium alloy [6, 7, 21], it is likely that, in the refining of magnesium grains, different factors act in the regions with the total reduction ratio greater than or less than 42%. The intermetallic compound suppresses the grain growth during the heating and holding periods.

Xu et al. showed that the size of the areas around the grain boundary decreased because of the fine dispersion of the beta phase by DRX [22]. del Valle et al. reported that, within the grains, shear deformation and the introduction of twins generate sites of DRX [21]. Although intermetallic compounds were not identified in this study, the optical images shown in Figure 4, the twinning microstructures in the IPF map, and the fine grain region within the magnesium region are consistent with these reports. Jain et al. concluded that the mean grain size has no significant influence on the k

value, based on their investigation of the Hall-Petch equation using an extruded AZ31 alloy, for which $d = 13$–$140\,\mu$m [15]. However, the influence of the mean grain size cannot be disregarded, because in case of a cast material such as AZX811 alloy, an intermetallic compound exists around the dendrite; the magnesium regions become small as the total reduction ratio increases, and the intermetallic compound aligns parallel to the rolling direction.

3.3. Influence of the Rolling Temperature on the Structure and Mechanical Properties.

Figure 2 suggests that the rate of strength improvement decreases as the difference in the mean grain size between the coarse and fine grains of the magnesium decreases with an increase in the total reduction ratio. Therefore, in order to fabricate a sheet thickness of 3 mm, a sheet thickness of 6 mm was used as a starting material for finish rolling and the influence of the rolling conditions shown in Table 3 on the structure and mechanical properties was investigated. Table 3 shows the mechanical properties of sheet thickness of 6 mm. Figure 5 summarizes YS, UTS, and the elongation for each rolling condition. Upon comparing cases 1 and 2, it was found that the UTS does not change at T_R of 250°C, while the elongation increases from 9.5 to 17.9% as a result of increasing T_S. A similar tendency is also observed at $T_R = 80$°C, and a UTS of 351 MPa and an elongation of 12.2% were obtained in case 4 ($T_S = 430$°C). When the results were compared to those obtained for the multipass rolling discussed in Section 3.2, in case 4 and in the region

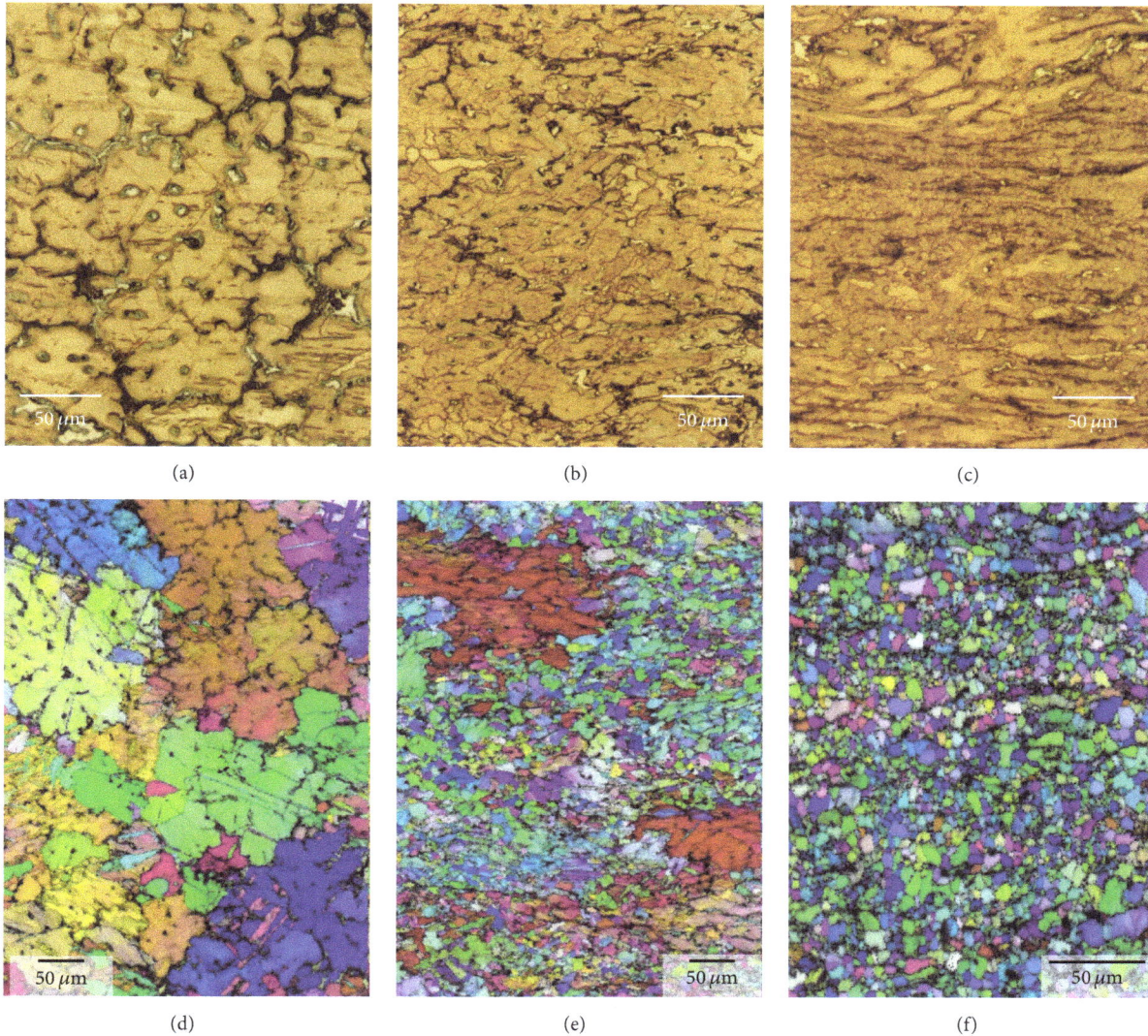

FIGURE 4: Optical micrographs and inverse pole figure maps of the AZX811 hot-rolled alloy. Total reduction ratios are as follows: (a), (d) 8.3%; (b), (e) 42%; and (c), (f) 75%.

TABLE 3: Variation in the rolling conditions for samples 6–3 mm in thickness. The mechanical properties and mean grain size of the 6 mm thick alloy are provided.

Case	Roll temp. (°C)	Sample temp. (°C)	Rolling speed (m/min)	$t = 6$ mm sample properties
The multipass rolling process	250	350		
1	250	250		$d = 17.7\ \mu$m YS 281 MPa
2	250	430	10	UTS 312 MPa
3	80	250		El 6.6%
4	80	430		

of $R = 42$–75%, the UTS increased from 22 MPa to 59 MPa in the multipass rolling, as a result of controlling T_R and T_S during the rolling.

For each rolling condition, Figure 6 shows the relationship between the area ratio (f) of the fine grain region, YS, UTS, and elongation, assuming that the plate is rolled to a thickness of 3 mm. The YS and UTS values increased while the elongation decreased from 18 to 2% as the area ratio of the fine grain area increased. Therefore, to compare the internal texture of the rolling materials between case 3, which does not show elongation because of the high mechanical strength, and case 4, in which the mechanical strength and elongation are balanced, Figures 7(a) and 7(b) show the IPF map and the distribution of the grain sizes.

The mean fine grain size and the area ratio are 2.7 μm ($f_{fine} = 29\%$) and 13.6 μm ($f_{coarse} = 71\%$) in case 4 and 2.9 μm ($f_{fine} = 65\%$) and 9.7 μm ($f_{coarse} = 35\%$) in case 3. In case 3, the area ratio of the fine grain area is 65%, but the IPF map shows residual shear deformation caused by the rolling. While both materials showed equivalent image-quality (IQ) levels, a residual processing strain was observed in case 3,

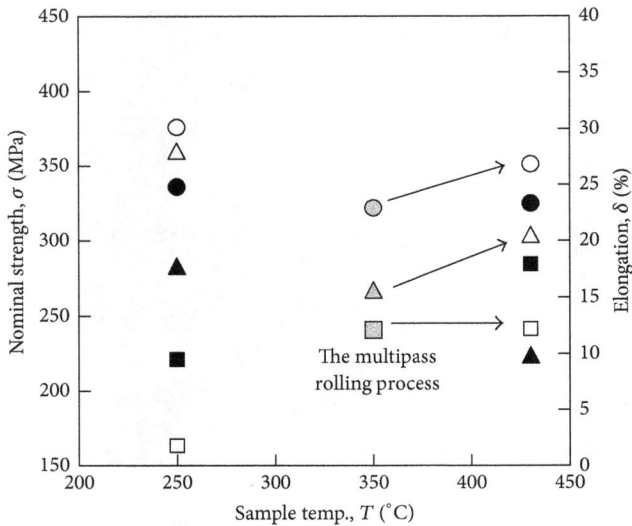

FIGURE 5: Relationship between the sample temperature and the mechanical properties of case 1 to case 4 and the multipass rolling process result (Section 3.2). The circle, triangle, and square symbols represent UTS, YS, and El, respectively.

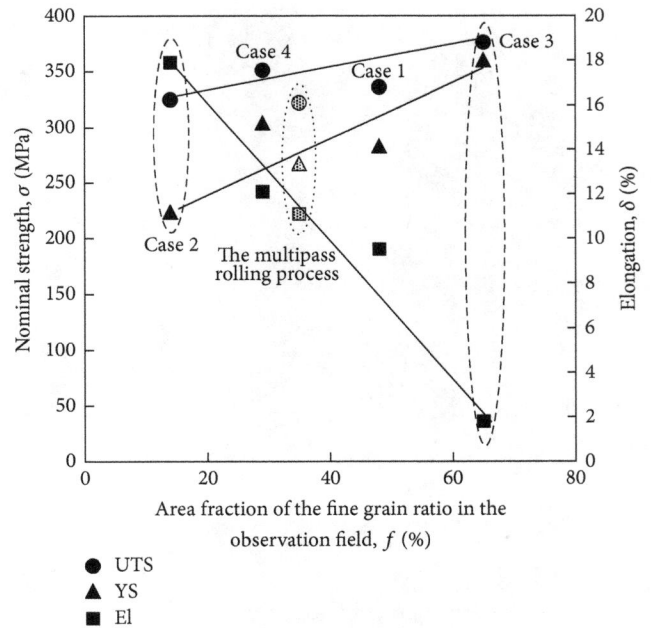

FIGURE 6: Relationship between the area fraction of the fine grain ratio in the observation field and the mechanical properties under rolling conditions of case 1 to case 4 and the results of the multipass rolling process (Section 3.2). The circle, triangle, and square symbols represent UTS, YS, and El, respectively.

where the black areas with a confidence index (CI) level of less than 0.1 remain. On the other hand, the grain size difference between the areas of the fine and coarse grains in case 4 is 1.6 times greater than that in case 3, and the structure indicates that coarse grains surround the fine grains. The grain size distribution shows that, in case 3, the first peak appears at a mean grain size of 1.3 μm; however, the second peak does not appear subsequently. In case 4, the first peak appears at a mean grain size of 1.3 μm and the second peak appears at a mean grain size of 10–13 μm. Subsequently (up to $d = 30\ \mu$m), the area ratio decreases slowly.

Kato et al. reported that, for homogeneous materials, the strength and ductility can be improved by the presence of connecting fine grains (0.32 μm) around the coarse grains (29.2 μm), although this was only reported for pristine copper [23]. Park and Yanagimoto reported that a warm or hot compression test of 0.2% carbon steel with a bimodal structure, which shows two peaks in the grain distribution, reveals balanced strength and ductility characteristics [24]. The characteristics of the internal structure of the material hot rolled at $T_R = 80°$C and $T_S = 430°$C are in agreement with those in the reports discussed above. In other words, to improve both the strength and elongation of the AZX811 alloy exclusively by rolling, it is important for the structure to show a bimodal distribution.

3.4. Rolling Feasibility and Macrorolling Ability. It is well-known that an edge crack develops actively in high strength Mg alloy during the hot rolling process. In this study, for the AZX811 alloy with 1 mass% of added Ca, cracks are likely to occur through the formation of Al-Ca compounds [25].

Figure 8 presents the appearance of the AZX811 rolled sheet at each roll temperature (T_R), sample temperature (T_S), and sheet thickness. For the sheet thickness of 6 mm, edge cracks could not be observed on the rolled sheet. We investigated the edge cracks and the total crack length for rolled sheet thicknesses of 5 to 3 mm, as shown in Figure 8. At the T_S of 250°C, large edge cracks propagated at the T_R of 80°C (case 3) and 250°C (case 1). On the other hand, as T_S was increased to 430°C, the number and length of the edge cracks decreased regardless of the T_R (case 2 and case 4), indicating that the AZX811 alloy has a good roll-ability. In case 2 and case 4, only minor edge cracks were observed on the rolled sheet. By increasing T_R from 80°C to 250°C, the sheets were produced without any obvious edge cracks for each sheet thickness.

The number of edge cracks and the total crack length were measured along the area of 100 mm length, in each sheet thickness [26]. The result is depicted in Figure 9. The number of edge cracks increased with decreasing sheet thickness for each rolling condition. When rolling is performed at a sample temperature of 250°C (case 3), the total crack length and the edge cracks reach 200 mm and 20 pieces, respectively. On the other hand, when rolling was performed at a T_S of 430°C (case 2 and case 4), the number of edge cracks increased with decreasing sheet thickness, but the total crack length did not indicate significant propagation of the cracks. In the rolling process of the AZX811 alloy, it was necessary to set T_S to a high enough temperature to produce a rolled sheet with high strength and elongation and less edge cracks.

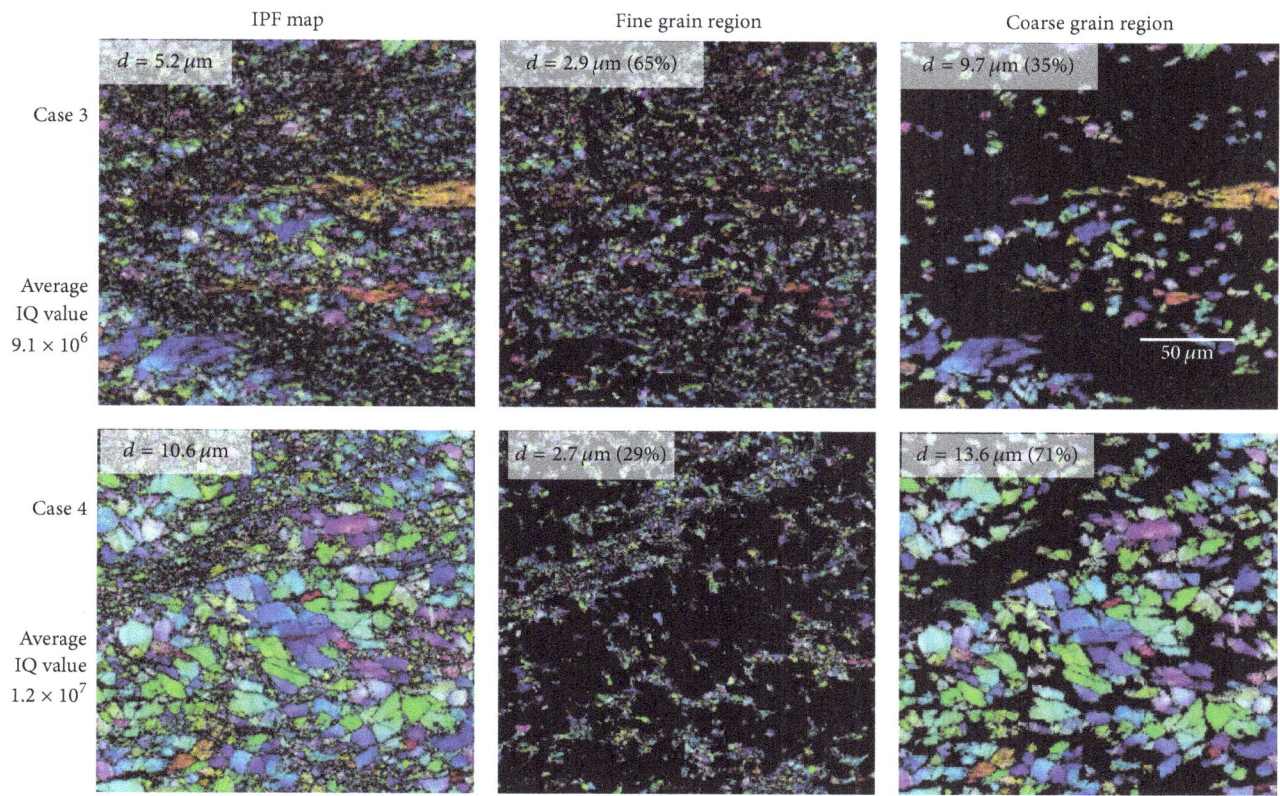

(a)

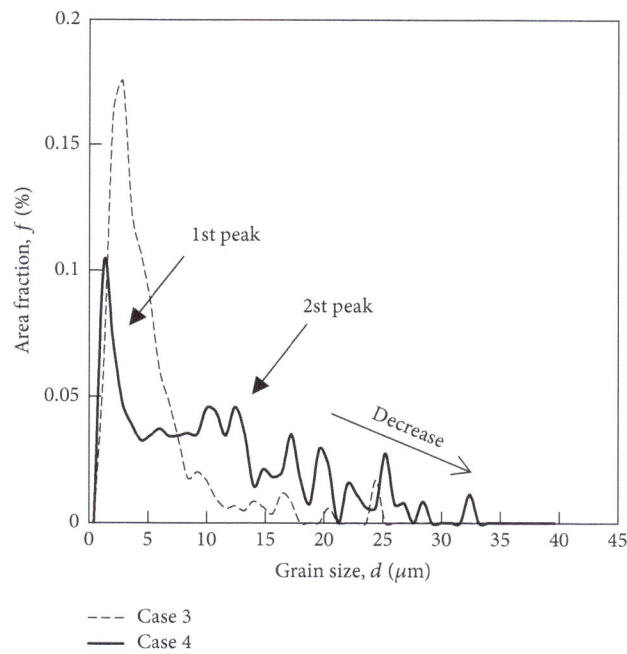

(b)

FIGURE 7: (a) Inverse pole figure maps of the 3 mm thick rolled sample and (b) the grain size distribution diagram. The grain size and area fraction are shown in the IPF maps.

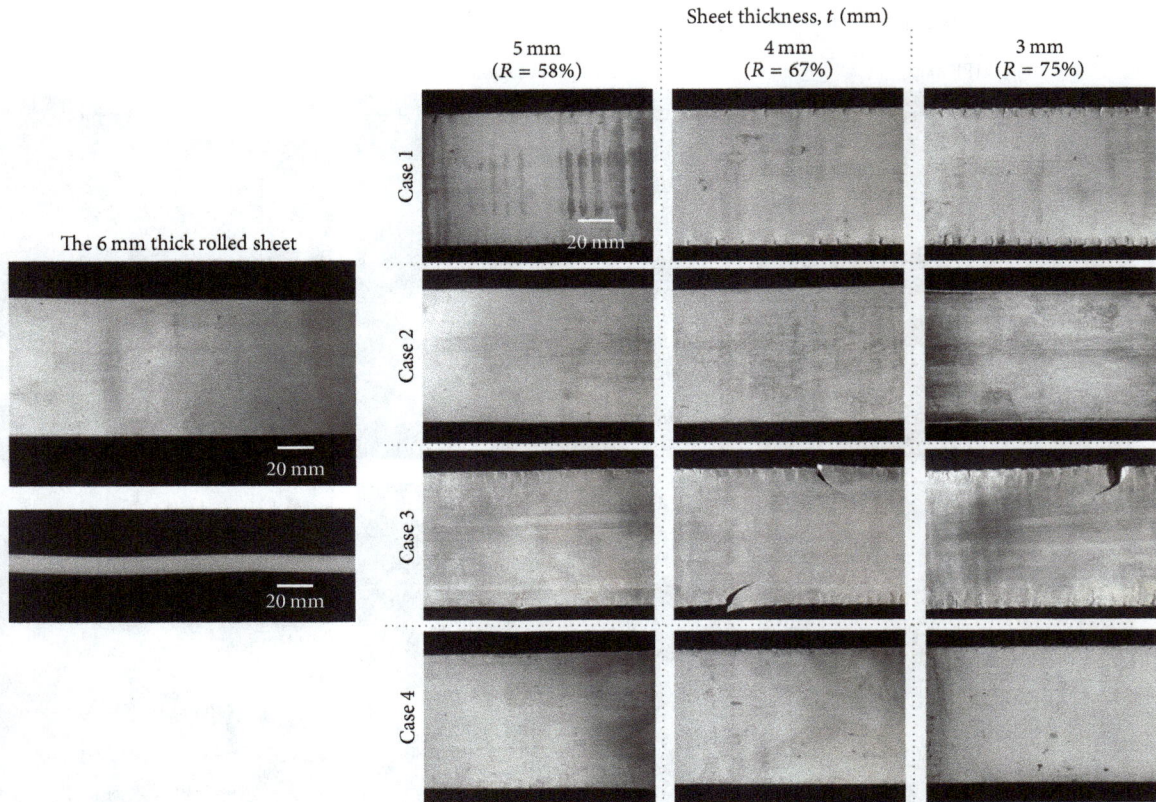

Sheet thickness, *t* (mm)

The 6 mm thick rolled sheet

FIGURE 8: The appearance of the AZX811 rolled sheet at each rolling temperature, sample temperature, and sheet thickness.

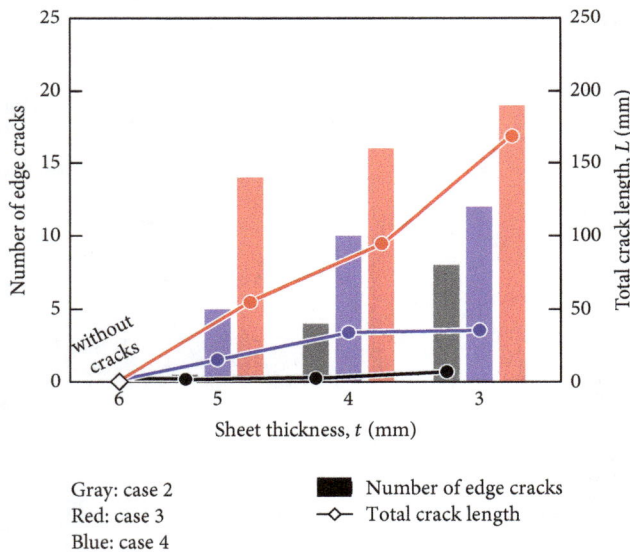

Gray: case 2
Red: case 3
Blue: case 4

■ Number of edge cracks
◇ Total crack length

FIGURE 9: The statistical data on the edge cracks of the rolled sheet at each rolling temperature, sample temperature, and sheet thickness.

4. Conclusion

The influence of the rolling rate, temperature, and intermediate heat treatment was investigated for the fabrication of metal plates with high strength and high ductility using incombustible cast AZX811 alloy as the starting material. In the rolling of cast materials, an improvement in the intensity of the materials was found to depend on the fineness of the magnesium grains up to 42% of the total reduction ratio. On the other hand, at 42% or more of the total reduction ratio, the linear gradient of the Hall-Petch equation changed partially because of the mechanism of grain refining, as affected by the intermetallic compound. Both the strength and ductility are improved by the formation of a bimodal structure, and the material with a plate thickness of 3 mm showed the greatest tensile strength at 351 MPa and an elongation of 12.2%.

Conflicts of Interest

The authors declare that they have no conflicts of interest.

Acknowledgments

This paper is based on results obtained from the Future Pioneering Program (Innovative Structural Materials Project) commissioned by the New Energy and Industrial Technology Development Organization (NEDO).

References

[1] A. A. Luo, "Magnesium: Current and potential automotive applications," *The Minerals, Metals & Materials Society*, vol. 54, pp. 42–48, 2002.

[2] H. Mori, K. Fujino, K. Kurita et al., "Application of the flame-retardant magnesium alloy to high speed rail vehicles," *Materia Japan*, vol. 52, no. 10, pp. 484–490, 2013.

[3] M. Sakamoto, S. Akiyama, T. Hagio, and K. Ogi, "Control of oxidation surface film and suppression of ignition of molten Mg-Ca Alloy by Ca addition," *Journal of Japan Foundry Engineering Society*, vol. 69, pp. 227–233, 1997.

[4] Y. Kawamura and M. Yamasaki, "Formation and mechanical properties of Mg97Zn1RE 2 alloys with long-period stacking ordered structure," *Materials Transactions*, vol. 48, no. 11, pp. 2986–2992, 2007.

[5] S. W. Xu, K. Oh-Ishi, S. Kamado, F. Uchida, T. Homma, and K. Hono, "High-strength extruded Mg-Al-Ca-Mn alloy," *Scripta Materialia*, vol. 65, no. 3, pp. 269–272, 2011.

[6] X. Huang, K. Suzuki, Y. Chino, and M. Mabuchi, "Texture and stretch formability of AZ61 and AM60 magnesium alloy sheets processed by higherature rolling," *Journal of Alloys and Compounds*, vol. 632, pp. 94–102, 2015.

[7] W. J. Kim, J. D. Park, and W. Y. Kim, "Effect of differential speed rolling on microstructure and mechanical properties of an AZ91 magnesium alloy," *Journal of Alloys and Compounds*, vol. 460, no. 1-2, pp. 289–293, 2008.

[8] W. Z. Chen, X. Wang, M. N. Kyalo, E. D. Wang, and Z. Y. Liu, "Yield strength behavior for rolled magnesium alloy sheets with texture variation," *Materials Science and Engineering A*, vol. 580, pp. 77–82, 2013.

[9] J. A. del Valle and O. A. Ruano, "Effect of annealing treatments on the anisotropy of a magnesium alloy sheet processed by severe rolling," *Materials Letters*, vol. 63, no. 17, pp. 1551–1554, 2009.

[10] S. H. Park, S.-H. Kim, Y. M. Kim, and B. S. You, "Improving mechanical properties of extruded Mg-Al alloy with a bimodal grain structure through alloying addition," *Journal of Alloys and Compounds*, vol. 646, Article ID 34372, pp. 932–936, 2015.

[11] M. Hirano, M. Yamasaki, K. Hagihara, K. Higashida, and Y. Kawamura, "Effect of extrusion parameters on mechanical properties of Mg$_{97}$Zn$_1$Y$_2$ alloys at room and elevated temperatures," *Materials Transactions*, vol. 51, no. 9, pp. 1640–1647, 2010.

[12] T. Sakai, "Microstructure and texture control of magnesium alloy sheets by rolling," *Journal of the Japan Society for Technology of Plasticity*, vol. 50, no. 578, pp. 201–205, 2009.

[13] M. Noda, K. Funami, H. Mori, Y. Gonda, and K. Fujino, "Thermal stability, formability, and mechanical properties of a high-strength rolled flame-resistant magnesium alloy," in *Light Metal Alloys Applications*, A. M. Waldemar, Ed., pp. 125–144, InTech, Rijeka, Croatia, 2014.

[14] W. Yuan, S. K. Panigrahi, J. Q. Su, and R. S. Mishra, "Influence of grain size and texture on Hall-Petch relationship for a magnesium alloy," *Scripta Materialia*, vol. 65, no. 11, pp. 994–997, 2011.

[15] A. Jain, O. Duygulu, D. W. Brown, C. N. Tomé, and S. R. Agnew, "Grain size effects on the tensile properties and deformation mechanisms of a magnesium alloy, AZ31B, sheet," *Materials Science and Engineering A*, vol. 486, no. 1-2, pp. 545–555, 2008.

[16] S. Kleiner, O. Beffort, A. Wahlen, and P. J. Uggowitzer, "Microstructure and mechanical properties of squeeze cast and semisolid cast Mg-Al alloys," *Journal of Light Metals*, vol. 2, no. 4, pp. 277–280, 2002.

[17] T. Ito, M. Noda, H. Mori, Y. Gonda, Y. Fukuda, and S. Yanagihara, "Effect of antigravity-suction-casting parameters on microstructure and mechanical properties of Mg-10Al-0.2Mn-1Ca cast alloy," *Materials Transactions*, vol. 55, no. 8, pp. 1184–1189, 2014.

[18] Y. Yamamoto, N. Sakate, and K. Sakamoto, "Cast-forge process for Al-Ca series magnesium alloy mold by semi solid injection molding," *Transactions of the Japan Society of Mechanical Engineers, Part A*, vol. 77, no. 780, pp. 1388–1397, 2011.

[19] T. Ito, S. Yanagihara, M. Noda, and H. Mori, "Effect of cast structure and forging conditions on upset forgeability of a flame-resistant magnesium alloys," *Journal of Japan Institute of Light Metals*, vol. 65, no. 12, pp. 611–616, 2015.

[20] Y. N. Wang, C. I. Chang, C. J. Lee, H. K. Lin, and J. C. Huang, "Texture and weak grain size dependence in friction stir processed Mg-Al-Zn alloy," *Scripta Materialia*, vol. 55, no. 7, pp. 637–640, 2006.

[21] J. A. del Valle, M. T. Pérez-Prado, and O. A. Ruano, "Texture evolution during large-strain hot rolling of the Mg AZ61 alloy," *Materials Science and Engineering A*, vol. 355, no. 1-2, pp. 68–78, 2003.

[22] S. W. Xu, N. Matsumoto, S. Kamado, T. Honma, and Y. Kojima, "Dynamic microstructural changes in Mg-9Al-1Zn alloy during hot compression," *Scripta Materialia*, vol. 61, no. 3, pp. 249–252, 2009.

[23] S. Kato, C. Swangrat, O. Yamaguchi, Y. Sudo, D. Orlov, and K. Ameyama, "Microstructure and mechanical properties of harmonic structure designed pure copper," in *Proceedings of the Collected Abstracts of the 2013 Autumn Meeting of the Japan Institute of Metals and Materials*, 2013, J26.

[24] H.-W. Park and J. Yanagimoto, "Formation process and mechanical properties of 0.2% carbon steel with bimodal microstructures subjected to heavy-reduction single-pass hot/warm compression," *Materials Science and Engineering A*, vol. 567, pp. 29–37, 2013.

[25] C. D. Yim, B. S. You, J. S. Lee, and W. C. Kim, "Optimization of hot rolling process of gravity cast AZ31-xCa ($x = 0$–2.0 mass%) alloys," *Materials Transactions*, vol. 45, no. 10, pp. 3018–3022, 2004.

[26] F. Guo, D. Zhang, X. Yang, L. Jiang, S. Chai, and F. Pan, "Influence of rolling speed on microstructure and mechanical properties of AZ31 Mg alloy rolled by large strain hot rolling," *Materials Science and Engineering A*, vol. 607, pp. 383–389, 2014.

6

Plasticity Improvement of Ball-Spun Magnesium Alloy Tube based on Stress Triaxiality

Chunjiang Zhao ⓘ,[1,2] **Feitao Zhang,**[1] **Jie Xiong,**[3] **Zhengyi Jiang** ⓘ,[1,2] **Xiaorong Yang,**[1,2] **and Hailong Cui**[2,4]

[1]School of Mechanical Engineering, Coordinative Innovation Center of Taiyuan Heavy Machinery Equipment, Taiyuan University of Science and Technology, Taiyuan 030024, China
[2]The School of Mechanical, Materials and Mechatronic Engineering, University of Wollongong, Wollongong, NSW 2522, Australia
[3]Sichuan Aerospace Special Power Research Institute, Chengdu 610100, China
[4]Institute of Machinery Manufacturing Technology, China Academy of Engineering Physics, Mianyang 621000, China

Correspondence should be addressed to Zhengyi Jiang; jiang@uow.edu.au

Academic Editor: Pavel Lejcek

The effects of thickness reduction, feed ratio, and ball diameter, and their coupling effects, on the average relative stress triaxiality during spinning are discussed via simulation results. The relationships among the parameters and the average value of relative stress triaxiality (AVRST) are fitted with multiple nonlinear functions to calculate the optimal process parameters. According to the trend of stress triaxiality, the corresponding process parameters are calculated for the minimum average value of relative stress triaxiality (AVRST). Room temperature experiments performed on an AZ31 magnesium alloy thin-walled tube with the optimal parameters reveal an improvement of cracking of the tube surface. The study reveals changes in the minimum AVRST and aids in selecting the process parameters to improve plastic performance.

1. Introduction

The ball-spinning process (Figure 1) employs a support ring, conical ring, screw tube, and numerous balls that collectively constitute the ball-spinning mold. The ball-spinning mold is present on the outer wall of the workpiece. The mold and the workpiece rotate relative to each other, and the mold moves along the axis of the workpiece to produce the axial feed. Then, the workpiece placed outside the mandrel comes into contact with the balls, and the workpiece is compressed to produce plastic deformation. The main parameters for the ball-spinning process are shown in Figure 2, where R is the ball radius, Δt is the thickness reduction, f is the feed ratio, and α is the spinning angle.

Rotarescu [1] performed a theoretical derivation and finite-element simulation to establish the relationship between the parameters for ball spinning. Abd-Eltwab et al. [2] studied the effects of processing variables pertaining to ball

spinning on the forming load and the quality of the formed sleeves and determined the optimum values of these variables. Li et al. [3] obtained a formula for calculating the ball-spinning pressure under the assumption of a plane strain state. Zhang et al. [4] analyzed the folding defects formed by ball spinning at the bottom of the inner grooves of copper tubes according to the results of finite-element analysis. Jiang et al. [5, 6] simulated the ball spinning of a nickel-titanium shape memory alloy tube by the rigid-viscoplastic finite-element method and investigated the interface compatibility of the composite tube of copper and aluminum during ball spinning. In [7], the finite-element method was used to simulate the thin-walled tube ball spinning, and the reasonable process parameters were obtained. Kuss and Buchmayr [8, 9] carried out a finite-element simulation and an experiment on the surface cracking phenomenon, which affects the spinning of the workpiece. Jiang et al. [10, 11] simulated multipass backward ball spinning and carried out

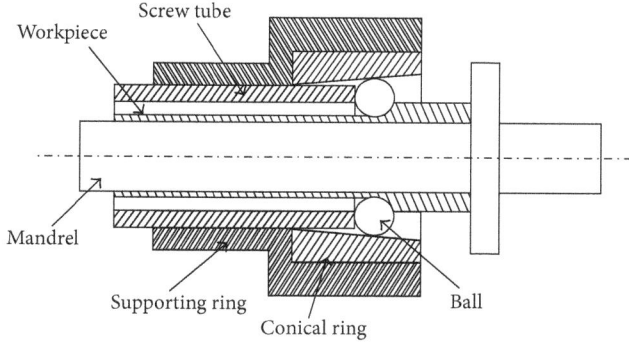

FIGURE 1: Schematic of ball spinning.

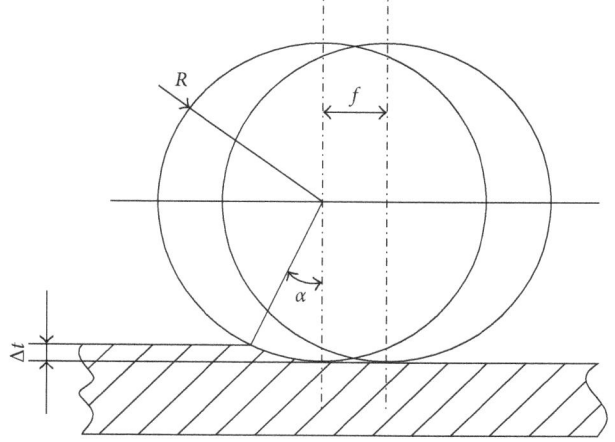

FIGURE 2: Process parameters for ball spinning.

a study on the influence of the ball size on deformability of thin-walled tubular part with longitudinal inner ribs.

As mentioned above, previous research on the ball-spinning process parameters mostly considered the influence of single-process parameters on the spinning tube, without taking into account the coupling effects of various parameters. As a result, when a process parameter changes, the remaining process parameters cannot be correspondingly adjusted.

2. Theoretical Basis and Related Hypotheses

Because of the close-packed hexagonal structure of the metal atom, the magnesium alloy shows poor plasticity and can be easily broken during spinning. Therefore, it is important to select appropriate process parameters to improve the plastic-forming ability and thus ensure surface quality.

Internal factors such as deformation temperature, deformation speed, and deformation methods as well as other external factors affect the deformation behavior of magnesium alloys. At present, a large number of studies on the mechanical properties of magnesium alloys are gradually transferred from normal temperature and quasi-static conditions to different temperatures and different strain rates, including fracture strength and fracture ductility [12].

Rod parameter, soft coefficient, and stress triaxiality are the commonly used stress state parameters for studying the deformation and fracture of a metal. From multidirectional tension to multidirectional compression, the stress triaxiality and different stress states show a significant monotonic change; hence, it is imperative to describe the stress state of the material.

The research results show that ductile fracture caused by plastic deformation is affected by parameters such as strain rate and temperature as well as the stress triaxiality [13, 14]. With an increase in stress triaxiality, the equivalent elastic modulus and equivalent yield stress of a magnesium alloy increase, but its fracture strain gradually decreases [15]. At present, a single stress or strain fracture criterion cannot explain the failure fracture behavior under the complex stress state of a magnesium alloy material. Considering the relationship between the stress triaxiality and the fracture strain as the core of the fracture criterion can help explain the magnesium alloy failure behavior in different stress states.

The stress triaxiality σ^* force is given by

$$\sigma^* = \frac{\sigma_m}{\sigma},$$

$$\sigma_m = \frac{\sigma_1 + \sigma_2 + \sigma_3}{3}, \qquad (1)$$

$$\sigma = \sqrt{\frac{1}{2}\left[(\sigma_1 - \sigma_2)^2 + (\sigma_2 - \sigma_3)^2 + (\sigma_3 - \sigma_1)^2\right]},$$

where σ_m is the spherical stress; σ_1, σ_2, and σ_3 are maximum, intermediate, and minimum principal stresses, respectively; and σ is the von Mises equivalent stress.

Generally, the smaller the σ^* value, the larger is the plastic deformation limit of the material and the better is the plastic-forming ability. El-Magd and Abouridouane [16] studied magnesium alloys and found that, under dynamic loading conditions ($\dot{\varepsilon} > 10^{-3}$), there was an increase in deformation when the strain rate increased.

From the aspect of cracking of the material surface, the fracture failure of the metal is related to the strain rate and temperature in addition to the stress triaxiality. The most widely accepted and used fracture failure criterion is the Johnson–Cook fracture failure model, which is expressed as follows [17]:

$$\varepsilon_f = \left[D_1 + D_2 \exp\left(D_3 \sigma^*\right)\right]\left(1 + D_4 \ln \dot{\varepsilon}\right)\left(1 + D_5 T^*\right), \qquad (2)$$

where ε_f is the fracture strain; σ^* is the stress triaxiality; σ_e is the Mises equivalent stress; D_1, D_2, D_3, D_4, and D_5 are the material constants; $\dot{\varepsilon}$ is the strain rate; and T^* is a temperature parameter.

According to the literature [17], in formula (2), stress triaxiality is the most important factor affecting the fracture strain; when the hydrostatic pressure increases, the fracture strain decreases rapidly. The fracture strain mainly depends on the hydrostatic pressure state and is less dependent on the strain rate and temperature.

Thus, stress triaxiality is the decisive factor for the fracture strain of a given material at medium and low strain rates. Although stress triaxiality and equivalent fracture strain can be calculated based on tested data, the material failure strain is not the same as the equivalent fracture strain. Hence, the actual relationship between equivalent strain and stress triaxiality cannot be determined experimentally. For this reason, a numerical simulation must be performed to obtain the accurate stress triaxiality of the specimen.

This study analyzes the change rule for the average value of relative strain triaxiality in the deformation influence zone during the ball spinning of an AZ31 magnesium alloy thin-walled tube. A method for selecting the process parameters based on the stress triaxiality is presented.

Ball spinning is a complex stress-strain process, and the material stress-strain curve changes with the stress state; hence, calculation of the real stress triaxiality is very difficult. Based on the above analysis, the finite-element calculation in this paper has been carried out with the following conservative processing: the strain rate is in the medium-low range and has little effect on the fracture strain; the simulation and experiment are carried out at room temperature, so the effect of temperature on the fracture strain is neglected; a bilinear model of the stress-strain relationship of the material is used in the finite-element model.

Thus, the stress triaxiality value at each point is not the true stress triaxiality but a relative representation of the stress triaxiality. The main purpose is to explore the change in stress triaxiality with different parameters and to provide a qualitative reference for the selection of process parameters toward a small stress triaxiality.

3. Finite-Element Simulation of Ball Spinning

3.1. Model Establishment. In this study, the commercial finite-element software ABAQUS is used to simulate the spinning process. The model is simplified accordingly. The support ring, screw tube, and conical ring are ignored, and ball movement is directly defined. The ball, thrust ring, and mandrel are defined as analytical rigid bodies, and only the tube is defined as the elastoplastic body. The eight-node linear hexahedral element C3D8R is used, and the plastic deformation region is remeshed. As the local deformation is large, an enhanced hourglass control is set up. The finite-element model is shown in Figure 3.

To compare the effects of different process parameters on the stress state of the workpiece (a thin-walled tube), multiple simulations must be conducted. Based on the above discussion, the elastic modulus and yield stress of the workpiece-magnesium alloy tube are given in a simple bilinear model [18] in Table 1. The material properties and process parameters of the tube are shown in Table 1.

3.2. Boundary Condition Settings. In order to maximize the fit of the actual spinning conditions, the boundary conditions for the simulation process are set as follows:

(1) During spinning, the ball rotates in a three-dimensional manner. Hence, the simulation limits its three directions

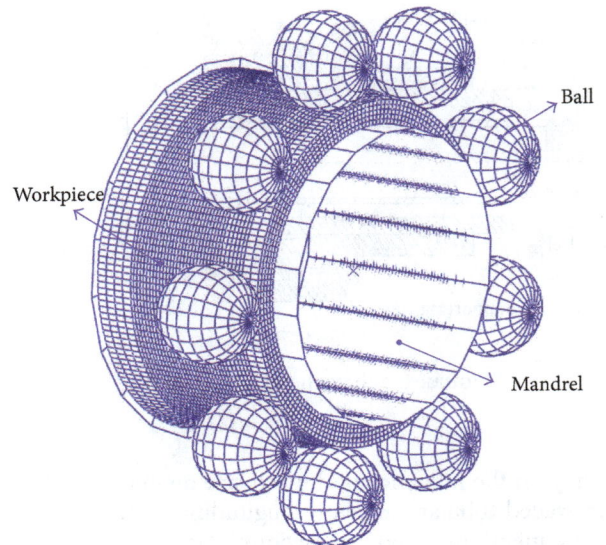

FIGURE 3: Finite-element model.

of translational freedom to retain the rotation freedom.

(2) The tube is in frictional contact with the mandrel and thrust ring at a friction coefficient of 0.08. The contact between the ball and the magnesium alloy material with lubrication corresponds to a friction coefficient of 0.1.

(3) The mandrel is fed axially with the workpiece, and the remaining directions of freedom are restricted.

3.3. Data Extraction from Simulation Results. In ball spinning, besides the metal extrusion by the ball just below the ball, the nearby area is also affected. Thus, this study considers the contact area between the ball and tube and the surrounding vicinity as a single ball-deformation-affected area (Figure 4).

The average value of relative stress triaxiality (AVRST) in the affected zone is taken as the basis for the selection of process parameters, which is mainly in the following considerations:

First, the ball and the workpiece are theoretically in the point contact state, so the actual deformation-affected area is very small. The location of the extreme value of stress triaxiality is usually not the position of the maximum position of the stress, and the AVRST can weaken the influence of fluctuations in the extreme value of stress triaxiality of an isolated unit.

Second, the balls are circumferentially distributed along the circumference of the workpiece, and the contact and noncontact states of the ball are continuously repeated at the same point on the workpiece. This repeated state is contained in the deformation-affected zone.

Therefore, it is more reasonable to use the change in the AVRST in the deformation-affected zone to investigate the plastic-forming ability of the deformation zone of the workpiece.

TABLE 1: Properties and process parameters of the blank tube.

Tube material	Elastic modulus (MPa)	Poisson's ratio	Yield stress (σ_s)	Outside diameter of tube (mm)	Tube-wall thickness (mm)	R (mm)	Δt (mm)	f (mm/r)
Magnesium alloy (AZ31B)	44800	0.31	180	18	1.5	2.5	0.1	0.1
						3	0.2	0.15
						3.5	0.3	0.2
						4	0.4	0.25
						4.5	0.5	0.3

Along the circumferential direction of the workpiece shell, the tension zone between two balls appears at intervals, immediately below the ball; eight units are taken from each side in the ball feeding direction to constitute the deformation-affected zone.

The stress triaxiality value of each element in the set is extracted, as shown in Table 2.

As mentioned above, the stress triaxial values are relative, but its change can be derived from multiple sets of process parameters; this can qualitatively guide the selection of the process parameters in favor of plasticity improvement.

4. Results of Finite-Element Calculation

The three main process parameters—ball diameter, thickness reduction, and feed ratio—affect the stress state of the workpiece during spinning, and the coupling effects between these parameters are also significant. Therefore, the relationships between one of these parameters and the other two parameters are studied.

The AVRST in the deformation-affected zone under different parameter configurations for each group in Table 3 is plotted as a graph. Cloud diagrams of relative stress triaxiality by the finite-element method, corresponding to each group of process parameters, are extracted. The areas in which the relative stress triaxiality is greater than zero are set in white color for significant distinction, as shown in Figures 5–10, for each graph and cloud diagram.

5. Discussion

According to the calculated data, the relative stress triaxiality for different ball diameters, amounts of thinning, and feed ratios is analyzed and discussed as follows.

5.1. Effect of Ball Diameter. As seen in Figure 5, as the ball diameter gradually increases, the AVRST in the deformation-affected zone decreases first and then increases. This observation indicates that excessively small or excessively large ball diameters are not suitable for the plastic deformation capacity.

As can be seen from curves 1 and 3 in Figure 5, the minimum AVRST in the deformation-affected zone appears at $R = 3$ mm, while the spinning angle is

$$\alpha = \arccos \frac{R - \Delta t}{R} = \arccos \frac{3 - 0.2}{3} = 21^\circ. \quad (3)$$

Curve 2 shows the minimum value when $R = 4.5$ mm, and the corresponding spinning angle is

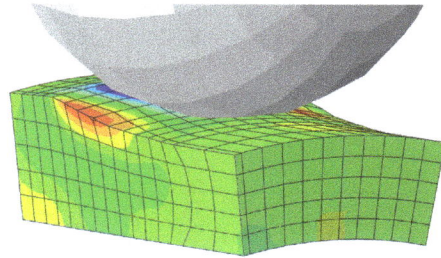

FIGURE 4: Deformation-affected area.

$$\alpha = \arccos \frac{R - \Delta t}{R} = \arccos \frac{4.5 - 0.3}{4.5} = 21.04^\circ. \quad (4)$$

This angle is consistent with the best spinning angle obtained by the production practice mentioned in the literature [19].

From the contrasting trend for curves 1 and 3 in Figure 6, it is seen that with an increase in the ball diameter, the difference in AVRST increases. The corresponding AVRST plotted on curves 1 and 3 increases rapidly, but curve 2 is relatively flat. This indicates that when a larger ball diameter is used, a smaller feed ratio and larger thickness reduction should be adopted.

To analyze the distribution of stress triaxiality in Figure 6, a nodal flow vector diagram of the section of the contact area between the ball and the workpiece is extracted, as shown in Figure 11.

Notably, the contact area of the ball is squeezed during spinning. In this case, the relative stress triaxiality is small. During the movement of the ball along the circumference of the workpiece, the material flow velocity is lower on the adjacent front and rear areas of the ball than in the ball contact area. Thus, the frontal pressure and rear tensile stress states are formed.

Moreover, a band-like tensile stress region is generated on the workpiece surface in the direction of about 45° because of the large shearing stress.

When the ball diameter is small, the deformation area is also small. In this case, the relative stress triaxiality in most areas is small and negative. With an increase in ball diameter, the area of plastic deformation and the area in which the relative stress triaxiality is positive increase, but the relative stress triaxiality pole value decreases from 5.16 to 4.71.

Moreover, when the ball diameter is $R = 4$ mm, the minimum value of relative stress triaxiality is larger than that at $R = 3$ mm, and this minimum value generally appears immediately below the ball. This indicates that as the ball diameter

TABLE 2: Average stress triaxial value for different process parameters.

No.	R	Δt	f	TRIAX	No.	R	Δt	f	TRIAX	No.	R	Δt	f	TRIAX
1	2.5	0.2	0.2	−0.7413	13	2.5	0.3	0.1	−0.1733	25	4.5	0.3	0.2	−0.8001
2	3.5	0.2	0.2	−0.9252	14	2.5	0.3	0.15	−0.4531	26	4.5	0.4	0.2	−0.6585
3	4	0.2	0.2	−0.8034	15	2.5	0.3	0.25	−0.4681	27	4.5	0.5	0.2	−0.6274
4	4.5	0.2	0.2	−0.7574	16	2.5	0.3	0.3	−0.2836	28	3	0.1	0.2	−0.6487
5	2.5	0.2	0.3	−0.6020	17	2.5	0.2	0.1	−0.4014	29	3	0.2	0.2	−1.0312
6	3.5	0.2	0.3	−0.7252	18	2.5	0.2	0.15	−0.6340	30	3	0.4	0.2	−0.5260
7	4	0.2	0.3	−0.5834	19	2.5	0.2	0.25	−0.7329	31	3	0.5	0.2	−0.4594
8	4.5	0.2	0.3	−0.5074	20	3	0.2	0.1	−0.5552	32	3	0.1	0.3	−0.5487
9	2.5	0.3	0.2	−0.5699	21	3	0.2	0.15	−0.8883	33	3	0.3	0.3	−0.5886
10	3	0.3	0.2	−0.8416	22	3	0.2	0.25	−0.9686	34	3	0.4	0.3	−0.3063
11	3.5	0.3	0.2	−0.7552	23	3	0.2	0.3	−0.8766	35	3	0.5	0.3	−0.2050
12	4	0.3	0.2	−0.7634	24	4.5	0.1	0.2	−0.4416					

TABLE 3: Process parameters.

Process parameters	Different ball diameters R (mm)		
Δt (mm)	0.2	0.2	0.3
f/mm/r	0.2	0.3	0.2
R (mm)	2.5, 3.0, 3.5, 4.0, 4.5		
Process parameters	Different amounts of thinning Δt (mm)		
f/mm/r	0.2	0.3	0.2
R (mm)	3	3	4.5
Δt (mm)	0.1, 0.2, 0.3, 0.4, 0.5		
Process parameters	Different feed ratios f (mm)		
R (mm)	2.5	2.5	3
Δt (mm)	0.2	0.3	0.2
f/mm/r	0.1, 0.15, 0.2, 0.25, 0.3		

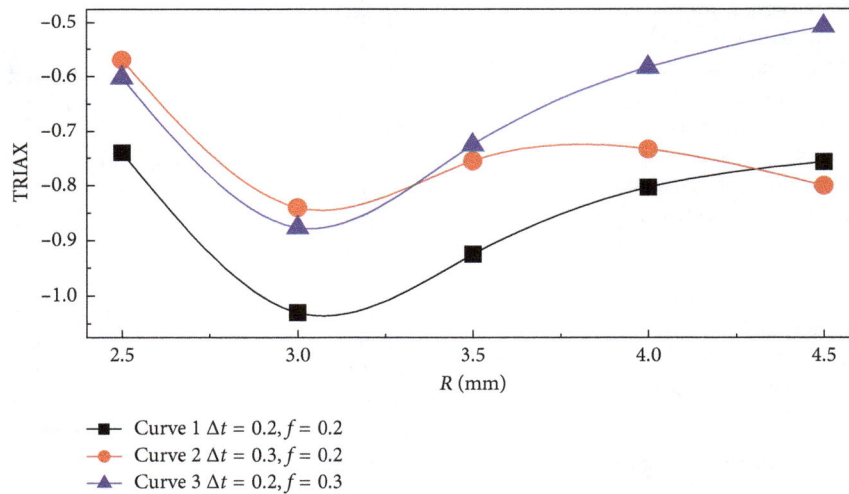

FIGURE 5: Graph of change in AVRST with ball diameter.

increases, the plastic limit of the material decreases, and particularly, the extent of the thickness reduction is diminished. Moreover, when the ball radius increases, the extremum of relative stress triaxiality in the tension region increases, so excessively small ball diameters are highly undesirable.

5.2. *Effect of Thickness Reduction.* In Figure 7, the AVRST decreases first and then increases with increasing thickness reduction. This observation indicates that excessively high or low thickness reductions are not conducive for ductile-forming ability. From the three curves in Figure 7, when the ball diameter is $R = 3$ mm, the thickness reduction corresponding to the minimum AVRST is 0.2. When the ball diameter is $R = 4.5$ mm, the thickness reduction corresponding to the minimum AVRST is 0.3. These two values satisfy the following relation:

$$\Delta t = R\left(1 - \cos 21°\right). \tag{5}$$

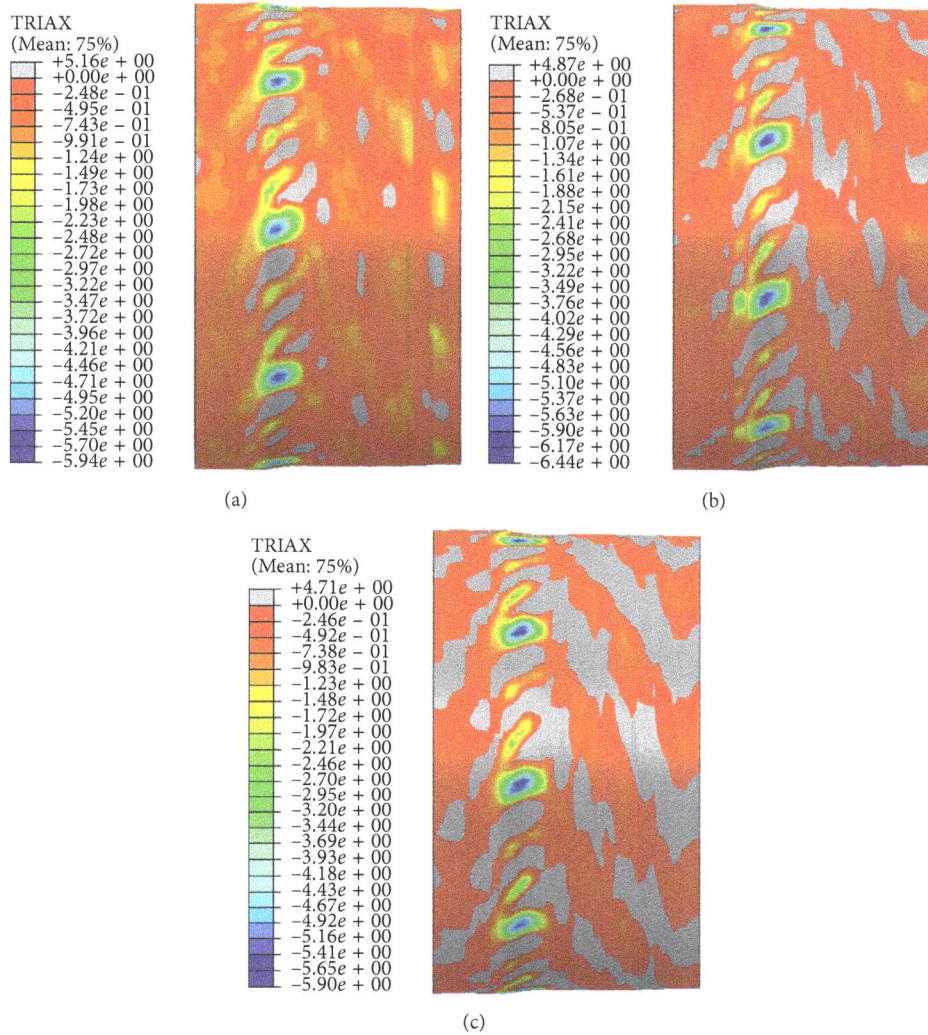

FIGURE 6: Cloud diagram of relative stress triaxiality for different ball diameters. (a) $R = 2.5\,\text{mm}$, $f = 0.2\,\text{mm/r}$, and $\Delta t = 0.2\,\text{mm}$. (b) $R = 3\,\text{mm}$, $f = 0.2\,\text{mm/r}$, and $\Delta t = 0.2\,\text{mm}$. (c) $R = 4\,\text{mm}$, $f = 0.2\,\text{mm/r}$, and $\Delta t = 0.2\,\text{mm}$.

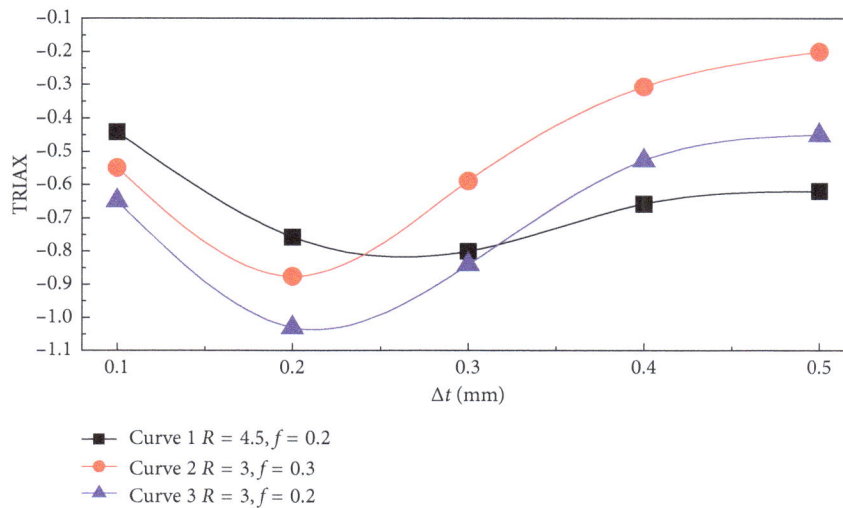

FIGURE 7: Graph of change in AVRST with thickness reduction.

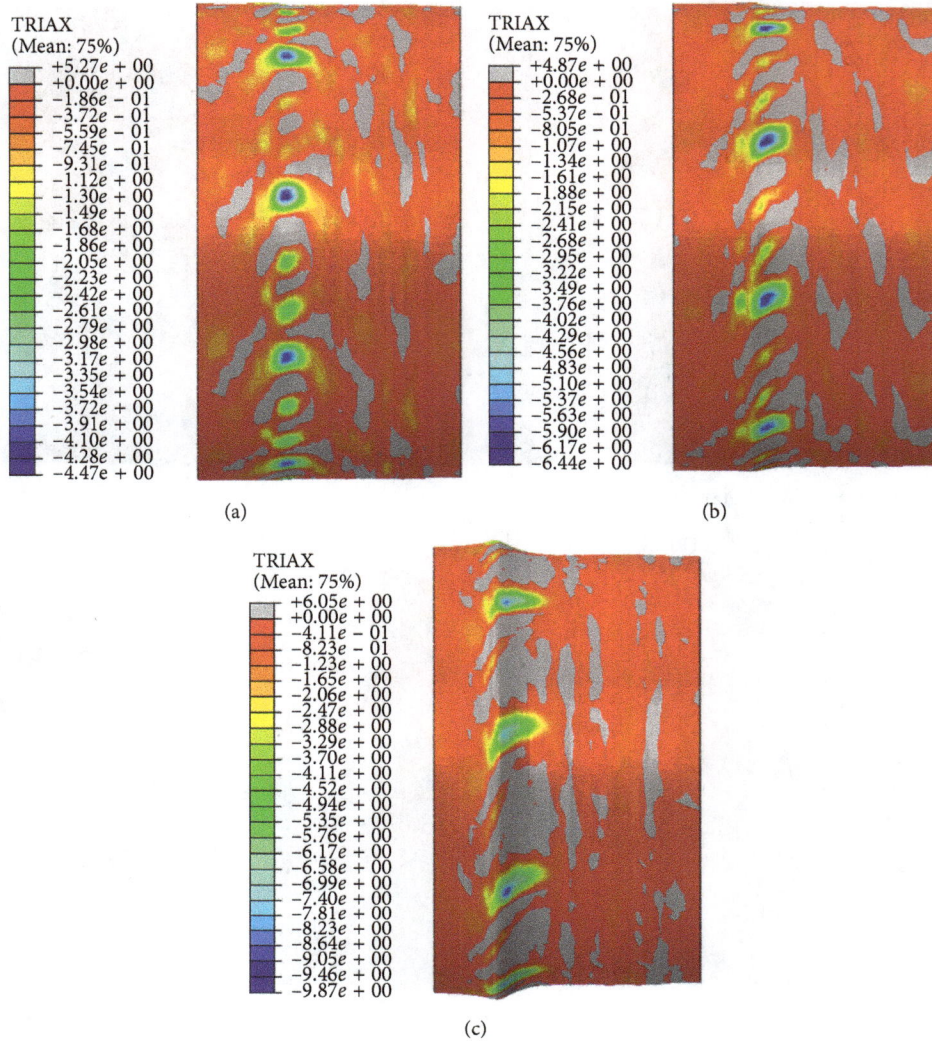

(a)

(b)

(c)

FIGURE 8: Cloud diagram of relative stress triaxiality for different amounts of thinning. (a) $R = 3$ mm, $f = 0.2$ mm/r, and $\Delta t = 0.1$ mm. (b) $R = 3$ mm, $f = 0.2$ mm/r, and $\Delta t = 0.2$ mm. (c) $R = 3$ mm, $f = 0.2$ mm/r, and $\Delta t = 0.4$ mm.

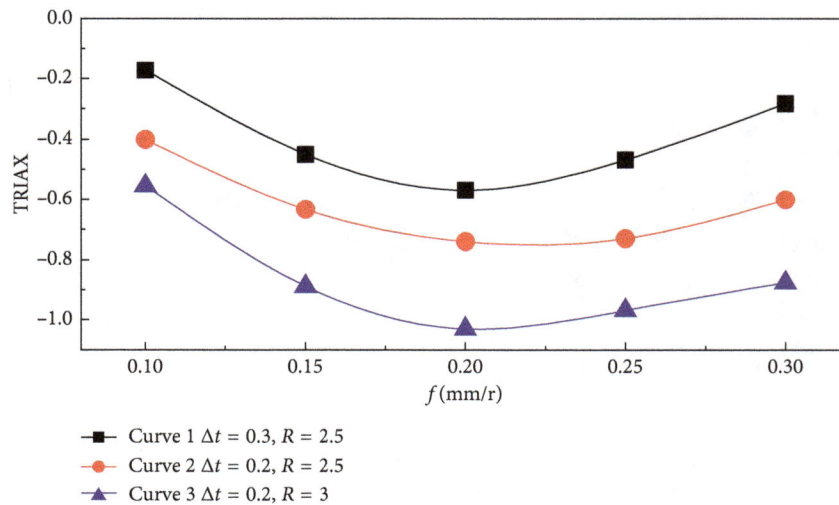

FIGURE 9: Graph of change in AVRST with feed ratio.

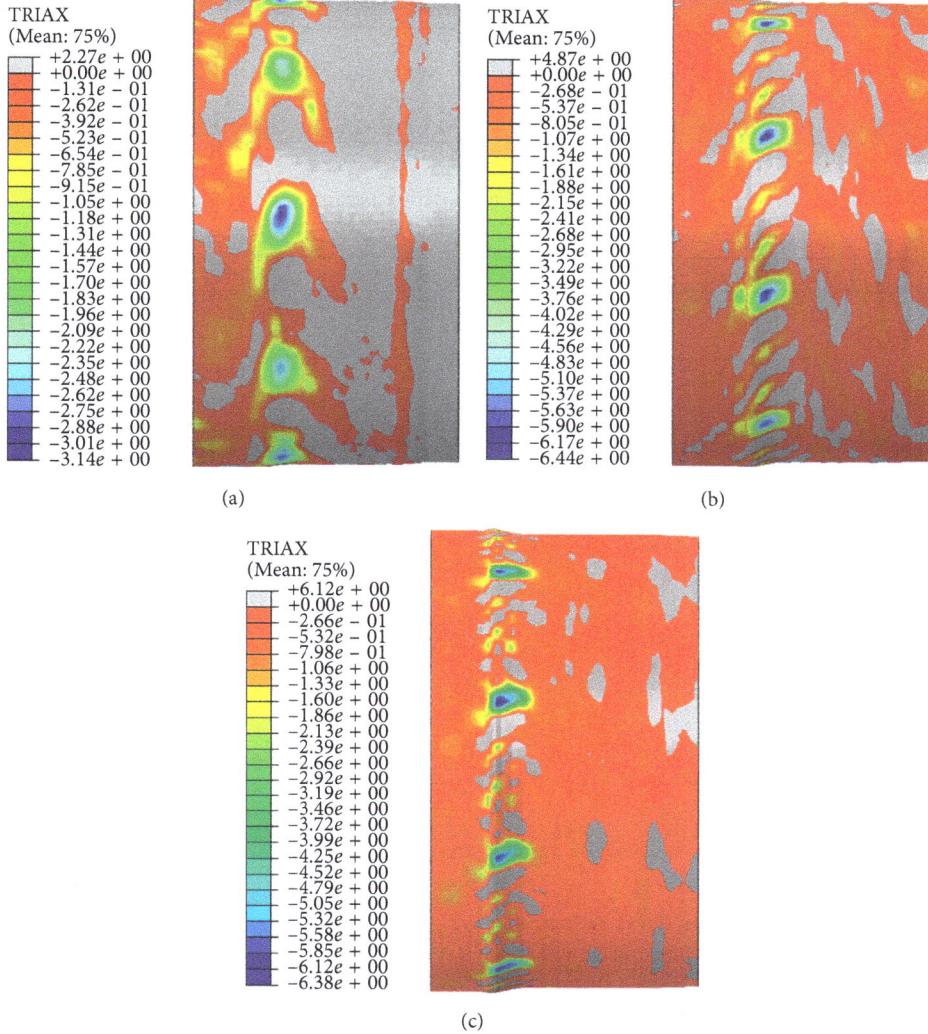

FIGURE 10: Cloud diagram of relative stress triaxiality under different feed ratios. (a) $R = 3$ mm, $f = 0.1$ mm/r, and $\Delta t = 0.2$ mm. (b) $R = 3$ mm, $f = 0.2$ mm/r, and $\Delta t = 0.2$ mm. (c) $R = 3$ mm, $f = 0.3$ mm/r, and $\Delta t = 0.2$ mm.

This correspondence implies that the optimum spinning angle is always about 21°, which is consistent with the analysis results in Section 5.1.

When the thickness reduction exceeds the optimum value, the growth of curves 2 and 3 is faster than that of curve 1. The smaller the ball diameter, the more sensitive is the change in the AVRST to the thickness reduction. Since there are intersections between curve 1 and curves 2 and 3, the influence of ball diameter on the AVRST exceeds the influence of feed ratio when the thickness reduction exceeds that corresponding to the intersection. Therefore, when the thickness reduction is large, the ball diameter match should be first considered. As the thickness reduction increases, curve 2 grows more rapidly than curve 3; that is, as the thickness reduction increases, a larger feed ratio leads to a poor stress state. Therefore, when the ball diameter is the same, the feed ratio should be reduced accordingly when the thickness reduction increases.

Figure 8 shows that the minimum value of relative stress triaxiality decreases with an increase in the thickness reduction,

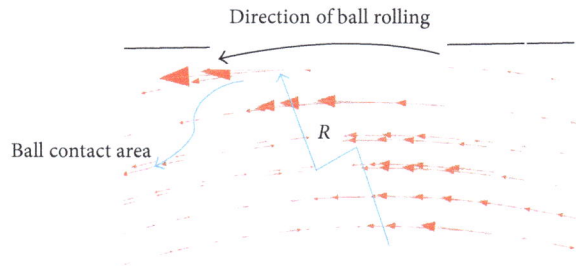

FIGURE 11: Flow vector diagram of nodes of cross section at the contact area center.

and that its maximum value decreases first and then decreases with an increase in the thickness reduction. With an increase in the thickness reduction, the area similar to an inclined strip, where the relative stress triaxiality is greater than 0 decreases and the inclination angle progressively decreases; however, the tensioned area between the two balls increases gradually.

TABLE 4: Fitting function coefficient.

a_1	a_2	a_3	a_4	a_5	a_6	a_7	a_8
$-2.84E-01$	$-5.49E+01$	$-4.49E+01$	$-7.78E-02$	$3.64E-01$	$-1.41E+01$	$-3.92E-01$	$-4.06E+00$
a_9	b_1	b_2	b_3	c_1	c_2	c_3	d
$2.42E+01$	$3.07E+00$	$6.02E+01$	$6.15E+01$	$-1.10E+01$	$-1.64E+01$	$-1.95E+01$	$1.52E+01$

TABLE 5: Fitness determination parameters.

RMSE	SSE	R	R^2	DC
0.032591414	0.037177010	0.987666548	0.975485211	0.975485211

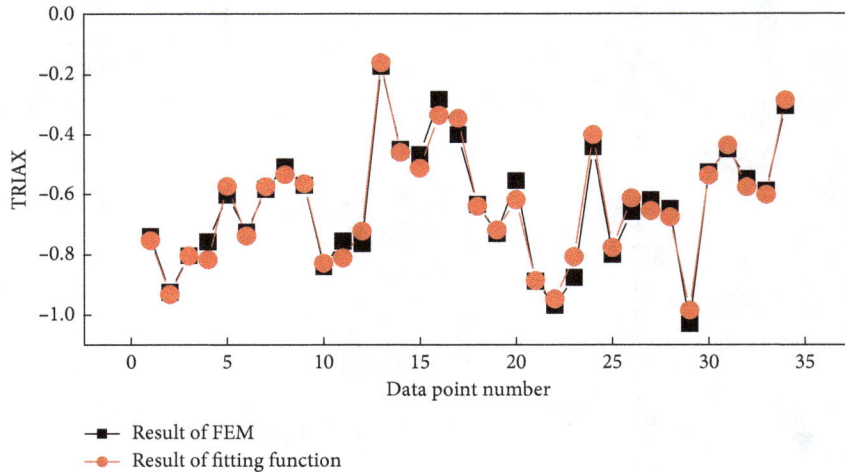

FIGURE 12: Comparison of the measured values of the average stress triaxiality and the calculated values of the fitting function.

In addition, with an increase in the thickness reduction, the area of the unspun section of the workpiece in which the relative stress triaxiality is greater than 0 shows a decreasing trend. This is because as the thickness reduction increases, the radial spinning force component increases faster than the axial force and tangential force component [20]; therefore, a larger thickness reduction is advantageous for reducing circumferential torsional failure and axial pressure buckling.

FIGURE 13: Experimental equipment.

5.3. *Effect of Feed Ratio.* As seen in the three curves in Figure 9, the AVRST first decreases and then increases with increasing feed ratio. This plot indicates that an excessively large or small feed ratio is not conducive for improving the plastic-forming ability of the tube, and all the feed ratios corresponding to the minimum AVRST is about 0.2. By comparing curve 1 and curve 2, it can be seen that, at a larger thickness reduction, we must use a smaller feed ratio to achieve better stress states. By comparing curve 2 and curve 3, it can be seen that when the ball diameter increases, the feed ratio used should also be high for a smaller AVRST.

Figure 10 shows that when the feed ratio is small, the AVRST of the deformation-affected zone is large. The area mainly distributed in the spinned region, where the relative stress triaxiality is greater than 0 is large, but the maximum relative stress triaxiality is 2.27, which is smaller than that for the other cases, indicating that it is difficult to break the material under these conditions.

6. Multivariate Nonlinear Function Fitting

From the above analysis, the trend of AVRST with the change of process parameters is obtained, so the nonlinear function is fitted according to the existing calculation data in the following text. So that when a process parameter changes, it is easy to match the remaining process parameters.

According to the simulation results, the three-variable cubic polynomial is selected as fitting function. During the fitting analysis using the standard ternary cubic polynomial model, it is found that a significant collinearity relationship exists among the four items of $R\Delta tf$, $R\Delta t$, Rf, and Δtf in the polynomial. However, when these four items are applied into the fitting function model, the model becomes distorted, and the fitting results are not estimated. Therefore, these four items on the standard ternary cubic polynomial model are eliminated, and the final fitting function model is attained consequently as follows:

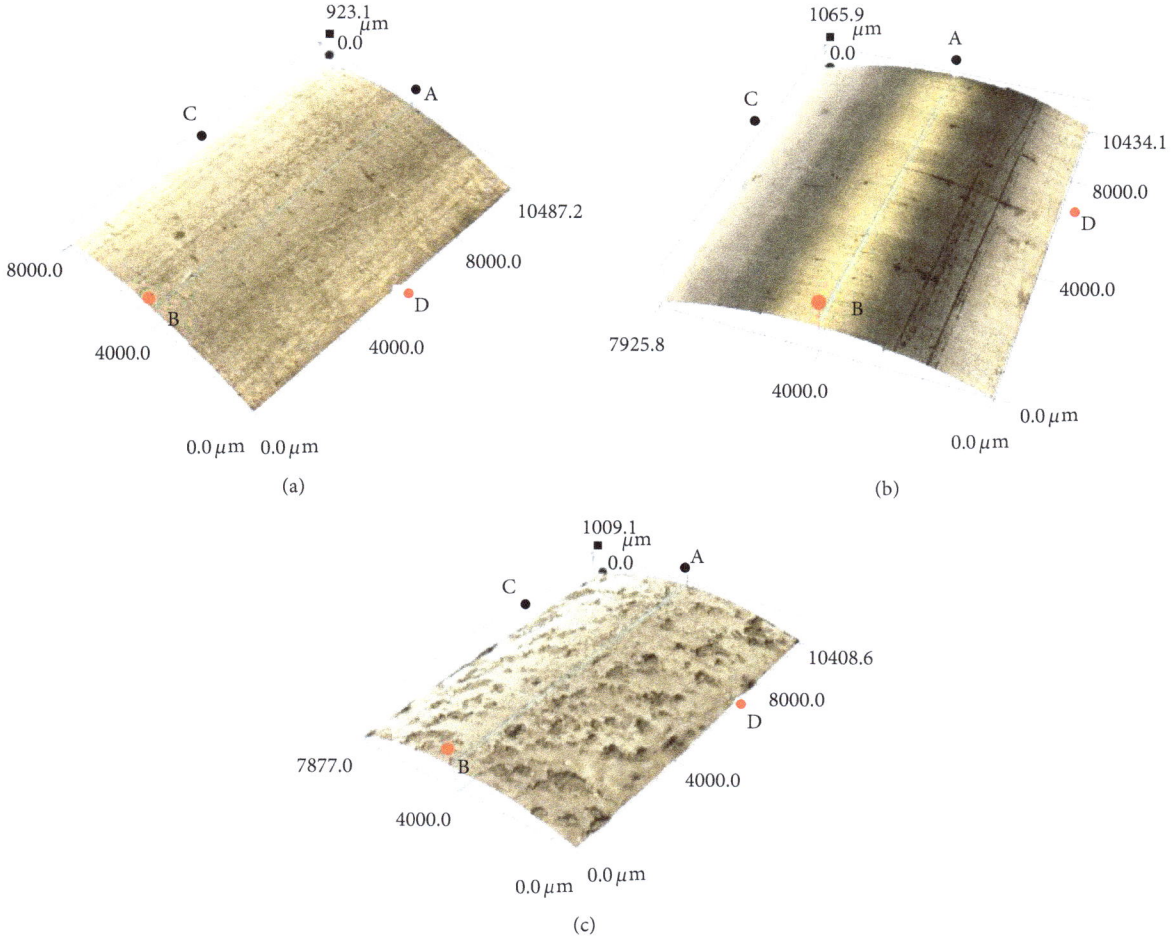

FIGURE 14: Tube after spinning at different thickness reductions. (a) $R = 3$ mm, $f = 0.2$ mm/r, and $\Delta t = 0.1$ mm. (b) $R = 3$ mm, $f = 0.2$ mm/r, and $\Delta t = 0.2$ mm. (c) $R = 3$ mm, $f = 0.2$ mm/r, and $\Delta t = 0.4$ mm.

$$R_d = a_1 R^3 + a_2 \Delta t^3 + a_3 f^3 + a_4 R^2 \Delta t + a_5 R^2 f$$
$$+ a_6 \Delta t^2 f + a_7 R \Delta t^2 + a_8 R f^2 + a_9 \Delta t f^2 + b_1 R^2 \quad (6)$$
$$+ b_2 \Delta t^2 + b_3 f^2 + c_1 R + c_2 \Delta t + c_3 f + d.$$

The data in Table 2 are used, and the results are shown in Table 4.

The fitting degree of the fitting function is also considered, and the determination parameters are shown in Table 5.

The plot in Figure 12 compares the compatibility between the results of FEM and fitting function.

In Figure 12, the compatibility between the measured value of the AVRST and the calculated value of the fitting function is high with no point of complete deviation, so the fitting function model given in this paper is reliable.

At the given range of ball diameter of 2.5 mm $\leq R \leq$ 4.5 mm, thickness reduction of 0.1 mm $\leq \Delta t \leq$ 0.5 mm, and the feed ratio of 0.1 mm/r $\leq f \leq$ 0.3 mm/r, the optimal process parameters that correspond to the minimum AVRST are obtained as follows: $R = 3.01$, $\Delta t = 0.205$, and $f = 0.208$.

7. Experimental Verification

The material used in the experiment is a magnesium alloy AZ31B extruded tube. The horizontal spinning machine used in the experiment is shown in Figure 13, and it can achieve feed ratios of 0.1, 0.2, and 0.3 mm/r.

However, the inner diameter of the conical ring is limited, so the ball diameter cannot be changed arbitrarily to adjust the range of thickness reductions. Therefore, the experimental ball diameter is fixed $R = 3.0$ mm, and the experiment only explores the changes of thickness reduction and feed ratio. In line with the previous finite-element analysis, the number of balls used in the experiment is 9, and the spinning mold is filled with grease.

To clearly observe the tube surface after spinning for comparative analysis, the spinned tube surface is examined by an ultradepth microscope.

Spinning experiments are carried out for different thickness reductions and feed ratios. The experimental results are shown in Figures 14 and 15.

Figure 14(a) shows that the pipe surface is smoother and shows minor cracks. In Figure 14(b), the surface finish is the highest, and there are no obvious cracks except for the

(a)

(b)

(c)

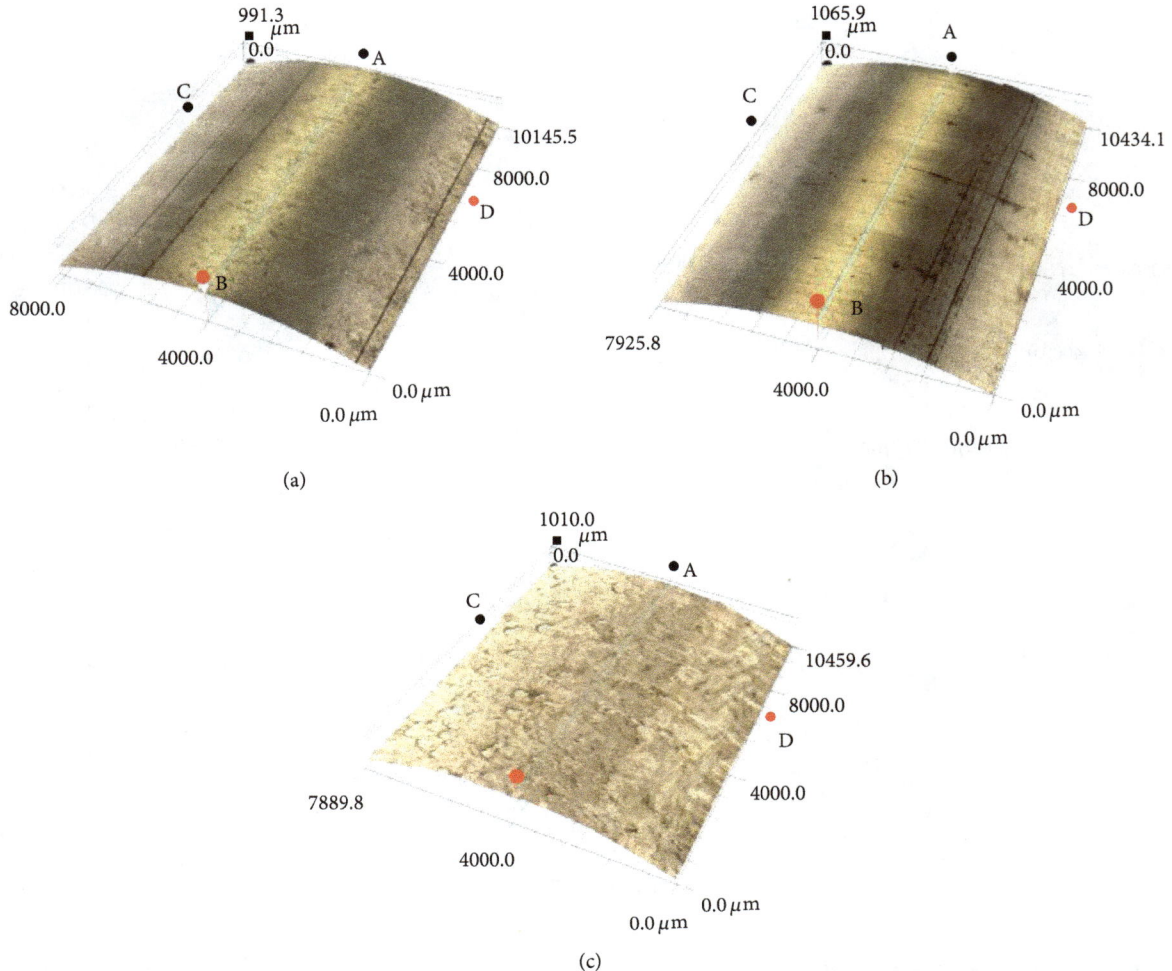

FIGURE 15: Tube after spinning under different feed ratios. (a) $R = 3$ mm, $f = 0.1$ mm/r, and $\Delta t = 0.2$ mm. (b) $R = 3$ mm, $f = 0.2$ mm/r, and $\Delta t = 0.2$ mm. (c) $R = 3$ mm, $f = 0.3$ mm/r, and $\Delta t = 0.2$ mm.

original scratches on the surface of the tube. The tube surface in Figure 14(c) is seriously damaged, and deep cracks are visible along the tube circumference.

In Figure 15(a), the pipe surface shows no obvious cracks and debris but displays a poor and dim finish. Figure 15(b) is the same as Figure 14(b). In Figure 15(c), the surface shows visible cracks and a rolled skin, and the micrographs reveal a stack of layers on the surface.

It can be seen from the experimental results that the quality of the spinned tube is closely related to the AVRST, and the failure of the tube after spinning is consistent with the simulation results. It is thus demonstrated that the method for using the AVRST to characterize the plastic-forming ability of the material is feasible.

8. Conclusion

In this paper, the influence of the process parameters on the stress state of the spinning deformation zone during ball spinning is described by finite-element simulation. The relationship among the three parameters—ball diameter, feed ratio, and thickness reduction—and the average stress triaxiality are discussed. Finally, spinning experiments are carried out, and the following conclusions are drawn.

The AVRST for the ball-spinning deformation first decreases and then increases with changes in the three main process parameters. Excessively large or small values of the ball diameter, feed ratio, and thickness reduction are not conducive for improving the plastic-forming ability of the tube. When a large thickness reduction is used, a large ball diameter can improve the stress state. When the feed ratio is large, the ball diameter is reduced, and the stress state in the deformation-affected zone is improved; increasing the ball diameter and reducing the feed ratio is beneficial for improving the plastic-forming capacity of the tube. The fitting formula used in this paper can predict the AVRST of the deformation-affected zone of the workpiece accurately within a certain range of process parameters.

Conflicts of Interest

The authors declare that there are no conflicts of interest.

Acknowledgments

The authors would like to thank the National Natural Science Foundation of China (no. 51375325), the Shanxi Coal Based Low Carbon Joint Fund (U1610118 and U1510131), the Shanxi Provincial Special Fund for Coordinative Innovation Center of Taiyuan Heavy Machinery Equipment, and the Fund for Shanxi "1331 Project" Key Subjects Construction for their support to this research.

References

[1] M. I. Rotarescu, "A theoretical analysis of tube spinning using balls," *Journal of Materials Processing Technology*, vol. 54, no. 1–4, pp. 224–229, 1995.

[2] A. A. Abd-Eltwab, S. Z. El-Abden, K. I. E. Ahmed, M. N. El-Sheikh, and R. K. Abdel-Magied, "An investigation into forming internally-spline sleeves by ball spinning," *International Journal of Mechanical Sciences*, vol. 134, pp. 399–410, 2017.

[3] M. Li, S. Zhang, D. Kang et al., "Ball diameter selection in ball spinning process," *Materials Science and Technology*, vol. 13, no. 6, pp. 594–597, 2005.

[4] G. Zhang, S. Zhang, B. Li, and H. Zhang, "Analysis on folding defects of inner grooved copper tubes during ball spin forming," *Journal of Materials Processing Technology*, vol. 184, no. 1–3, pp. 393–400, 2007.

[5] S. Jiang, Y. Q. Zhang, Y. N. Zhao et al., "Finite element simulation of ball spinning of NiTi shape memory alloy tube based on variable temperature field," *Transactions of Nonferrous Metals Society of China*, vol. 23, no. 3, pp. 781–787, 2013.

[6] S. Jiang, Y. Q. Zhang, Y. Zhao, X. Zhu, D. Sun, and M. Wang, "Investigation of interface compatibility during ball spinning of composite tube of copper and aluminum," *International Journal of Advanced Manufacturing Technology*, vol. 88, no. 1–4, pp. 683–690, 2017.

[7] F. A. Hua, Y. S. Yang, Y. N. Zhang et al., "Three-dimensional finite element analysis of tube spinning," *Journal of Materials Processing Technology*, vol. 168, no. 1, pp. 68–74, 2005.

[8] M. Kuss and B. Buchmayr, "Damage minimised ball spinning process design," *Journal of Materials Processing Technology*, vol. 234, pp. 10–17, 2016.

[9] M. Kuss and B. Buchmayr, "Analytical, numerical and experimental investigations of a ball spinning expansion process," *Journal of Materials Processing Technology*, vol. 224, pp. 213–221, 2015.

[10] S. Jiang, Z. Ren, C. Li, and K. Xue, "Role of ball size in backward ball spinning of thin-walled tubular part with longitudinal inner ribs," *Journal of Materials Processing Technology*, vol. 209, no. 4, pp. 2167–2174, 2009.

[11] S. Jiang, Y. Zheng, Z. Ren, and C. Li, "Multi-pass spinning of thin-walled tubular part with longitudinal inner ribs," *Transactions of Nonferrous Metals Society of China*, vol. 19, no. 1, pp. 215–221, 2009.

[12] N. Gupta, D. Dung, P. Luong, and K. Rohatgi, "A method for intermediate strain rate compression testing and study of compressive failure mechanism of Mg-Al-Zn alloy," *Journal of Applied Physics*, vol. 109, no. 10, p. 103512, 2011.

[13] C. Lou, X. Zhang, G. Duan et al., "Characteristics of twin lamellar structure in magnesium alloy during room temperature dynamic plastic deformation," *Journal of Materials Science and Technology*, vol. 30, no. 1, pp. 41–46, 2014.

[14] M. R. Barnett, K. Zohreh, A. G. Beer, and D. Atwell, "Influence of grain size on the compressive deformation of wrought Mg-3Al-1Zn," *Acta Materialia*, vol. 52, no. 17, pp. 5093–5103, 2004.

[15] Q. Li, F. Ji, and F. Li, "Relationship between stress state and stress state parameters in plastic deformation," *Forging and Stamping Technology*, vol. 39, no. 3, pp. 122–126, 2014.

[16] E. El-Magd and M. Abouridouane, "Characterization, modelling and simulation of deformation and fracture behaviour of the light-weight wrought alloys under high strain rate loading," *International Journal of Impact Engineering*, vol. 32, no. 5, pp. 741–758, 2006.

[17] J. W. Hancock and D. K. Brown, "On the role of strain and stress state in ductile failure," *Journal of Mechanics and Physics of Solids*, vol. 31, no. 1, pp. 1–24, 1983.

[18] B. Wang, D. Q. Yi, W. Gu, and X. Fang, "Thermal simulation on hot deformation behavior of ZK60 and ZK60 (0.9Y) magnesium alloys," *Rare Metal Material and Engineering*, vol. 39, no. 1, pp. 106–122, 2010.

[19] D. Kang, M. Li, and Y. Chen, "A study of force and power in ball spinning," *Materials Science and Technology*, vol. 10, no. 2, pp. 179–182, 2002.

[20] Y. Liu, X. Gao, C. Zhao, and N. Wang, "Research on numerical simulation of ball spinning for thin-wall tube," *Chinese Journal of Engineering Design*, vol. 22, no. 6, pp. 562–568, 2015.

7

Effect of Casting Conditions on the Fracture Strength of Al-5 Mg Alloy Castings

Fawzia Hamed Basuny,[1] Mootaz Ghazy,[2]
Abdel-Razik Y. Kandeil,[2] and Mahmoud Ahmed El-Sayed[2]

[1]Industry Service Complex, Arab Academy for Science, Technology & Maritime Transport, P.O. Box 1029,
 Abu Qir, Alexandria 21599, Egypt
[2]Department of Industrial and Management Engineering, Arab Academy for Science, Technology & Maritime Transport,
 P.O. Box 1029, Abu Qir, Alexandria 21599, Egypt

Correspondence should be addressed to Mahmoud Ahmed El-Sayed; m_elsayed@aast.edu

Academic Editor: Fawzy H. Samuel

During the transient phase of filling a casting running system, surface turbulence can cause the entrainment of oxide films into the bulk liquid. Previous research has suggested that the entrained oxide film would have a deleterious effect on the reproducibility of the mechanical properties of Al cast alloys. In this work, the Weibull moduli for the ultimate tensile strength (UTS) and % elongation of sand cast bars produced under different casting conditions were compared as indicators of casting reliability which was expected to be a function of the oxide film content. The results showed that the use of a thin runner along with the use of filters can significantly eliminate the surface turbulence of the melt during mould filling which would lead to the avoidance of the generation and entrainment of surface oxide films and in turn produce castings with more reliable and reproducible mechanical properties compared to the castings produced using conventional running systems.

1. Introduction

Due to their unique properties, the usage of aluminium alloys in different industrial sectors has grown dramatically in the last decades. Their high elasticity, high electrical and thermal conductivity, and high strength-to-weight ratio allowed them to be widely adopted in the automotive and aerospace industries [1, 2]. However, the mechanical properties of Al castings were found to be greatly affected by the presence of double oxide film defects (or bifilms) which were reported to not only reduce the tensile strength and fatigue limit of the castings but also increase their variability [3–6].

Bifilms were suggested to result from the surface disturbance during metal flow which causes the surface of the liquid metal to fold over onto itself. This causes the upper and lower oxidized surfaces of the folded-over metal to come together and trap a layer of the mould atmosphere between them, creating a double oxide film defect [7–12]. This defect is then incorporated into the bulk liquid in an entrainment action,

which typically constitutes a crack in the solidified casting, as shown in Figure 1 [4]. Oxide films were also shown to act as nucleation sites for hydrogen porosity and iron intermetallics [13, 14]. Such consequences were found to have detrimental effects on the mechanical properties of the castings produced [15]. Results of research performed by Green and Campbell [16] and Nyahumwa et al. [17] suggested that aluminium alloys usually achieve just a small fraction of their intended properties while these defects are present.

Several researchers had explained the mechanism responsible for the entrainment of bifilms in Al castings that usually occurs during pouring of the metal into the mould by introducing the concept of the critical velocity. The critical velocity (V_c) is the flow velocity at the mould entrance (the ingate) above which entrainment of surface oxide films would occur [18] and is commonly written as the following:

$$V_c = 2 \left(\frac{\gamma g}{\rho} \right)^{1/4}, \tag{1}$$

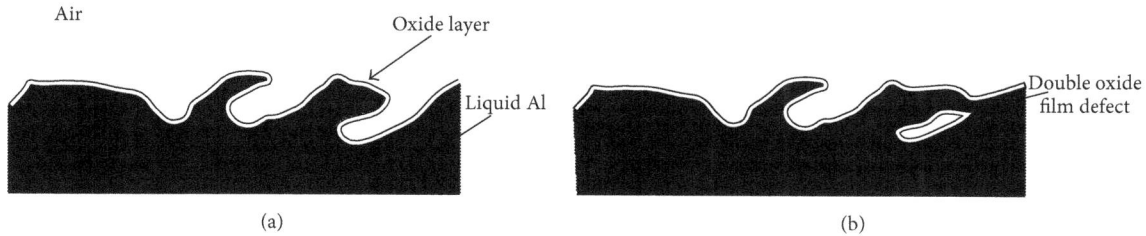

FIGURE 1: The formation of a double oxide film defect. (a) Surface turbulence leads to a breaking wave on the metal surface, and (b) the two unwetted sides of the oxide films come into contact with each other as the bifilm is entrained into the bulk liquid metal [4].

TABLE 1: Chemical composition of the alloy used.

Element	Si	Fe	Cu	Mn	Mg	Cr	Ni	Zn	Pb	Sn	Ti	Al
%	0.11	0.10	0.01	0.01	5.00	0.01	0.01	0.01	0.01	0.01	0.02	Bal

where ρ is the density of the melt in kg/m^3, γ is the surface tension in N/m, and g is the gravity acceleration in m/s^2. For liquid aluminium, $\gamma = 1\,\mathrm{Nm}^{-1}$ and $\rho = 2400\,\mathrm{kgm}^{-3}$; hence, the critical velocity can be estimated to be about 0.5 ms^{-1}. If the mould-entry velocity exceeds the critical velocity, the surface of the melt would be forced to propel upwards, achieving a height sufficient to enfold its oxide surface as it falls back under gravity [13, 18]. Results of the experiments by Runyoro and coauthors [19], Halvaee and Campbell [20], and Bahreinian et al. [21] for different Al and Mg alloys suggested that the critical velocity would be between 0.4 and 0.6 ms^{-1}.

It was suggested that only bottom-gated filling systems can produce reliable castings if the ingate velocities are to be kept below the critical velocity which might prevent the surface turbulence of the molten metal during pouring and reduce the possibility of entraining the oxidized surface inside the bulk liquid [22]. In this work, two different parameters were considered: the height of the runner and the use of filter. The effect of these parameters on the creation of oxide films in Al castings and by implication on the tensile properties of the resulting castings was determined. Understanding these issues could lead to the development of techniques by which oxide film defects might be reduced or eliminated in aluminum castings.

2. Experimental Work

In this study, castings of Al-5 wt.% Mg alloy were produced via gravity casting. Chemical composition of the alloy used is given in Table 1. Four different casting experiments were carried out. In each experiment, two resin-bonded sand moulds were prepared, with the shape and dimensions shown in Figure 2, each producing 10 test bars. The moulds were then held under partial vacuum of about 0.5 bar for 2 weeks before casting, which was suggested in an earlier work to remove the solvent of the resin binder from the moulds and in turn minimize the hydrogen pick-up by the liquid metal from the mould during casting [23]. Two different heights of the runner (thin (10 mm) and thick (25 mm)) were considered. For each runner height, castings were produced with and

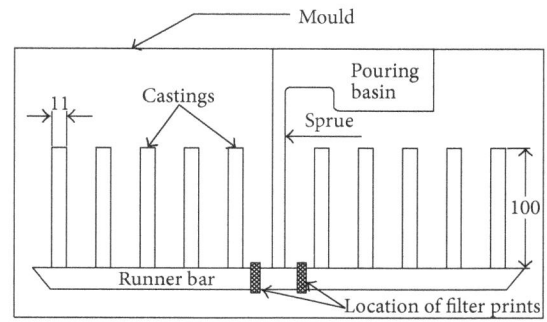

FIGURE 2: Shape and dimensions of the sand mould used in the experiment (dimensions are in mm).

without the use of filters. The experimental plan is shown in Table 2.

In each experiment, about 6 kg of the aluminium alloy was melted in an induction furnace. Once the temperature of the melt reached 850°C, the crucible containing the liquid metal was placed in a vacuum chamber, where the melt was held at about 800°C, under vacuum of about 0.2 bar for two hours, a procedure intended to remove previously introduced bifilms from the melt [24, 25]. The melt was then argon-degassed using a lance for 1 h before pouring from a height of about 1 meter into the sand moulds. In Experiments 2 and 4, two 10 PPI (pores per linear inch) ceramic filters, of dimensions 50 × 50 × 20 mm, were placed in the filter prints at the locations shown in Figure 2. The adopting of thin runner along with the use of filters during pouring was intended to reduce the melt velocity at the ingate which might minimize the possibility of having surface turbulence of the molten metal during mould filling with the corresponding entrainment of the surface oxide films. The filters were expected to play an additional role which was the removal of inclusions from the melt.

After solidification, the castings were machined into tensile test bars of a cylindrical cross section of 10 mm diameter with a gauge length of 100 mm, which were pulled to fracture with a strain rate of 1 mm/min. The fracture

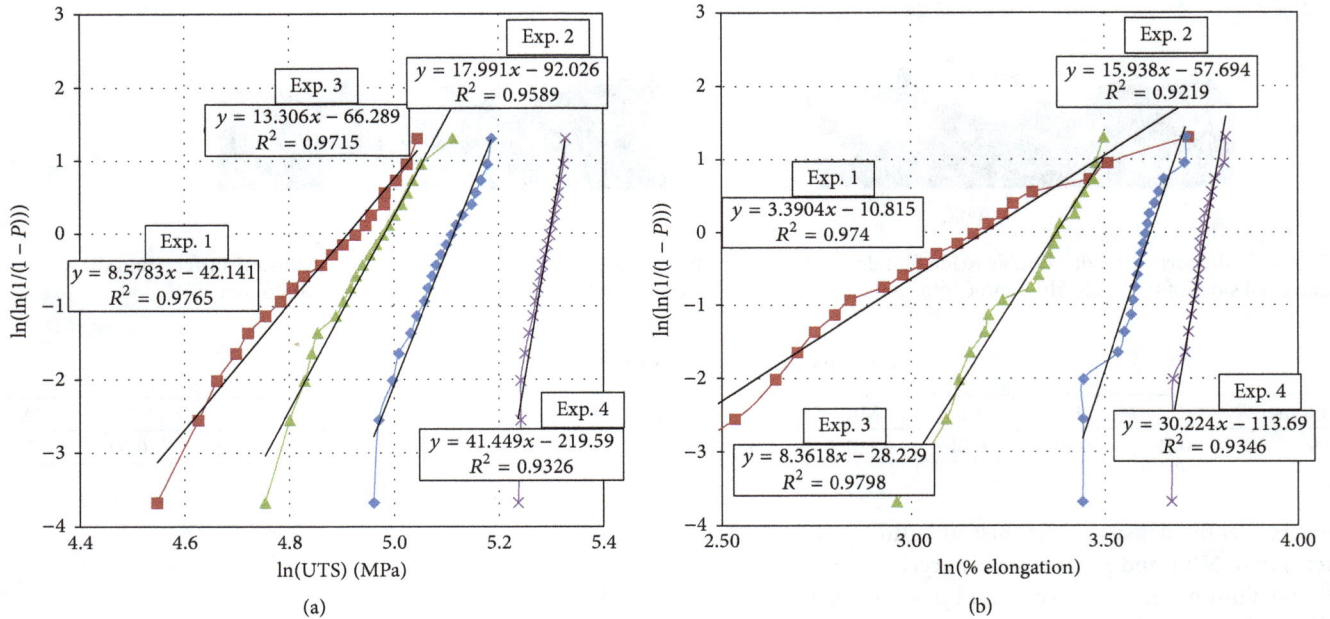

FIGURE 3: Weibull distribution of (a) ultimate tensile strength and (b) % elongation of Al-5 Mg alloy specimens from different experiments listed in Table 1.

TABLE 2: Experimental plan.

Experiment number	Runner height (mm)	Filters used
1	25	No
2	25	Yes
3	10	No
4	10	Yes

surfaces of the samples were subsequently examined using scanning electron microscopy (SEM), equipped with energy dispersive X-ray (EDX) analysis, for the evidence of oxide film. The tensile results were evaluated using a Weibull statistical analysis approach to assess the influence of different casting parameters, typically the runner height and the use of filters, on the variability of the mechanical properties of the castings.

It should be emphasized that, in this work, the melt was kept under partial vacuum for two hours before pouring to remove most, or perhaps all, previously existing oxide films. In addition, the amount of hydrogen in the final casting was minimized by keeping the sand moulds under partial vacuum for two weeks before being used which would allow them to lose most of the solvent and therefore minimize the amount of hydrogen picked up by the melt from the mould. Finally, the melt was degassed before pouring to reduce its hydrogen content. These arrangements were considered to eliminate the effect of any other parameter rather than the casting conditions on the mechanical properties of the resulting casting.

3. Results

The two-parameter Weibull distribution was used to analyze the scatter in the mechanical properties of the Al-5 Mg alloy castings produced under different casting conditions, as it was suggested to be more appropriate than a normal distribution [26, 27]. The Weibull modulus (the slope of the line fitted to the log-log Weibull cumulative distribution data) is a single value that shows the spread of properties; a higher Weibull modulus reveals less variability among the studied properties.

Weibull plots of the UTS and % elongation of the test bars cut from all castings (see Table 2) are represented in Figures 3(a) and 3(b), respectively. The values of the correlation coefficients (R^2) suggested that the data points expressing both the UTS and % elongation values were linearly distributed. It was noted that, for both the UTS and % elongation, the Weibull moduli (the slope of the trend line) of the castings from Experiment 4, where filters and the thin runner were used, were the highest among all castings. Figures 4(a) and 4(b) show plots of the Weibull moduli of the UTS and % elongation of the Al alloy versus the height of the runner with and without the use of filters. Weibull modulus of the UTS, when the runner height was 25 mm and without the use of filters, was 8.58. Decreasing the runner height to 10 mm increased the modulus to 13.41, while the use of filters increases the modulus to 17.99. Nevertheless, the use of 10 mm thick runner accompanied by the use of 10 PPI filters caused the Weibull modulus to increase to 41.45. Also, an elongation modulus of 3.39 was obtained for the casting produced using a runner height of 25 mm and without the use of filters. Castings with Weibull moduli of 8.36, 15.94, or

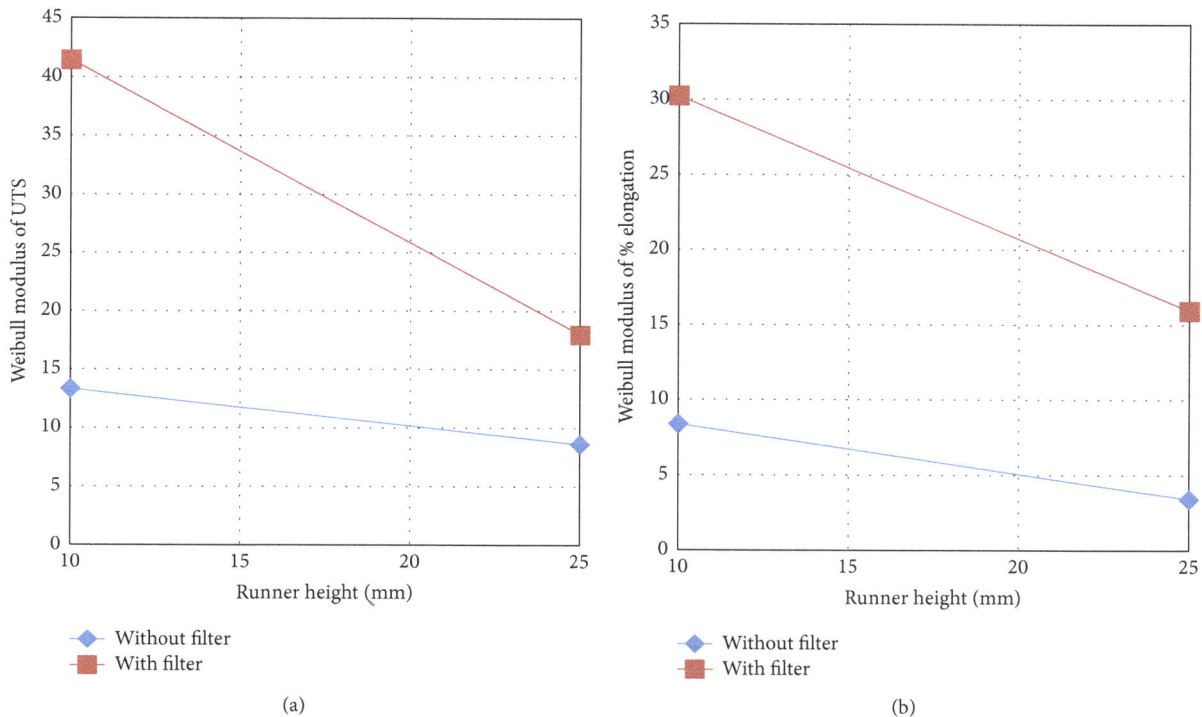

FIGURE 4: Plot of runner height versus (a) Weibull modulus of UTS and (b) Weibull modulus of % elongation.

30.22 were obtained when a 10 mm thick runner was used, when filters were implemented, or when both the thin runner and the filters were adopted.

Plots of the position parameter (the characteristic stress at which $1/e$ of the samples survived) [26] of both the UTS and % elongation of the Al alloy versus the height of the runner with and without the use of filters are represented in Figures 5(a) and 5(b), respectively. The position parameter showed a similar behavior to that of the Weibull moduli of both the UTS and % elongation. The use of thin runner together with the utilization of filters caused the position parameter of both the UTS and % elongation to increase from 134 to 200 MPa and from 24% to 43%, respectively.

Oxide film defects were found on all the fracture surfaces of test bars investigated from Experiments 1, 2, and 3. Only the fracture surfaces of the specimens from Experiment 4 were found to be free of oxide films. Figures 6–8 show SEM images of bifilm defects found on the fracture surfaces of test bars from Experiments 1–3, respectively, while an SEM micrograph of the fracture surface of a specimen from Experiment 4 is presented in Figure 9. The fracture surfaces were always selected from test bars that showed the lowest tensile strengths. Analysis by EDX was carried out at locations marked with "X" where it is suggested that MgO existed on the surfaces.

4. Discussion

In earlier studies of the effect of oxide films on the mechanical properties of different Al castings, the mould design shown in Figure 2 (using a 25 mm thick runner and without the use of

filters) was suggested to cause severe surface turbulence of the melt during mould filling which resulted in the creation of a significant amount of oxide films [23]. In the present work, the poor mould design, shown in Figure 2, was deliberately used in Experiment 1. This was expected to cause the velocity of the molten metal at mould entrance (the ingate velocity) to firmly exceed the critical velocity. This would lead to oxide film entrainment, which was subsequently confirmed by the SEM examination of the fracture surfaces (see Figure 6). This caused a significant reduction in the UTS and % elongation of the casting produced in Experiment 1 (position parameters of 134 MPa and 24%, resp.) and also increased the scatter of both properties (Weibull moduli of 8.6 and 3.4, resp.), as shown in Figures 4 and 5.

In an attempt to reduce the ingate velocity, two different methodologies were considered in this study: the use of filters and decreasing the runner height (Experiments 2 and 3, resp.). Each of the two approaches showed a noticeable reduction of the amount of oxide films on the fracture surfaces of the specimens from castings in these experiments, as shown in Figures 7 and 8, respectively. This resulted also in a perceptible improvement of both the Weibull moduli and position parameters of the UTS and % elongation, as shown in Figures 4 and 5.

However, the use of 10 PPI filters seems to have a stronger effect on enhancing the mechanical properties than the effect of reducing the runner height, especially for the Weibull moduli. This could be due to their secondary role in the removal of inclusions out from the melt. For instance, the reduction of runner height from 25 to 10 mm without the use of filters (Experiment 3) increased the Weibull modulus of

(a)

(b)

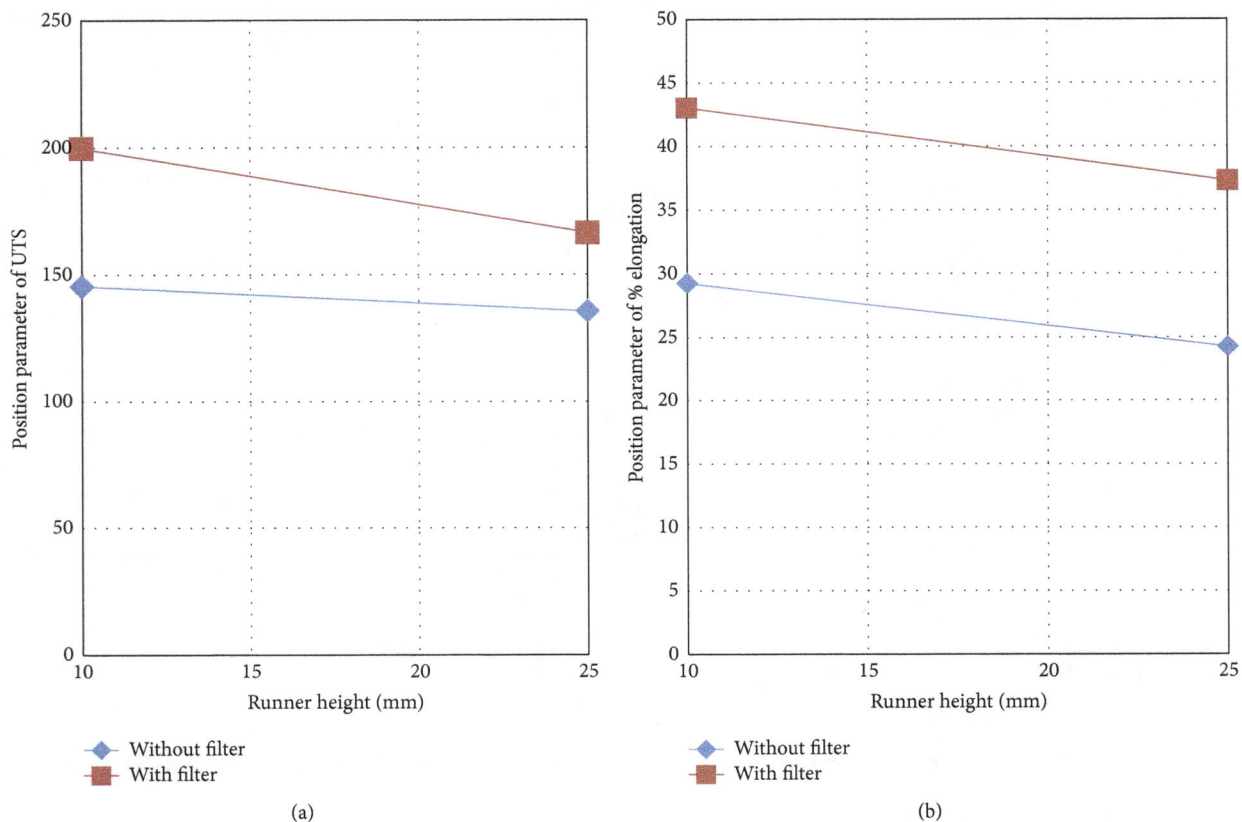

FIGURE 5: Plot of runner height versus (a) position parameter of UTS and (b) position parameter of % elongation.

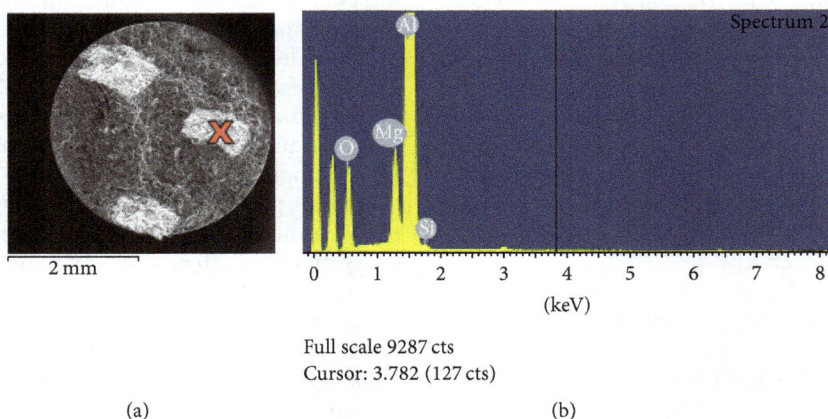

(a)

(b)

Full scale 9287 cts
Cursor: 3.782 (127 cts)

FIGURE 6: (a) An SEM image of the fracture surface of a specimen from Experiment 1; (b) EDX analysis at the location marked "X" in (a).

the UTS by about 56%. Conversely, a rise of such modulus of about 109% was achieved through the use of filters while keeping the runner height at 25 mm.

Combination of the two methodologies was found to considerably improve the mechanical properties. The Weibull moduli of the UTS and % elongation experienced a remarkable boost of about 380% and 790%, respectively. Also, the position parameter of the UTS increased by about 50%, while that of the % elongation was almost doubled. It could be

suggested that the use of thin runner could eliminate the jetting of the molten metal during its journey through the runner. In addition, the use of filters seems to help in reducing the acceleration of the incoming flow of liquid metal inside the runner before entering the mould cavity. This allowed for more quiescent filling regime of the mould cavity and in turn led to a reduction of the ingate velocity to less than 0.5 m/s, which minimized the amount of entrained oxide films and correspondingly enhanced the mechanical properties. This

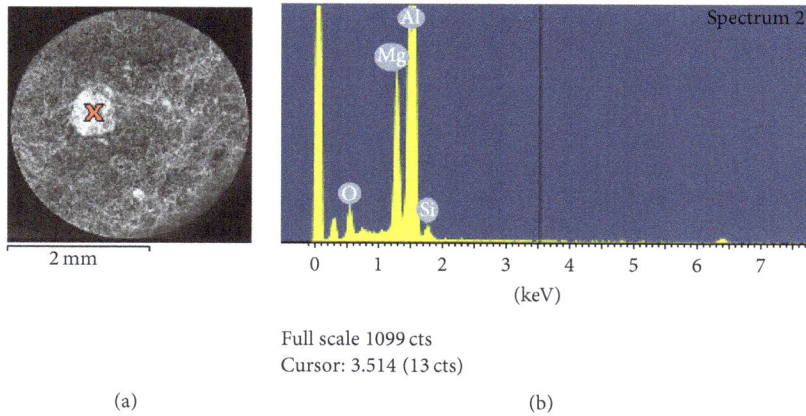

FIGURE 7: (a) An SEM image of the fracture surface of a specimen from Experiment 2; (b) EDX analysis at the location marked "X" in (a).

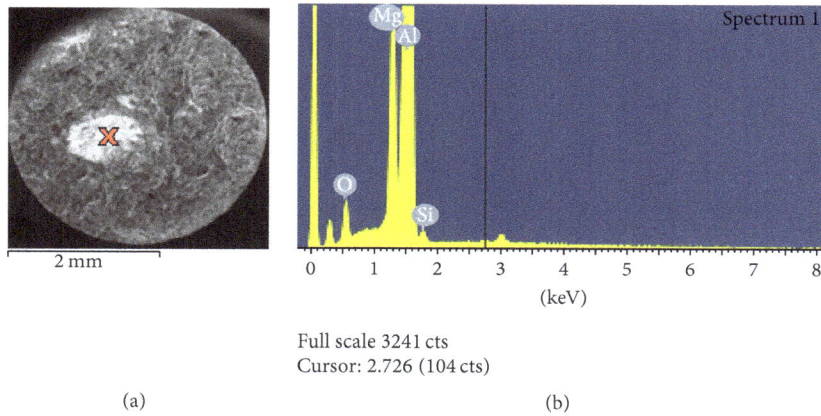

FIGURE 8: (a) An SEM image of the fracture surface of a specimen from Experiment 3; (b) EDX analysis at the location marked "X" in (a).

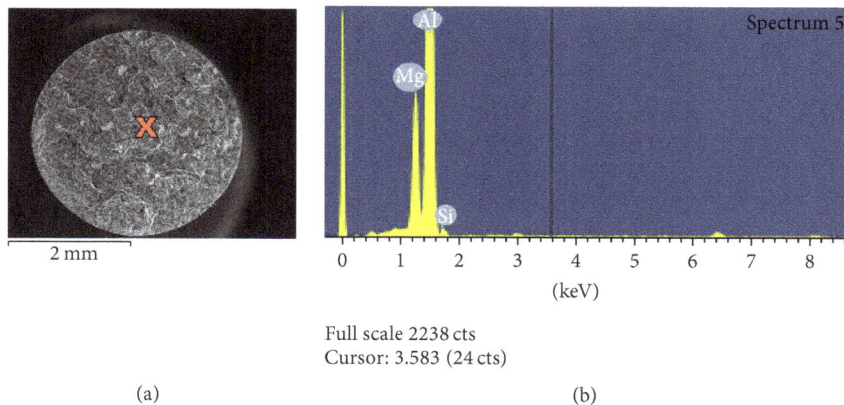

FIGURE 9: (a) An SEM image of the fracture surface of a specimen from Experiment 4; (b) EDX analysis at the location marked "X" in (a).

could be demonstrated via the SEM image in Figure 9 that did not show any oxide fragment on the fracture surface, which was also confirmed by the accompanied EDX analysis results. These results were in agreement with the results obtained by Dai et al. [28], Eisaabadi Bozchaloei et al. [29], and Nyahumwa et al. [17], who suggested that optimization of

the design of the running system and the use of filters in the running system might help keep the ingate velocity below the critical velocity which could eliminate the possibility of oxide film entrainment. The implication of these results is that the optimization of the runner system design and improving the flow behavior during mould filling could significantly reduce the production of oxide films so that the mechanical strength and reliability of aluminium alloy castings can be enhanced.

5. Conclusions

(1) Entrained bifilm defects reduce the mechanical properties of Al-5 Mg alloy castings.

(2) The use of a 10 mm thick runner increased the Weibull moduli of the UTS and % elongation by about 56% and 147%, respectively, while the use of 10 PPI filters increased the moduli by 109% and 368%, respectively.

(3) Adopting the 10 mm thick runner along with the use of 10 PPI filters resulted in a substantial improvement of the Weibull moduli of the UTS and % elongation by about 380% and 790%, respectively, perhaps due to the improved mould filling conditions that eliminated the chance of oxide film entrainment.

(4) The more careful and quiescent the mould filling practice, the higher the quality and reliability of the castings produced.

Competing Interests

The authors declare that they have no competing interests.

References

[1] M. El-Sayed, H. Salem, A. Kandeil, and W. D. Griffiths, "A study of the behaviour of double oxide films in Al alloy melts," *Materials Science Forum*, vol. 765, pp. 260–265, 2013.

[2] M. A. El-Sayed, "Effect of welding conditions on the mechanical properties of friction stir welded 1050 aluminum alloy," *International Review of Mechanical Engineering*, vol. 9, no. 3, pp. 252–256, 2015.

[3] W. D. Griffiths, A. J. Caden, and M. A. El-Sayed, "An investigation into double oxide film defects in aluminium alloys," *Materials Science Forum*, vol. 783–786, pp. 142–147, 2014.

[4] M. A. M. El-Sayed, *Double Oxide Film Defects and Mechanical Properties in Aluminium Alloys*, University of Birmingham, Birmingham, UK, 2012.

[5] M. A. El-Sayed, H. A. G. Salem, A. Y. Kandeil, and W. D. Griffiths, "Determination of the lifetime of a double-oxide film in al castings," *Metallurgical and Materials Transactions B: Process Metallurgy and Materials Processing Science*, vol. 45, no. 4, pp. 1398–1406, 2014.

[6] W. D. Griffiths, A. J. Caden, and M. A. El-Sayed, "The behaviour of entrainment defects in aluminium alloy castings," in *Proceedings of the 2013 International Symposium on Liquid Metal Processing and Casting*, pp. 187–192, John Wiley & Sons, Hoboken, NJ, USA, 2013.

[7] W. D. Griffiths and N.-W. Lai, "Double oxide film defects in cast magnesium alloy," *Metallurgical and Materials Transactions A: Physical Metallurgy and Materials Science*, vol. 38, no. 1, pp. 190–196, 2007.

[8] J. Campbell, "An overview of the effects of bifilms on the structure and properties of cast alloys," *Metallurgical and Materials Transactions B: Process Metallurgy and Materials Processing Science*, vol. 37, no. 6, pp. 857–863, 2006.

[9] J. Campbel, "The origin of porosity in castings," in *Proceedings of the in 4th Asian Foundry Conference*, Australian Foundry Institute, Queensland, Australia, 1996.

[10] J. Campbel, *The Modeling of Entrainment Defects during Casting*, The Minerals, Metals & Materials Society, 2006.

[11] J. Campbell, "Invisible macro defects in castings," *Journal de Physique IV*, vol. 3, no. 7, pp. C7-861–C7-872, 1993.

[12] J. Campbell, "The entrainment defect: the new metallurgy," in *Proceedings of the Staley Honorary Symposium*, ASM, Indianapolis, Ind, USA, 2001.

[13] J. Campbell, "Entrainment defects," *Materials Science and Technology*, vol. 22, no. 2, pp. 127–145, 2006.

[14] M. A. El-Sayed, H. Hassanin, and K. Essa, "Bifilm defects and porosity in Al cast alloys," *The International Journal of Advanced Manufacturing Technology*, vol. 86, no. 5, pp. 1173–1179, 2016.

[15] L. Liu and F. H. Samuel, "Effect of inclusions on the tensile properties of Al-7% Si-0.35% Mg (A356.2) aluminium casting alloy," *Journal of Materials Science*, vol. 33, no. 9, pp. 2269–2281, 1998.

[16] N. Green and J. Campbell, "Influence of oxide film filling defects on the strength of Al-7Si-Mg alloy castings," *AFS Transactions*, vol. 102, pp. 341–347, 1994.

[17] C. Nyahumwa, N. R. Green, and J. Campbell, "The concept of the fatigue potential of cast alloys," *Journal of the Mechanical Behavior of Materials*, vol. 9, no. 4, pp. 227–235, 1998.

[18] J. Campbell, *Castings*, Butterworth-Heinemann, Oxford, UK, 2nd edition, 2003.

[19] J. Runyoro, S. M. A. Boutorabi, and J. Campbell, "Critical gate velocities for film-forming casting alloys: a basic for process specification," *AFS Transactions*, vol. 100, pp. 225–234, 1992.

[20] A. Halvaee and J. Campbell, "Critical mold entry velocity for aluminum bronze castings," *AFS Transactions*, vol. 105, pp. 35–46, 1997.

[21] F. Bahreinian, S. M. A. Boutorabi, and J. Campbell, "Critical gate velocity for magnesium casting alloy (ZK51A)," *International Journal of Cast Metals Research*, vol. 19, no. 1, pp. 45–51, 2006.

[22] M. Cox, M. Wickins, J. P. Kuang, R. A. Harding, and J. Campbell, "Effect of top and bottom filling on reliability of investment castings in Al, Fe, and Ni based alloys," *Materials Science and Technology*, vol. 16, no. 11-12, pp. 1445–1452, 2000.

[23] M. A. El-Sayed and W. D. Griffiths, "Hydrogen, bifilms and mechanical properties of al castings," *International Journal of Cast Metals Research*, vol. 27, no. 5, pp. 282–287, 2014.

[24] M. A. El-Sayed, H. A. G. Salem, A. Y. Kandeil, and W. D. Griffiths, "Effect of holding time before solidification on double-oxide film defects and mechanical properties of aluminum alloys," *Metallurgical and Materials Transactions B: Process Metallurgy and Materials Processing Science*, vol. 42, no. 6, pp. 1104–1109, 2011.

[25] M. A. El-Sayed, H. Hassanin, and K. Essa, "Effect of casting practice on the reliability of Al cast alloys," *International Journal of Cast Metals Research*, 2016.

[26] W. Weibull, "A statistical distribution function of wide applicability," *Journal of Applied Mechanics*, vol. 18, pp. 293–297, 1951.

[27] N. R. Green and J. Campbell, "Statistical distributions of fracture strengths of cast Al-7Si-Mg alloy," *Materials Science and Engineering A*, vol. 173, no. 1-2, pp. 261–266, 1993.

[28] X. Dai, X. Yang, J. Campbell, and J. Wood, "Effects of runner system design on the mechanical strength of Al-7Si-Mg alloy castings," *Materials Science and Engineering A*, vol. 354, no. 1-2, pp. 315–325, 2003.

[29] G. Eisaabadi Bozchaloei, N. Varahram, P. Davami, and S. K. Kim, "Effect of oxide bifilms on the mechanical properties of cast Al–7Si–0.3Mg alloy and the roll of runner height after filter on their formation," *Materials Science and Engineering A*, vol. 548, pp. 99–105, 2012.

8

Investigation on Surface Roughness of Inconel 718 in Photochemical Machining

Nitin D. Misal[1] and Mudigonda Sadaiah[2]

[1]*Department of Mechanical Engineering, SVERI's College of Engineering Pandharpur, Maharashtra 413304, India*
[2]*Department of Mechanical Engineering, Dr. Babasaheb Ambedkar Technological University, Lonere, Raigad,*
 Maharashtra 402103, India

Correspondence should be addressed to Nitin D. Misal; nitin.misal72@gmail.com

Academic Editor: Patrice Berthod

The present work is focused on estimating the optimal machining parameters required for photochemical machining (PCM) of an Inconel 718 and effects of these parameters on surface topology. An experimental analysis was carried out to identify optimal values of parameters using ferric chloride ($FeCl_3$) as an etchant. The parameters considered in this analysis are concentration of etchant, etching time, and etchant temperature. The experimental analysis shows that etching performance as well as surface topology improved by appropriate selection of etching process parameters. Temperature of the etchant found to be dominant parameter in the PCM of Inconel 718 for surface roughness. At optimal etching conditions, surface roughness was found to be 0.201 μm.

1. Introduction

The Inconel 718 superalloy has a variety of industrial applications like aircraft engines, submarines, space vehicles, and so forth due to its distinctive properties. However, machining of this material becomes one of the key issues due to its hard nature. Photochemical machining is nontraditional micromachining processes. It produces flat metallic components which are free from burr and stress. In PCM, metal removal takes place through dissolution of metal atoms in chemical solution. It involves etching of metal from the restricted area using photo tool assisted photoresist. In present days, the PCM is acting a vital role in the precision parts manufacturing in industries like aerospace, automobile, electronics, ornament, medical, and so forth. The products made by using PCM are useful in the microelectromechanical system (MEMS) and nanoelectromechanical system (NEMS) [1, 2].

Inconel 718 is an important nickel based super alloy and it is used for various engineering applications. It especially is used in aerospace, marine, nuclear, and food processing applications [3]. This nickel alloy resists spalling during temperature fluctuations by developing a tightly sticking oxide scale. Inconel 718 has applications in high temperature and high pressure condition [4].

Saraf and Sadaiah studied effect of magnetic field on the etch rate of SS316L [5]. Patil and Mudigonda conducted experiments on Inconel 718 for understanding the effects of control variables such as speed of cutting, rate of feed, and depth of cut on surface finish and residual stresses at different level of parameters [6]. Qu et al. carried out the study on Monel 400 using PCM. Study shows that rolling direction affects etch rate [7]. For the first time Bruzzone and Reverberi introduced simulation in PCM. For this, a 2D Monte Carlo simulation model was used and experimental value was performed [8]. Çakir studied the effect of ferric chloride etchant on Aluminum as Al etching is the critical issue in the PCM industry [9]. Ho et al. performed the chemical machining, analysis of nanocrystalline nickel and predicted the effect of etchant concentration on etching quality [10]. Çakir et al. carried out comparative analysis of the copper etching process with ferric chloride and cupric chloride etchant. Etching rate was observed more with ferric chloride etchant while smooth surface quality was produced with cupric chloride etchant [11]. Allen and Almond discussed

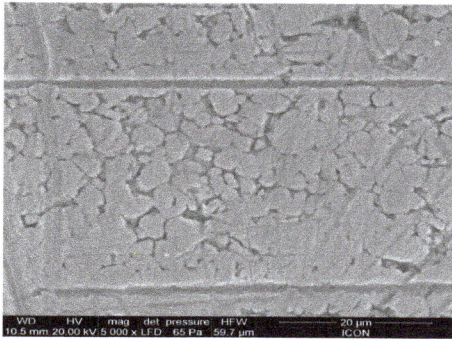

FIGURE 1: SEM image of Inconel 718 before machining.

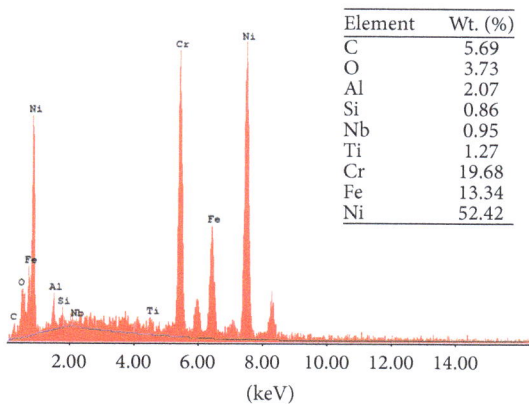

Element	Wt. (%)
C	5.69
O	3.73
Al	2.07
Si	0.86
Nb	0.95
Ti	1.27
Cr	19.68
Fe	13.34
Ni	52.42

FIGURE 2: EDAX profile of Inconel 718.

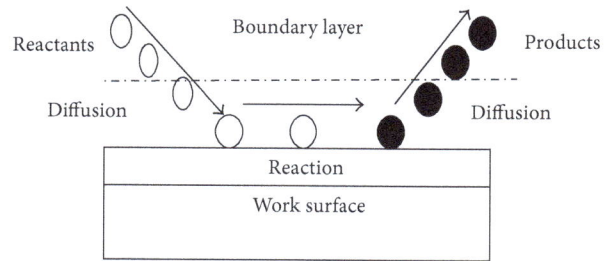

FIGURE 3: Photochemical machining mechanism.

TABLE 1: Input parameters.

Parameter	Levels for experimentation		
	Level 1	Level 2	Level 3
Temperature (°C)	45	50	55
Etchant conc. (g/L)	500	550	600
Time (min)	20	30	40

issues in Quality Control (QC) for ferric chloride etchant in PCM Industries [12].

Literature shows that no investigation has been reported on surface topology in PCM. Inconel 718 is a difficult to cut alloy and it is having wide range of applications. It can be machined using Electro Discharge Machining, laser beam machining, and so forth where stresses are induced during machining. The PCM is stress and burr free process in which no significant study has been observed on Inconel alloys. The effect of process parameter on surface topology is a critical issue in PCM. This study investigates the photochemical machining of Inconel 718 for the first time.

2. Materials and Methods

2.1. Material. The material selected for experimentation was Inconel 718. The plate of $250 \times 250 \times 1$ mm was taken initially for measurement of surface roughness value. The initial values of surface roughness were taken at different places on the workpiece and the average value was observed as 1.98 μm. The specimens were prepared in size of 20×20 mm. The scanning electron microscopy (SEM) image of unmachined Inconel 718 is as shown in Figure 1. Energy Dispersive X-Ray Analysis (EDAX) profile as presented in Figure 2 provides the chemical composition of Inconel 718.

2.2. Machining Mechanism. In photochemical machining, metal removal takes place by etching. There are three major stages observed (Figure 3).

(a) Ions or molecules from etchant solution diffused towards the exposed film on the work surface through the boundary layer.

(b) Due to chemical reaction between etchant exposed film soluble and gaseous byproducts formation takes place.

(c) A byproduct from the surface of work piece gets diffused through the boundary layer into the etchant solution.

2.3. Experimental Procedure. The different steps followed in PCM experimentation are given in Figure 4.

The specimens were cleaned by using ultrasonic cleaner. Solution used for cleaning contains deionizer water with the 1% of hydrochloric acid. The cleaned surface was observed under SEM for surface alterations and EDAX for the chemical composition. The specimens were machined by using $FeCl_3$ (ferric chloride) as etchant. The design of experiments is based on full factorial ($3k$) methodology. From past literature, it has been observed that the parameters which have significant effect on the response parameter of PCM [13–18] are concentration of etchant, etching temperature, and etching time. The preliminary experimentation has been carried out by using one factor at a time method for deciding the ranges of process parameters. The process parameters selected are etchant concentration, etchant temperature, and etching time with three levels each as shown in Table 1. For this combination of process parameters, 27 experiments are required to be carried out using a full factorial method. Also, response parameter as surface roughness (R_a) values is included in Table 2.

TABLE 2: Experimental layout with coded and actual values.

Run	Coded values			Actual values			Surface finish R_a (µm)
	Temperature (°C)	Concentration (g/L)	Time (min)	Temperature (°C)	Concentration (g/L)	Time (min)	
01	1	1	1	45	500	20	0.660
02	1	1	2	45	500	30	0.535
03	1	1	3	45	500	40	0.480
04	1	2	1	45	550	20	0.479
05	1	2	2	45	550	30	0.462
06	1	2	3	45	550	40	0.457
07	1	3	1	45	600	20	0.450
08	1	3	2	45	600	30	0.418
09	1	3	3	45	600	40	0.399
10	2	1	1	50	500	20	0.390
11	2	1	2	50	500	30	0.375
12	2	1	3	50	500	40	0.339
13	2	2	1	50	550	20	0.298
14	2	2	2	50	550	30	0.297
15	2	2	3	50	550	40	0.296
16	2	3	1	50	600	20	0.281
17	2	3	2	50	600	30	0.278
18	2	3	3	50	600	40	0.271
19	3	1	1	55	500	20	0.268
20	3	1	2	55	500	30	0.257
21	3	1	3	55	500	40	0.252
22	3	2	1	55	550	20	0.210
23	3	2	2	55	550	30	0.207
24	3	2	3	55	550	40	0.207
25	3	3	1	55	600	20	0.205
26	3	3	2	55	600	30	0.203
27	3	3	3	55	600	40	0.201

2.4. Experimental Setup. Experimental set-up for PCM of Inconel 718 is shown in Figure 5. Etching bath is used for carrying out experiments. It consists of an insulated cover which maintains the temperature inside the bath. The temperature controller is used to control the temperature of the bath with accuracy of ±1°C. The measurement of surface finish (R_a) was carried out by using Taylor Hobson talysurf profilometer.

3. Results and Discussion

The statistical analysis for the influence of process variables on surface roughness was made with the help of Mean Effective Plots and Analysis of Variance (ANOVA). Thus, the effect of process variables on response parameters was analyzed to obtain optimized condition for low surface roughness and a high material removal rate.

3.1. Analysis of Surface Roughness (R_a). Table 3 shows the analysis of variance (ANOVA) for surface roughness. From ANOVA results, it was observed that the effects of process parameters are significant on surface finish values. For temperature and concentration it shows selected range is nearly 100% significant and for the time it is about 95.1% significant. The most statistically significant factor was temperature.

Figure 6 indicates the main effect plot for the surface roughness of the photochemically machined Inconel 718. The main effect plots showed that the surface roughness was decreased with an increase in levels of input parameters. The etching temperature, concentration, and time in decreasing order of importance are control variables having an effect on R_a.

At 45°C temperature, the reaction of ferric chloride with Inconel 718 was just initiated, so initially the surface roughness was observed more. As the temperature increases, the viscosity of etchant reduces. Therefore, it results in improved penetration of cation across the diffusion layer. At high temperature, etchant attack is not along the grain boundaries but rather distributed over grain areas, leading to

TABLE 3: Summary of ANOVA for surface roughness.

Source	Degree of freedom	Adj. sum of square	Adj. mean of square	F-value	P value
Temperature	2	0.310680	0.155340	167.80	0.000
Concentration	2	0.043659	0.021830	23.58	0.000
Time	2	0.006500	0.003250	3.51	0.049
Error	20	0.018515	0.000926		
Total	26	0.379354			

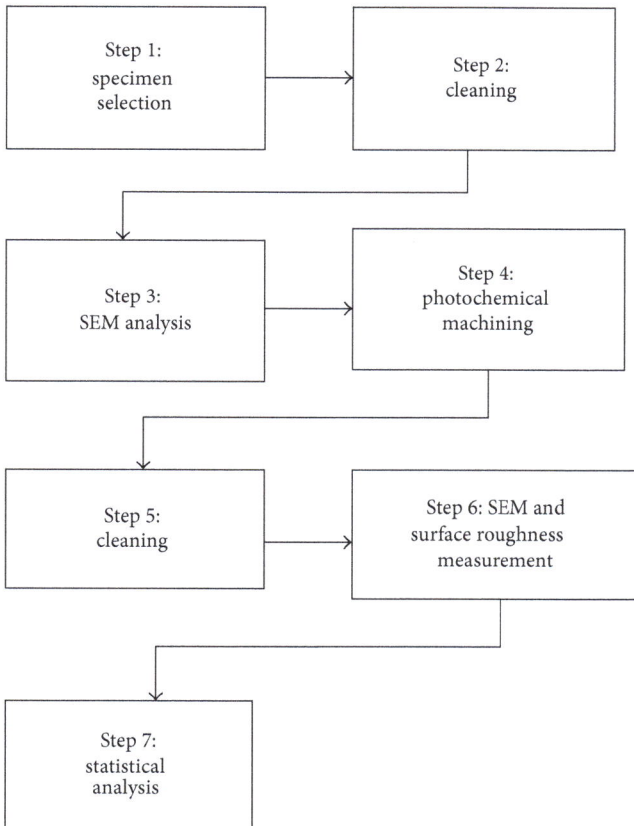

FIGURE 4: Flow chart of PCM experimentation.

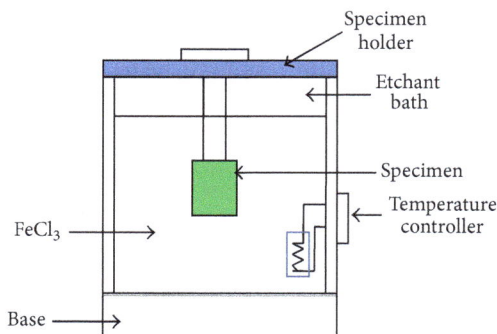

FIGURE 5: Experimental set-up.

TABLE 4: Summary of ANOVA for surface roughness.

Concentration in grams/liters	Sample weight before etching in grams	Sample weight after etching in grams	Weight loss in grams
(A) 500	0.977	0.958	0.019
(B) 550	0.971	0.956	0.015
(C) 600	0.982	0.970	0.012

a smoother surface. Hence it was seen from main effect plots as the temperature increases, the surface roughness decreases.

It was seen that as the concentration of etchant increases, etching rate decreases (as shown in Table 4 and Figure 7).

When etchant concentration is high, movement of cation across the diffusion layer becomes difficult as ferric chloride becomes more viscous. Thus, the reduced rate of diffusion leads to better surface finish. The time of reaction has a very less effect on surface roughness as temperature and concentration play a vital role for machining purposes.

Figure 8 shows the interaction plots between control parameter and the response variable. It can be seen that there is a significant decrease in the surface roughness value with increase in temperature, concentration, and time. As temperature increases the surface roughness value decreases; for 500 g/L concentration at temperature 45°C the average surface roughness value recorded was 0.56 μm and at 55°C it was 0.259 μm (refer Figure 8(a)). The time is less significant parameter for surface roughness as there is very less change in surface roughness values with increase in time for the constant temperature as shown in Figure 8(b). For 20-minute etching time, average 0.44 μm surface roughness was recorded for 500 g/L concentration and 0.31 μm for 600 g/L concentration (refer to Figure 8(c)).

3.2. Surface Morphology

3.2.1. SEM Analysis. Figure 9(a) shows the microstructure of unmachined Inconel 718 and Figure 9(b) shows the microstructure of Inconel 718 after machining. Using optimum etching parameter of 55°C temperature, 600 g/L, and 40 minutes time of etching the machining of Inconel 718 was carried out. The machined surface gives smaller grain size as compared to the unmachined surface. The machined surface also shows small clusters of alloying element chlorides which may adhere to the machined surface due to fusion.

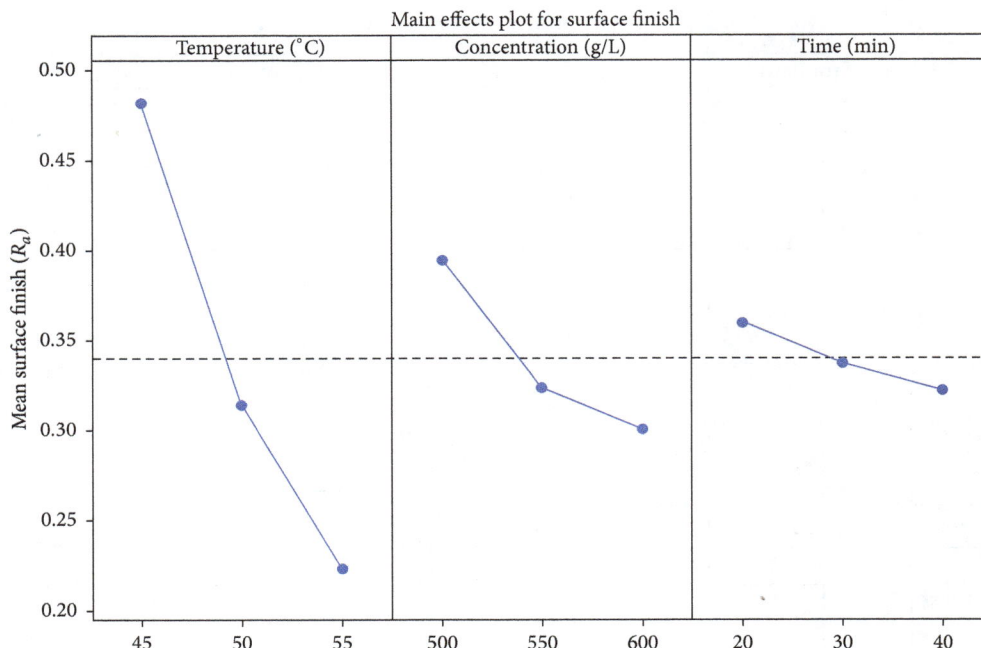

FIGURE 6: Main effect plot for roughness (R_a).

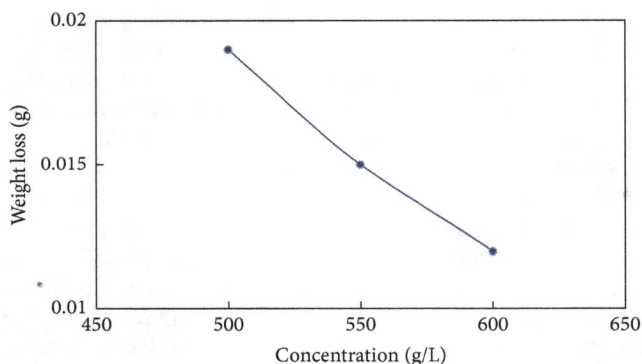

FIGURE 7: Graph of sample weight loss against concentration at 50°C and 20 Minutes of etching.

3.2.2. Sample Microstructure of Inconel 718. The microstructure of the Inconel 718 samples after machining was examined with Nikon microscope. The microstructure of specimens before machining is shown in Figure 10(a) and after machining is shown in Figures 10(b), 10(c), and 10(d) for constant concentration of etching solution at 600 g/L and time of etching at 40 min. with change in temperature as 45°C, 50°C, and 55°C, respectively. There is a significant effect of etchant temperature observed on the microstructure and the machining marks become smoother with an increase in the temperature.

As the temperature is increased from 45°C to 55°C the surface roughness decreases. At temperatures 45°C, 50°C, and 55°C, the roughness value R_a is found to be 0.399 μm, 0.271 μm, and 0.210 μm, respectively, and the roughness profile for the same is shown in Figures 11, 12, and 13.

The material removal initiated by the corrosion phenomenon in photochemical machining. The formation of passive layers is generally affected by chloride ions which eventually lead to corrosion. Very less corrosion sites are nucleated at low temperature and thus a rough and undesirable surface is produced. At higher etchant temperatures, corrosion sites nucleated are more which gives uniform corrosion and leave a better surface finish. This can be seen from Figure 14. At lower temperature (45°C) void formation is observed (Figure 14(a)) resulting in higher surface roughness. The good intergranular and intragranular corrosion during etching (Figure 14(b)) is observed which leads to a better surface roughness at higher temperature (55°C).

4. Conclusions

For machining of Inconel 718, photochemical machining is found to be a suitable machining process. The experimental investigation was carried out to analyze the influence of control variables on surface roughness in PCM. Findings of the above study are as follows:

(i) Higher temperature resulted in better surface finish as the etchant reacts with more grain area for uniform surface alterations.

(ii) As the etchant concentration increases, the surface roughness decreases.

(iii) The optimum surface finish, R_a as 0.201 μm was observed at temperature 55°C, etchant concentration 600 g/L, and time 40 min.

(iv) Time shows less effect on surface roughness as compared to temperature and concentration.

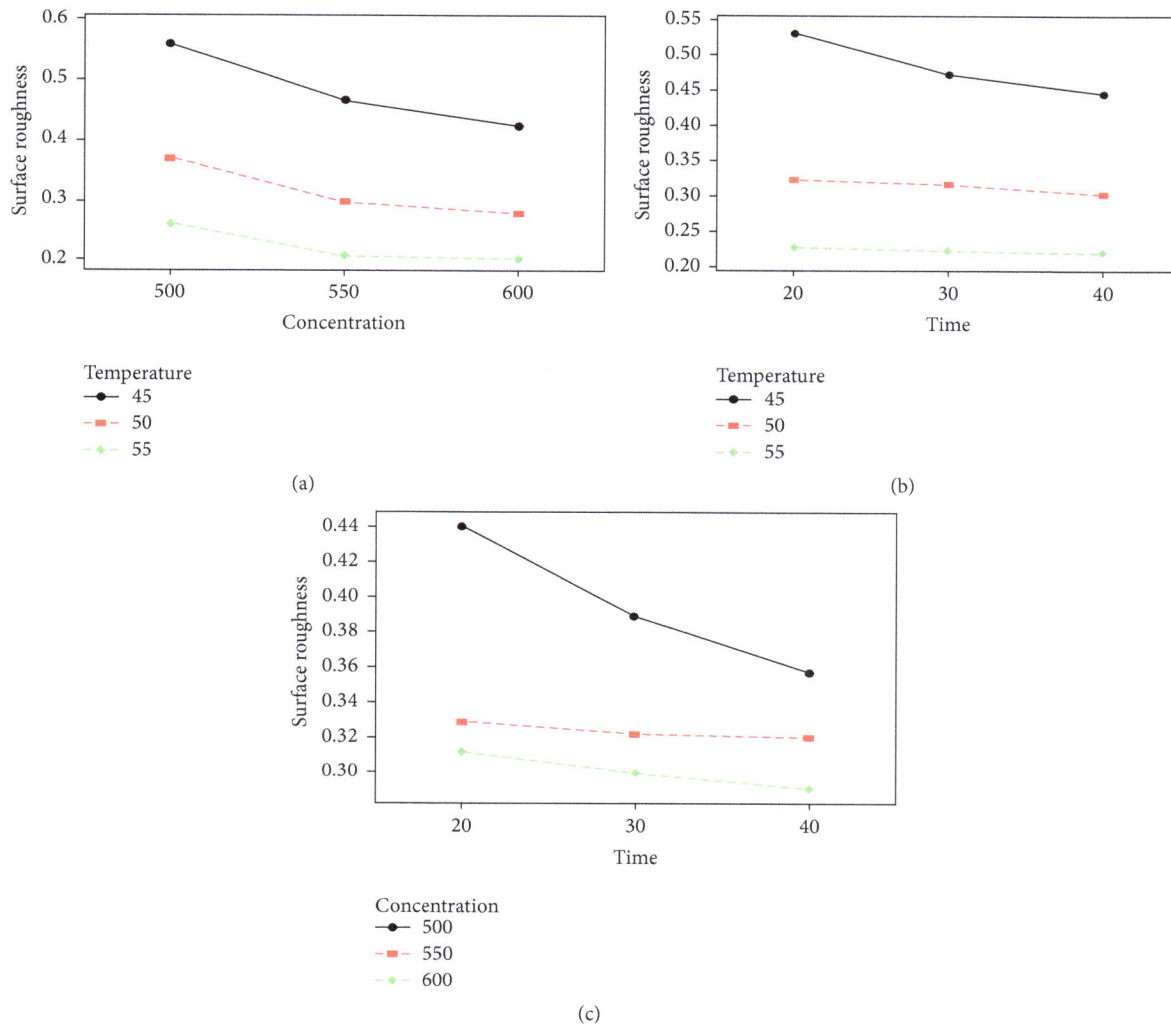

(a)

(b)

(c)

FIGURE 8: Interaction plots: (a) temperature and concentration, (b) temperature and time, (c) etchant concentration and time.

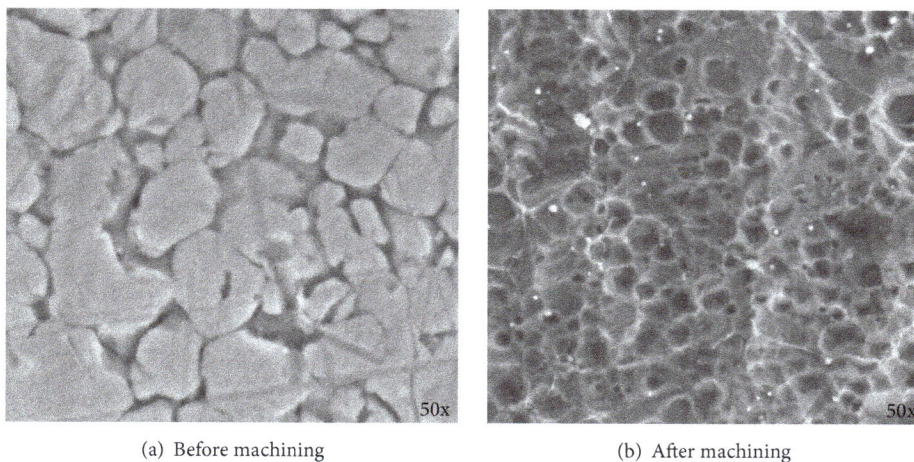

(a) Before machining

(b) After machining

FIGURE 9: SEM images showing the surface morphology of Inconel 718.

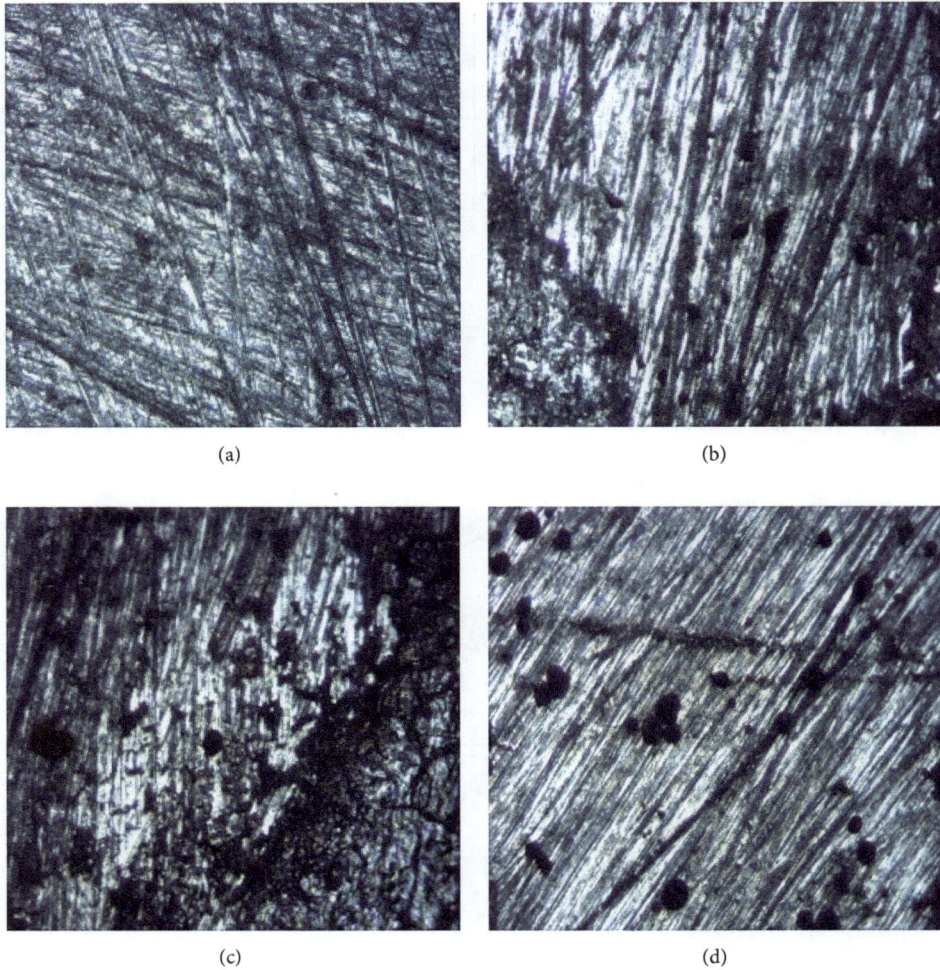

(a)

(b)

(c)

(d)

FIGURE 10: Microstructures of photochemically machined Inconel 718. (a) Before machining (b) at 45°C; after machining (c) at 50°C; after machining (d) at 55°C; after machining (etchant concentration: 600 g/L and etching time: 40 minute).

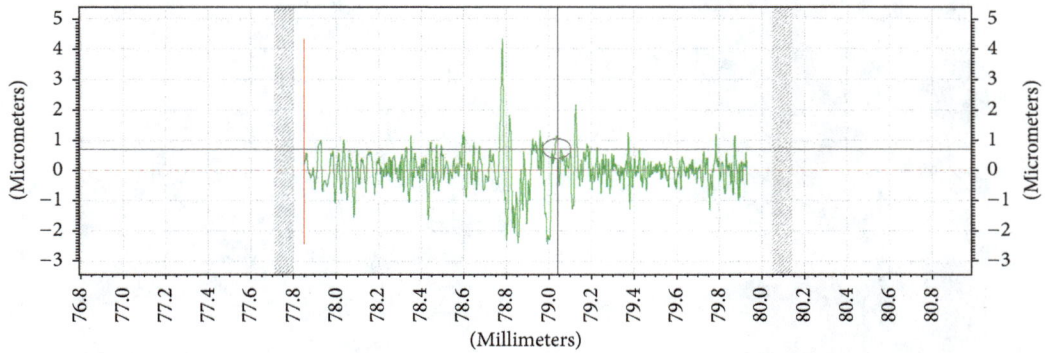

FIGURE 11: 2D roughness profile of Inconel 718 specimen (parameters: temperature: 45°C, concentration: 600 g/L, and time: 40 minutes).

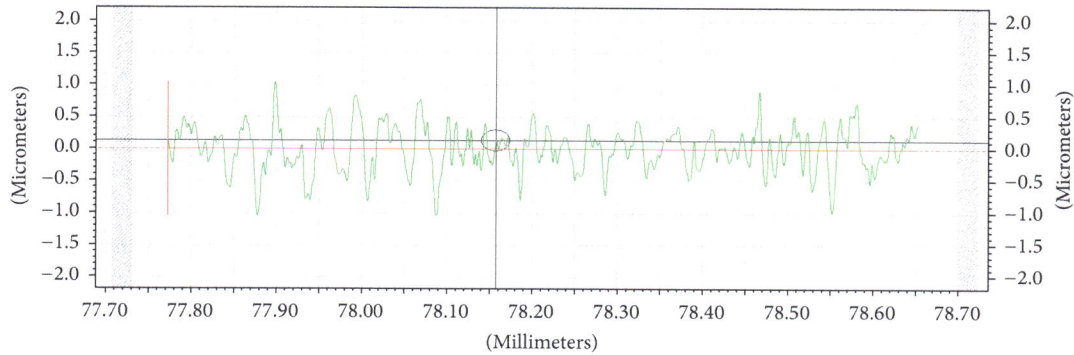

FIGURE 12: 2D roughness profile of Inconel 718 (parameters: temperature: 50°C, concentration: 600 g/L, and time: 40 minutes).

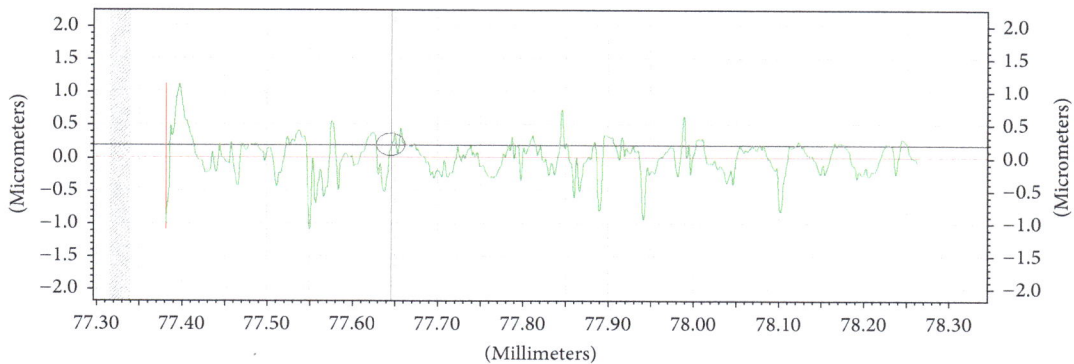

FIGURE 13: 2D surface roughness profile of machined Inconel 718 (parameters: temperature: 55°C, concentration: 600 g/L, and time: 40 minutes).

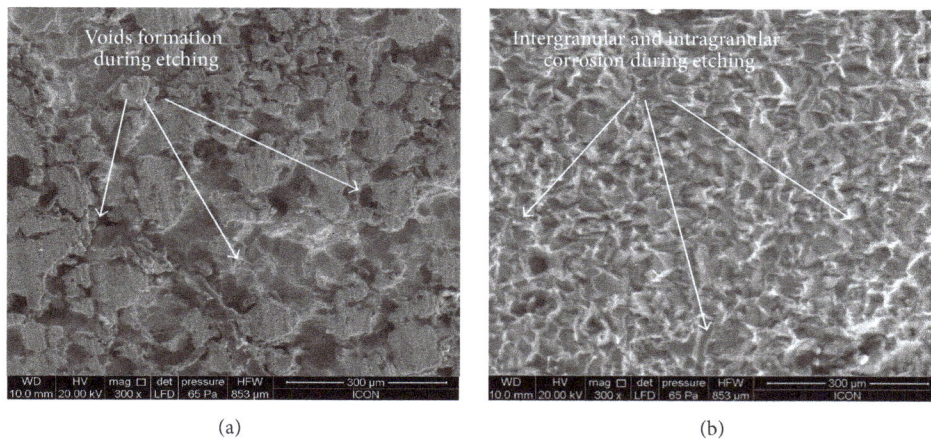

FIGURE 14: SEM image for a specimen photochemically machined at (a) 45°C and (b) 55°C.

Nomenclature

PCM: Photochemical machining
$FeCl_3$: Ferric chloride.

Competing Interests

The authors declare that there is no conflict of interests regarding the publication of this paper.

References

[1] D. M. Allen, *The Principles and Practice of Photochemical Machining and Photoetching*, Adam Hilger, IOP, Bristol, UK, 1986.

[2] D. M. Allen, "Photochemical machining: from 'manufacturing's best kept secret' to a $6 billion per annum, rapid manufacturing process," *CIRP Annals—Manufacturing Technology*, vol. 53, no. 2, pp. 559–572, 2004.

[3] F. Fareh, V. Demers, N. R. Demarquette, S. Turenne, and O. Scalzo, "Molding properties of inconel 718 feedstocks used in

low-pressure powder injection molding," *Advances in Materials Science and Engineering*, vol. 2016, Article ID 7078045, 7 pages, 2016.

[4] M. Dehmas, J. Lacaze, A. Niang, and B. Viguier, "TEM study of high-temperature precipitation of delta phase in inconel 718 alloy," *Advances in Materials Science and Engineering*, vol. 2011, Article ID 940634, 9 pages, 2011.

[5] A. R. Saraf and M. Sadaiah, "Magnetic field-assisted photo-chemical machining (MFAPCM) of SS316L," *Materials and Manufacturing Processes*, vol. 32, no. 3, pp. 327–332, 2016.

[6] D. H. Patil and S. Mudigonda, "The effect of the rolling direction, temperature, and etching time on the photochemical machining of monel 400 microchannels," *Advances in Materials Science and Engineering*, vol. 2016, Article ID 6751305, 9 pages, 2016.

[7] N. S. Qu, Y. Hu, D. Zhu, and Z. Y. Xu, "Electrochemical machining of blisk channels with progressive-pressure electrolyte flow," *Materials and Manufacturing Processes*, vol. 29, no. 5, pp. 572–578, 2014.

[8] A. A. G. Bruzzone and A. P. Reverberi, "An experimental evaluation of an etching simulation model for photochemical machining," *CIRP Annals—Manufacturing Technology*, vol. 59, no. 1, pp. 255–258, 2010.

[9] O. Çakir, "Chemical etching of aluminium," *Journal of Materials Processing Technology*, vol. 199, no. 1, pp. 337–340, 2008.

[10] S. Ho, T. Nakahara, and G. D. Hibbard, "Chemical machining of nanocrystalline Ni," *Journal of Materials Processing Technology*, vol. 208, no. 1–3, pp. 507–513, 2008.

[11] O. Çakir, H. Temel, and M. Kiyak, "Chemical etching of Cu-ETP copper," *Journal of Materials Processing Technology*, vol. 162-163, pp. 275–279, 2005.

[12] D. M. Allen and H. J. A. Almond, "Characterisation of aqueous ferric chloride etchants used in industrial photochemical machining," *Journal of Materials Processing Technology*, vol. 149, no. 1–3, pp. 238–245, 2004.

[13] R. P. Yadav and S. N. Teli, "A Review of issues in photochemical machining," *International Journal of Modern Engineering Research*, vol. 4, no. 7, pp. 49–53, 2014.

[14] O. Cakir, "Copper etching with cupric chloride and regeneration of waste etchant," *Journal of Materials Processing Technology*, vol. 175, no. 1–3, pp. 63–68, 2006.

[15] A. B. Bhasme and M. S. Kadam, "Parameter optimization by using grey relational analysis of photochemical machining," *International Research Journal of Engineering and Technology*, vol. 3, no. 3, pp. 992–997, 2016.

[16] A. B. Bhasme and M. S. Kadam, "Experimental investigation of PCM using response surface methodology on SS316L steel," *International Journal of Mechanical Engineering and Technology*, vol. 7, no. 2, pp. 25–32, 2016.

[17] P. Mumbare and A. J. Gujar, "Multi objective optimization of photoche-mical machining for ASME 316 steel using grey relational analysis," *International Journal of Innovative Research in Science, Engineering and Technology*, vol. 5, no. 7, pp. 12418–12425, 2016.

[18] S. S. Wangikar, P. K. Patowari, and R. D. Misra, "Effect of process parameters and optimization for photochemical machining of brass and german silver," *Materials and Manufacturing Processes*, pp. 1–9, 2016.

9

Uneven Precipitation Behavior during the Solutionizing Course of Al-Cu-Mn Alloys and their Contribution to High Temperature Strength

Jinlong Chen, Hengcheng Liao ⓘ, and Heting Xu

School of Materials Science and Engineering, Jiangsu Key Laboratory for Advanced Metallic Materials, Southeast University, Nanjing 211189, China

Correspondence should be addressed to Hengcheng Liao; hengchengliao@seu.edu.cn

Academic Editor: Pavel Lejcek

The dispersoid precipitation behavior during the solutionizing and aging of Al-$x$wt.%Cu-1.0 wt.% Mn alloys (x = 2.0, 4.5, and 7.5) and contribution to mechanical properties were investigated using tensile testing and microstructural characterization. A shell-core structure of primary α-Al dendrites is found in Al-Cu-Mn alloys, in which the Cu content in the shell is higher than that in the core. The area of shell zone (Cu-rich) increases with an increase in Cu content in the alloy. Large amounts of fine dispersoid Al-Cu-Mn particles precipitate in solution. An alloy with low Cu content results in only the T_{Mn} (Al$_{20}$Cu$_2$Mn$_3$) particles being precipitated. However, in an alloy with high Cu content, AlCu$_3$Mn$_2$ particles are first found to precipitate beside T_{Mn}. However, this precipitation behavior is uneven. The precipitation zones in the solution microstructure are consistent with the Cu-rich regions in the as-cast microstructure. A number of fine particles (dozens nanometer in size) are first found to precipitate on the rod-like T_{Mn} particles during the aging phase. The redissolution and granulation of the eutectic CuAl$_2$ phase during the solutionizing process result in the formation of particle-free bands between the precipitation zones. The tension test at 300°C demonstrates that the increase in high temperature strength is due to the dispersoid precipitation during solutionizing, and the precipitation behavior in the aging phase has little or no effect, however, largely improves the tensile strength at room temperature. High temperature strength is significantly increased with an increase in Cu content, which correlates to an increase in number and decrease in size of T_{Mn} and AlCu$_3$Mn$_2$ particles.

1. Introduction

Because of their good high temperature performance, Al-Cu-Mn alloys, such as 2519 and 2219 as well as A2219, have been used in structural parts for auto and space vehicles working at higher temperature environment [1–4]. In order to further improve the high-temperature mechanical properties, controlling the size and number of strengthening precipitates with high thermal stability is of great importance.

In Al-Cu-Mn alloys, the main strengthening phases at high temperature have been proposed to be the T_{Mn} and θ phases. For 2×24 alloys, the interaction between the dislocation motion and dispersoid precipitates of T_{Mn} (Al$_{20}$Cu$_2$Mn$_3$) and θ (CuAl$_2$) phases play key roles in strengthening at high temperature [5–8]. For the 2×24 alloys (Al-Cu-Mg-Mn alloys), large amounts of rod-like T-phase (Al$_{20}$Cu$_2$Mn$_3$) dispersoid particles are precipitated in ingot homogenization or solution treatment processes [9, 10]. Chen et al. [11] reported that the overwhelming majority of particles of the T_{Mn} phase take a shape of a rod in the Al-5Cu-1Mn alloy after solutionizing. Chen et al. [12] produced Al-4.6Cu-Mn ribbons using melt spinning, and during this process, rod-like T_{Mn} dispersoids with a diameter of 120–160 nm formed during the aging process at 190°C. The work of Wang et al. [13] was in agreement, reporting a rod-like T_{Mn} phase in Al-Cu-Mn alloys that have the lattice parameters a = 2.41 nm, b = 1.25 nm, c = 0.78 nm. Using CBED (convergent beam electron diffraction), Park and Kim [14] determined the structure parameters of the T_{Mn} phase of a = 2.345 nm,

$b = 1.183$ nm, and $c = 0.729$ nm. Recently, Shen et al. [10] proposed that the T_{Mn} phase belongs to a space group of BBMM (Brownian bridge movement model) with lattice parameters of $a = 23.98$ Å, $b = 12.54$ Å, and $c = 7.66$ Å using near-atomic resolution chemical mapping. Feng et al. [9] adopted HRTEM and high-angle annular dark-field scanning transmission electron microscopy (HAADF-STEM) to confirm that the T_{Mn} phase has an orthorhombic structure. Many studies report that the dispersoid Al-Cu-Mn compound particles precipitated during the solutionizing are the T_{Mn} phase, but another crystal structure of an Al-Cu-Mn compound with formula $AlCu_3Mn_2$ was proposed as a single-phase system using X-ray and neutron diffractometer traces [15]. There are many reports in the literature focusing on the crystal structure of Al-Cu-Mn precipitates. However, the conditions during precipitation from a supersaturated aluminum solution and strengthening effect are not understood and require further study.

PFZ (precipitation-free zone) often occurs at the grain boundaries in Al-Cu alloys. It is widely accepted that it is the escaping of both the vacancies and solute atoms of Cu into the grain boundary during aging course that leads to the formation of PFZ. The formation of PFZs has an important influence on mechanical properties, especially on antistress corrosion property. Li et al. [16] reported that the width of the PFZs became larger with increasing aging time of the Al-Zn-Mg-Sc-Zr alloy. Muntiz and Cotler [17] thought that the smaller the width of PFZs is, the higher the tensile strength is. Both the precipitation behaviors of T_{Mn} particles during solutionzing and θ particles during aging are formed from Al solution; however, it is not clear whether PFZs occur during solutionizing.

In this study, the microstructure and mechanical properties of three Al-$x$wt.%Cu-1.0 wt.% Mn alloys at ambient and elevated temperatures were investigated to find the precipitation behavior of T_{Mn} and $AlCu_3Mn_2$ phases in the Al-Cu-Mn system alloy and their contribution to the mechanical properties of the alloy.

2. Experimental Procedures

Three nominal Al-$x$wt.%Cu-1.0Mn alloys ($x = 2.0$, 4.5, and 7.5) were prepared by melting pure Al (99.7%) ingot and Al-20Cu and Al-10Mn master alloys in a resistance-heated furnace with the graphite crucible of 3 kg capacity. The prepared alloys are marked as A_1, A_2, and A_3, respectively, for 2, 4.5, and 7.5 wt. % of Cu contents. MAX×LMF15 spectrum was used to measure the chemical composition of the studied alloys, as listed in Table 1. After processing, the melts were poured into a metal mold with a cavity size of 170 mm × 100 mm × 20 mm which was preheated at 250°C for at least 5 h.

Parts of the obtained castings underwent a heat treatment: solutionized at 525°C for 6 h + 535°C for 6 h (a modified T6 state), followed by quenching in warm water and then aging at 170°C for 4 hrs. Metallography samples were cut from the castings. Optical microscope, SEM with GENESIS 60S X-ray EDS, and transmission electron

TABLE 1: Chemical composition of the studied alloys (wt.%).

Alloy	Cu	Mn	Si	Mg	Fe	Al
A_1	2.04	0.998	0.064	0.0009	0.130	Bal.
A_2	4.54	0.917	0.059	0.0004	0.131	Bal.
A_3	7.41	1.01	0.061	0.0006	0.138	Bal.

microscopy (TEM) were applied to characterize the microstructure of as-cast, as-solutionized, and T6 samples. Tensile test samples (based on Chinese standard: GB/T228-2002) were also cut from the castings with a gauge size of $35 \times 10 \times 3$ mm^3. The tensile testing was performed on a CMT4503 electronic universal testing machine with a rate of 1 mm/min at 25°C and 300°C. The data of mechanical properties are taken as an average of 3 samples.

3. Results and Discussion

3.1. Precipitation Behaviors during Solutionizing. Figure 1 shows the microstructure of the studied alloys as-solutionized. At the first sight, it seems that these optical photos are not qualified. Actually, it is a great number of precipitates that lead to this fuzziness. High-magnification photo (Figure 1(d)) clearly exhibits the fine dispersoid particles precipitated during the solutionizing course of the Al-Cu-Mn alloys. Here, the area where precipitation particles exist is designated as the precipitation zone (PZ). Of more importance, this precipitation behavior is uneven. The contour of the primary dendrites of the α-Al phase is dimly discernible, meaning that no or few particles are precipitated in the inner of dendrites (here, named as no particle zones, NPZs). It is seen that the area of NPZs is decreased with Cu content in the alloy, that is, the area of PZs is increased. And also there are a number of continuous or semicontinuous PFBs (precipitation-free bands) among the PZs. SEM image of the A_2 alloy as-solutionized more powerfully illustrates this precipitation characteristic (Figure 2(a)). The EDS results (Figure 2(b)) suggest these precipitated particles contain high level of Al, Cu, and Mn, being an Al-Cu-Mn compound. Figure 2 also shows a PFB (marked by Arrow 3) and $CuAl_2$ particle that are survived from the solutionizing course.

Figure 3 shows the TEM images and EDS results of the precipitates in the studied alloys as-solutionized and SAD patterns of representative precipitation particles. TEM images exhibit that most of them take a shape of the rod, long or short in size, consistent with [9–11, 18]. Usually, the composition detected by TEM-EDS is much more accurate than that by SEM-EDS. In the A_1 alloy with a low Cu content of about 2 wt. %, TEM-EDS result indicates the precipitation particles are Al-Cu-Mn compound with a Cu/Mn atomic ratio close to 2 : 3 (Figure 3(b)). Feng et al. [9], Shen et al. [10], and Chen et al. [11] thought the rod-like particles precipitated in the solutionized 2 × 24 alloy and Al-5Cu-1Mn alloy are T_{Mn} phase with a formula of $Al_{20}Cu_2Mn_3$. Liao et al. [18] also observed short rod-like dispersoid particles of the Al-Cu-Mn compound in the solutionized Al-Si-Cu-Mn alloy and identified it as T_{Mn} ($Al_{20}Cu_2Mn_3$) by calculating the SAD patterns. According to the phase diagram of the Al-Cu-Mn system [19],

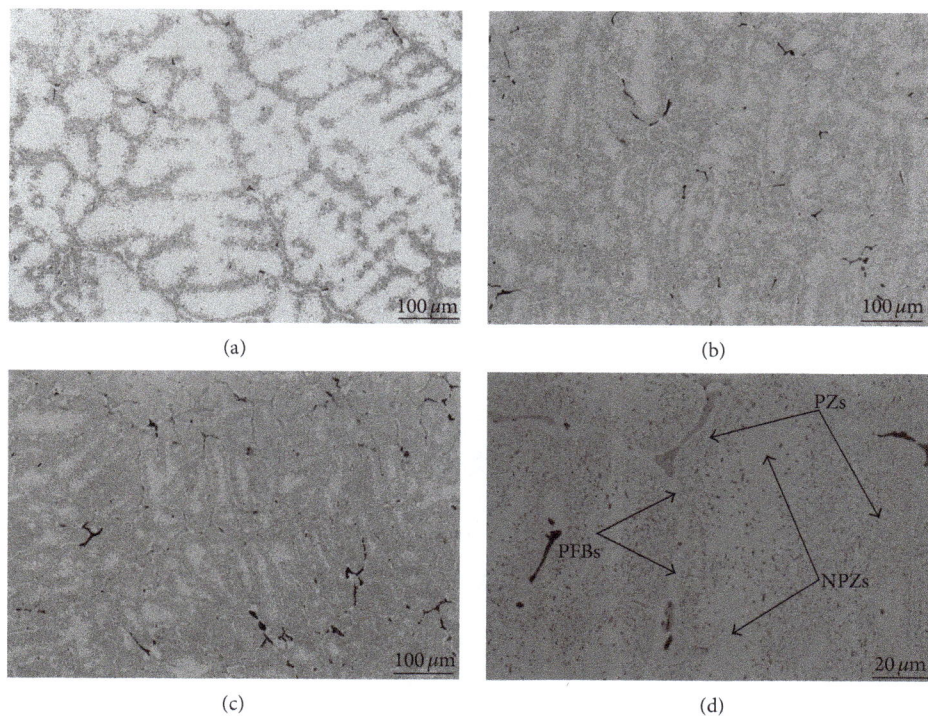

FIGURE 1: Optical microstructure of the studied alloys as-solutionized, showing PZs (precipitation zones), NPZs (no precipitation zones), and PFBs (precipitation-free bands): (a) A_1 alloy; (b) A_2 alloy; (c) A_3 alloy; (d) large magnification photo of the A_2 alloy.

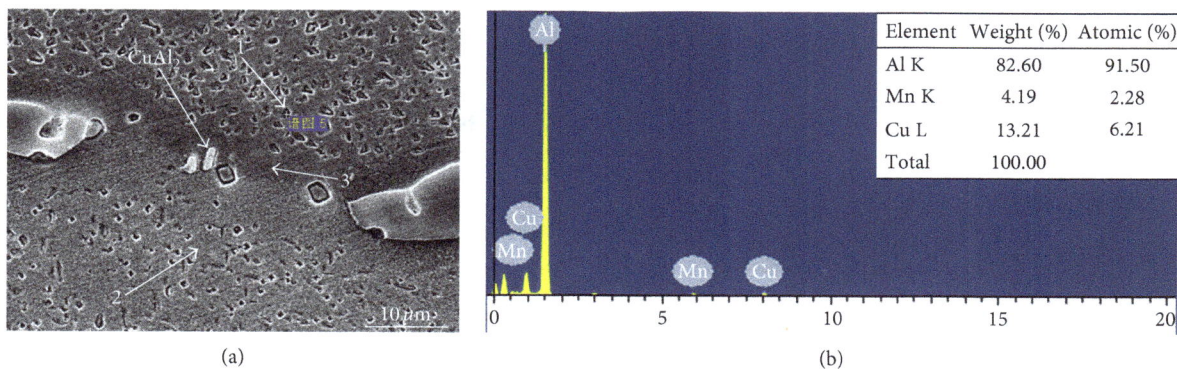

Element	Weight (%)	Atomic (%)
Al K	82.60	91.50
Mn K	4.19	2.28
Cu L	13.21	6.21
Total	100.00	

FIGURE 2: SEM image showing the precipitated particles (a) and its EDS results marked by Arrow 1 (b) in the A_2 alloy as-solutionized

in the Cu-rich corner, the most possible phase of Al-Cu-Mn compounds is T_{Mn}. Thus, the precipitates in the A_1 alloy are thought to be the T_{Mn} phase. Calculation of the SAD pattern in Figure 3(c) further demonstrates it, having an orthorhombic structure with lattice parameters $a = 2.420$ nm, $b = 1.250$ nm, and $c = 0.772$ nm that is consistent with Wang's et al. [13] and Park and Kim's [14] models. But A_2 and A_3 alloys with high Cu content are about 4.5 wt.% and 7.0 wt.%, respectively, besides T_{Mn} precipitates (such as Arrow 2 in Figure 3(e) and EDS in Figure 3(f) and Arrow 4 in Figure 3(i) and EDS in Figure 3(j)), and another Al-Cu-Mn precipitation particles are found (such as Arrow 1 in Figure 3(e) and Arrow 3 in Figure 3(i)), in which the EDS results (Figures 3(d) and 3(h)) indicate they have an almost same Cu/Mn atomic ratio of $3 : 2$. In 1968, Johnston and Hall [15] identified the

crystal structure of the Cu_3Mn_2Al compound in its single-phase system by X-ray and neutron diffractometer traces: it is a cubic Laves phase with the 8 manganese atoms ordered at the geometrically larger A sites, while the copper and aluminum atoms are randomized in the 16-fold B sites, and its parameter is 6.9046 Å. In this study, the calculation of the SAD pattern of this compound (Figure 3(g)) suggests that it has a simple cubic structure with $a = 0.6904$ nm, equal to $AlCu_3Mn_2$ (PDF card no.: 23-0760). Thus, these Al-Cu-Mn precipitation particles with a Cu/Mn atomic ratio of $3 : 2$ found in the as-solutionized A_2 and A_3 alloys are identified as $AlCu_3Mn_2$. Now, it is concluded that, in the Al-Cu-Mn alloy with a low Cu content, there are only T_{Mn} ($Al_{20}Cu_2Mn_3$) dispersoid particles precipitated during the solutionizing course, but in the alloy with a high Cu

(a)

Element	Weight (%)	Atomic (%)
Al K	88.33	94.27
Mn K	6.04	3.17
Fe K	0.23	0.12
Cu K	5.40	2.45
Total	100.00	

(b)

(c)

Element	Weight (%)	Atomic (%)
Al K	88.93	94.69
Mn K	4.22	2.20
Fe K	0.20	0.10
Cu K	6.65	3.01
Total	100.00	

(d)

(e)

Element	Weight (%)	Atomic (%)
Al K	78.49	88.81
Mn K	10.96	6.09
Fe K	0.49	0.27
Cu K	10.05	4.83
Total	100.00	

Full scale 26152 cts cursor: 0.000   (keV)

(f)

(g)

Element	Weight (%)	Atomic (%)
Al K	88.69	94.56
Mn K	4.18	2.19
Fe K	0.30	0.15
Cu K	6.83	3.09
Total	100.00	

(h)

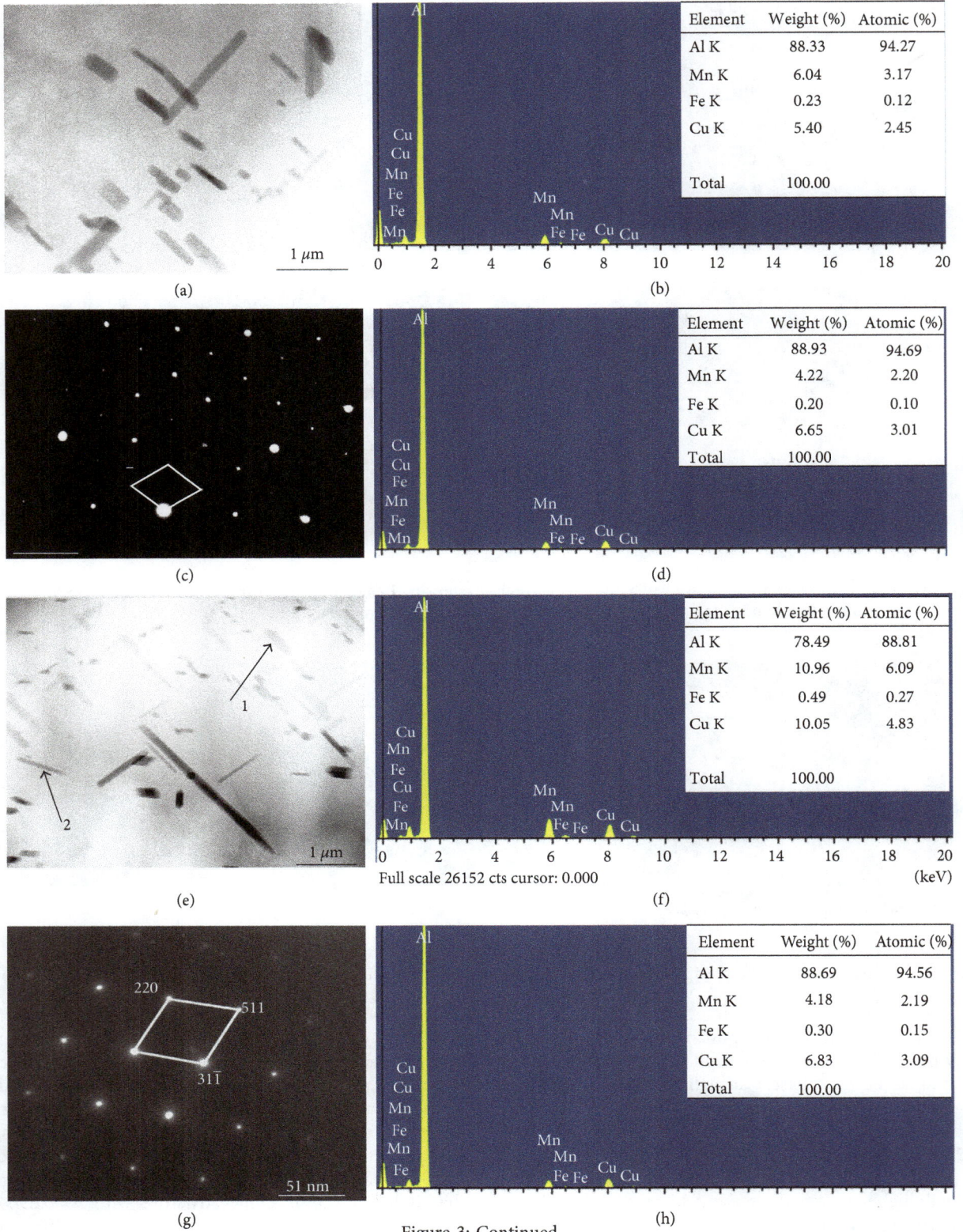

Figure 3: Continued.

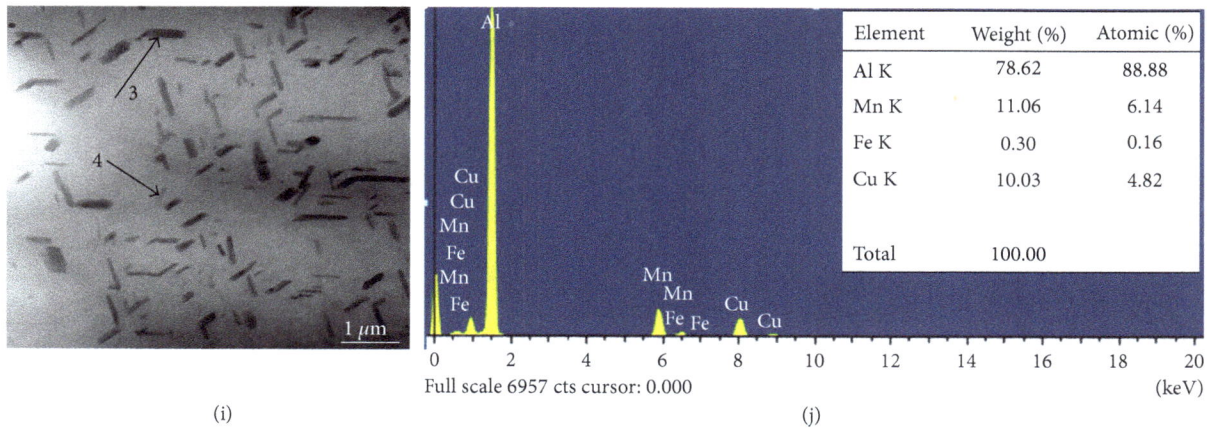

Element	Weight (%)	Atomic (%)
Al K	78.62	88.88
Mn K	11.06	6.14
Fe K	0.30	0.16
Cu K	10.03	4.82
Total	100.00	

Full scale 6957 cts cursor: 0.000

(i)   (j)

FIGURE 3: Transmission electron micrographs of precipitated particles in A_1 (a), A_2 (e), and A_3 (i) alloys as-solutionized; representative EDS (b) and SAD (c) results of the T_{Mn} phase in the A1 alloy; EDS (d) and SAD (g) results of the $AlCu_3Mn_2$ phase and EDS (f) of the T_{Mn} phase in the A2 alloy, respectively, marked by Arrows 1 and 2 in (e); EDS results (h, j) of the particles $AlCu_3Mn_2$ and T_{Mn} in the A3 alloy, respectively, marked by Arrows 3 and 4 in (i)

content, $AlCu_3Mn_2$ particles are demonstrated to precipitate beside T_{Mn} particles. From Figure 3, it is also worth to note that the number of precipitation particles is increased with Cu content in the alloys; however, the particle size is decreased simultaneously, which will exert impact to the tension properties.

To deeply understand this precipitation behavior during solutionizing, uneven precipitation of the T_{Mn} phase, and formation of PFBs among the PZs, it is necessary to carefully characterize the microstructure as-cast. Figure 4 shows the optical microstructure of the A_1, A_2, and A_3 alloys as-cast. There are particulate-like or continuous/semicontinuous network constitutes in the interdendritic area. They are eutectic resultants during the final stages of nonequilibrium solidification of the Al-Cu-Mn alloy: $L \rightarrow \alpha\text{-Al} + \theta$ and $L \rightarrow \alpha\text{-Al} + \theta + T_{Mn}$ [19–21]. SEM images of three constitute and their EDS results in Figure 5 demonstrate it. With Cu content in the alloy, the amount of network θ phase is increased in the eutectic resultants (Figure 5).

It is very interesting to note that, in Figure 4, there are two contrasts on primary α-Al dendrites, like a shell-core structure. The part in the core of dendrites is partly bright, and the shell around the core is partly grey. In SEM image (Figure 6(a)), the core of dendrites presents as dark-grey and the shell as light-grey. EDS results of two regions are illustrated in Figures 6(b) and 6(c), respectively, suggesting that Cu content in the shell is much higher than that in the core; however, Mn content is almost equal. EPMA diagrams in Figure 7 show the distribution of Cu, Fe, and Mn elements in the microstructure of the A_2 alloy as-cast. It demonstrates again that the shell of dendrites is much more Cu-rich than the core. And except for a few regions with the T_{Mn} phase, the other regions where $CuAl_2$ phase exists alone are highly Mn-poor. Now, the shell region of α-Al dendrites can be denoted as the Cu-rich region. The Al-rich corner of the Al-Cu-Mn phase diagram is rather complicated [19]. The physic process for the formation of this obvious shell-core structure of primary α-Al dendrites is still unknown.

Of more importance, the area of Cu-rich regions is increased with Cu content in the alloy (Figure 4). Carefully comparing Figure 1 with Figure 4, it is found in surprise that Cu-rich regions in as-cast microstructure are highly consistent with the precipitation zones in as-solutionized microstructure. It is just in the Cu-rich regions that precipitation of T_{Mn} and $AlCu_3Mn_2$ dispersoid particles occurs. Due to nonequilibrium solidification, a shell-core structure of primary α-Al dendrites is formed. In Cu-rich regions, Cu content is about 5 wt.% (Figure 6(c)), very close to the maximum solubility of Cu solute in Al solution in the Al-Cu binary system. In other words, these Cu-rich regions are supersaturated actually. In the course of solutionizing at 525°C for 6 hrs + 535°C for 6 hrs, part of the eutectic $CuAl_2$ phase formed during nonequilibrium solidification is forced to redissolve into the Al matrix. In the A1 alloy with only 2 wt.% Cu content, the eutectic $CuAl_2$ phase is dissolved completely (Figure 8(a)); however, in the A_2 and A_3 alloys, there are some eutectic $CuAl_2$ particles which survived (Figures 8(b) and 8(c), resp.). Eutectic $CuAl_2$ phase is located in the Cu-rich regions (Figure 4). The redissolution of these regions further enhances the degree of supersaturation. Even at 535°C of solutionizing temperature, it is supersaturated. Thus, it is the supersaturation of the Cu solute in the Al solution that drives the precipitation of T_{Mn} and $AlCu_3Mn_2$ to occur in the Cu-rich regions. Because the diffusion coefficient of manganese in the Al solution is very low ($1 \times 10^{-12} \cdot cm^2/s$), much less than that of copper ($5 \times 10^{-9} \cdot m^2/s$) [22], the precipitated particles of T_{Mn} and $AlCu_3Mn_2$ phase are fine and dispersed. With Cu content in the alloy, the area of the Cu-rich regions is increased, and hence, the area of PZs is also increased. In the alloy with high Cu content, the degree of supersaturation becomes larger, and thus, the driving force for precipitation becomes stronger that leads to an increase in amount and a decrease in size of the precipitated particles. But in the core part of dendrites, the Cu solute is much poor, so no precipitation occurs, forming NPZs in the original core parts of dendrites.

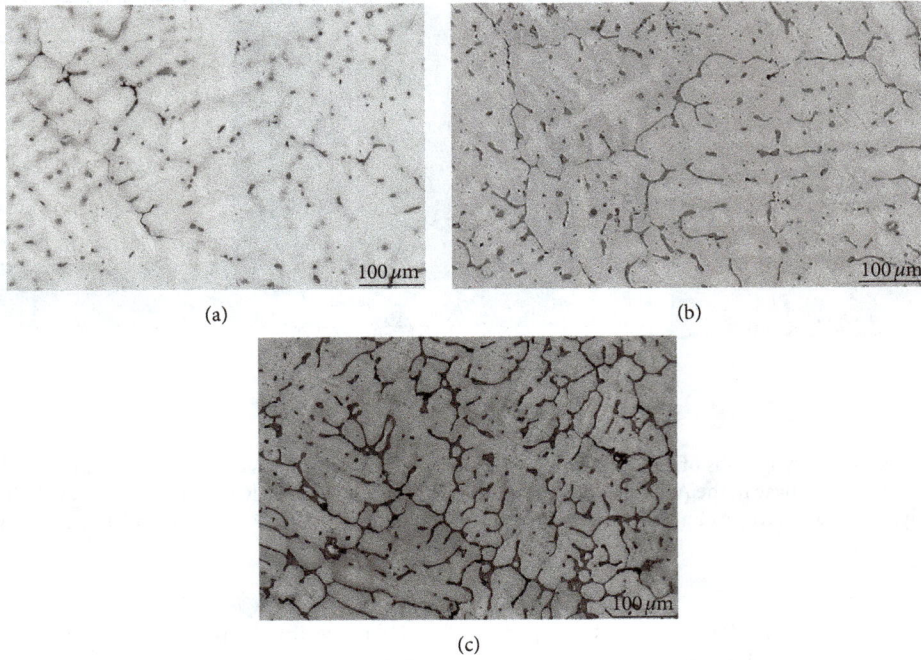

FIGURE 4: Optical microstructures of three Al-Cu-Mn alloys as-cast: (a) A_1 alloy; (b) A_2 alloy; (c) A_3 alloy.

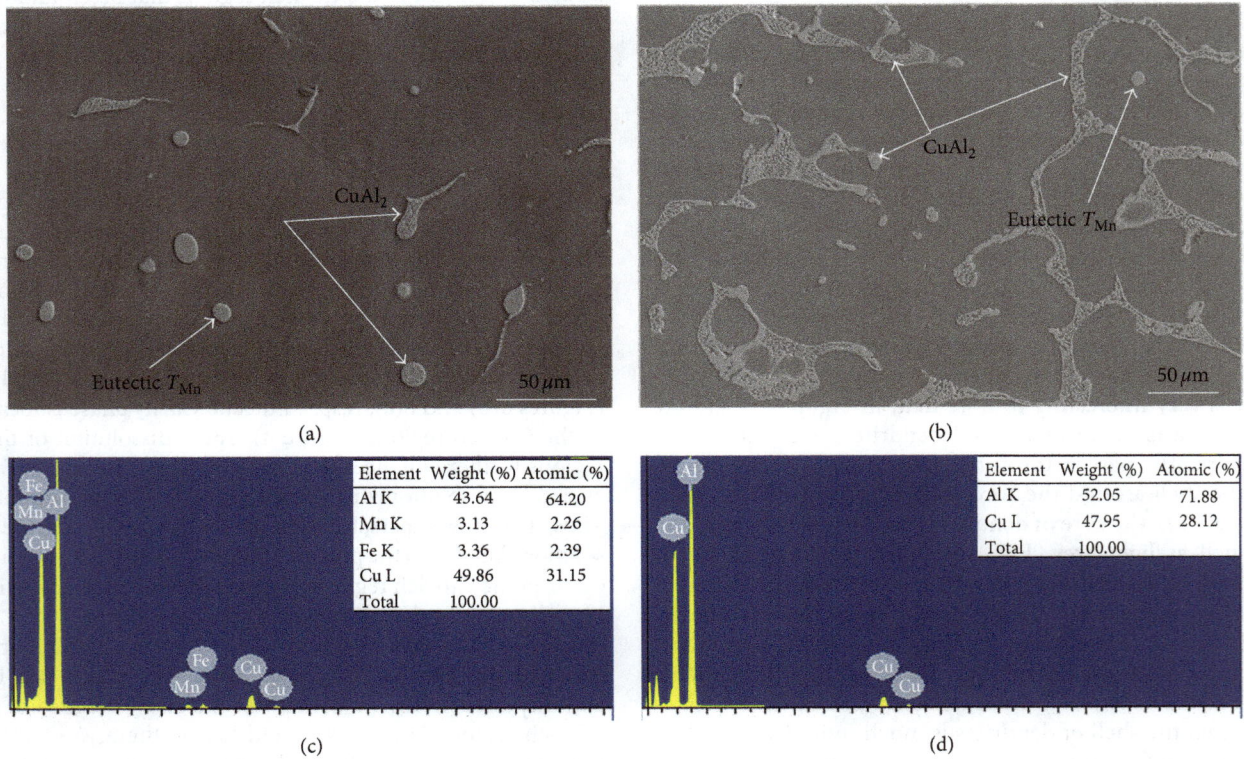

Element	Weight (%)	Atomic (%)
Al K	43.64	64.20
Mn K	3.13	2.26
Fe K	3.36	2.39
Cu L	49.86	31.15
Total	100.00	

Element	Weight (%)	Atomic (%)
Al K	52.05	71.88
Cu L	47.95	28.12
Total	100.00	

FIGURE 5: Continued.

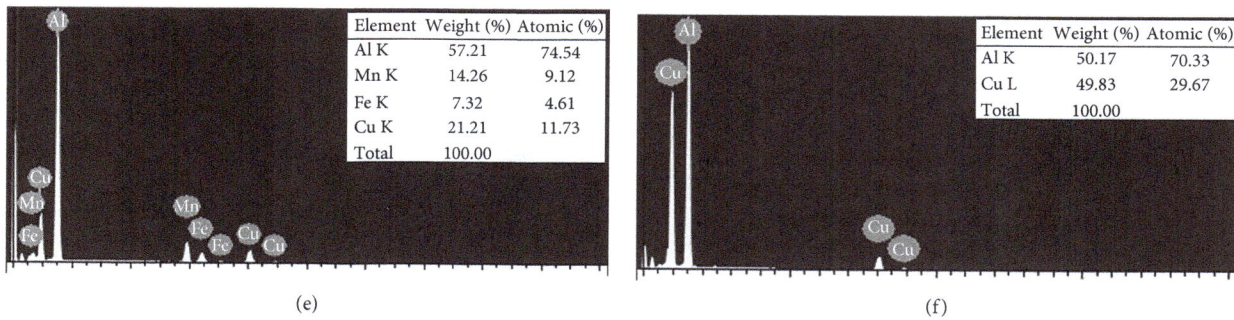

Element	Weight (%)	Atomic (%)
Al K	57.21	74.54
Mn K	14.26	9.12
Fe K	7.32	4.61
Cu K	21.21	11.73
Total	100.00	

(e)

Element	Weight (%)	Atomic (%)
Al K	50.17	70.33
Cu L	49.83	29.67
Total	100.00	

(f)

FIGURE 5: SEM images of Al-Cu-Mn alloys as-cast: (a) A_1 alloy; (b) A_3 alloy; (c, d) EDS results of the T_{Mn} and θ phases in the A_1 alloy, respectively; (e, f) EDS results of the T_{Mn} and θ phases in the A_3 alloy, respectively.

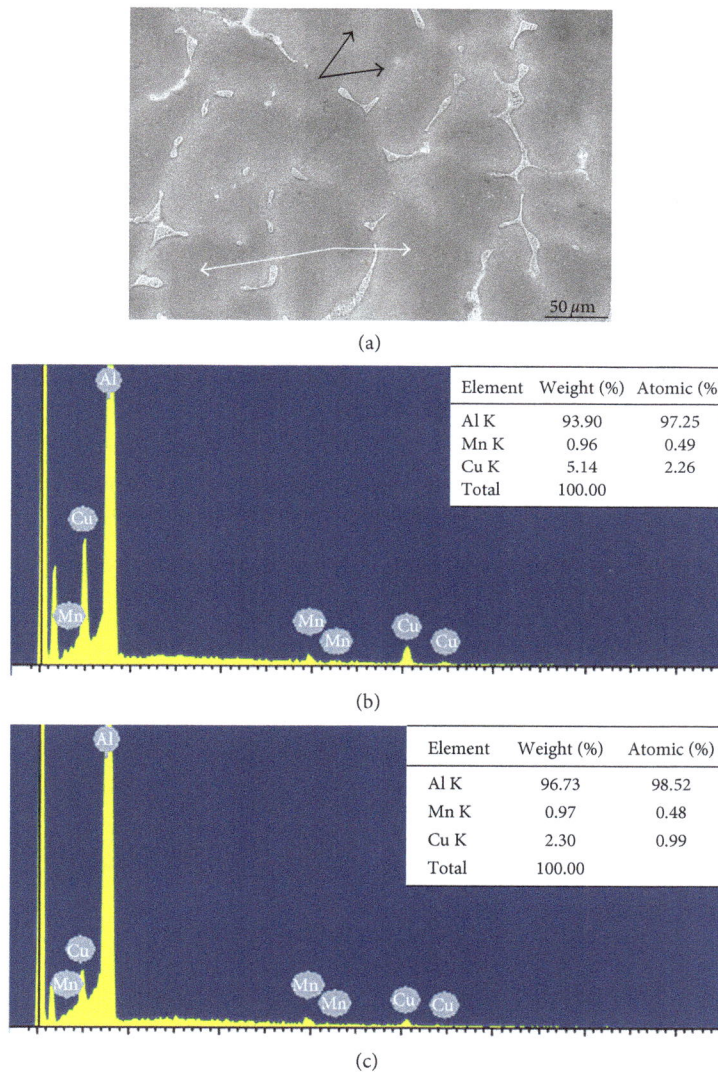

(a)

Element	Weight (%)	Atomic (%)
Al K	93.90	97.25
Mn K	0.96	0.49
Cu K	5.14	2.26
Total	100.00	

(b)

Element	Weight (%)	Atomic (%)
Al K	96.73	98.52
Mn K	0.97	0.48
Cu K	2.30	0.99
Total	100.00	

(c)

FIGURE 6: SEM image showing the light-grey and dark-grey areas of dendrites in the A_2 alloy as-cast (a) and the EDS results of the light-grey area (b) and dark-grey area (c).

As seen in Figure 1, a number of PFBs are formed, and it becomes more apparent with Cu content in the alloy. Figure 2 more clearly exhibits a PFB between PZs, marked by Arrow 3. In the PFB, there are some $CuAl_2$ particles which survived.

Arrow 2 in Figure 2 presents the blank region in PZs. Figure 9 illustrates the EDS results of the regions denoted by Arrows 2 and 3. In the blank region in PZs, the content of Mn is still high to about 0.7 wt.%, and even T_{Mn} and $AlCu_3Mn_2$ particles have

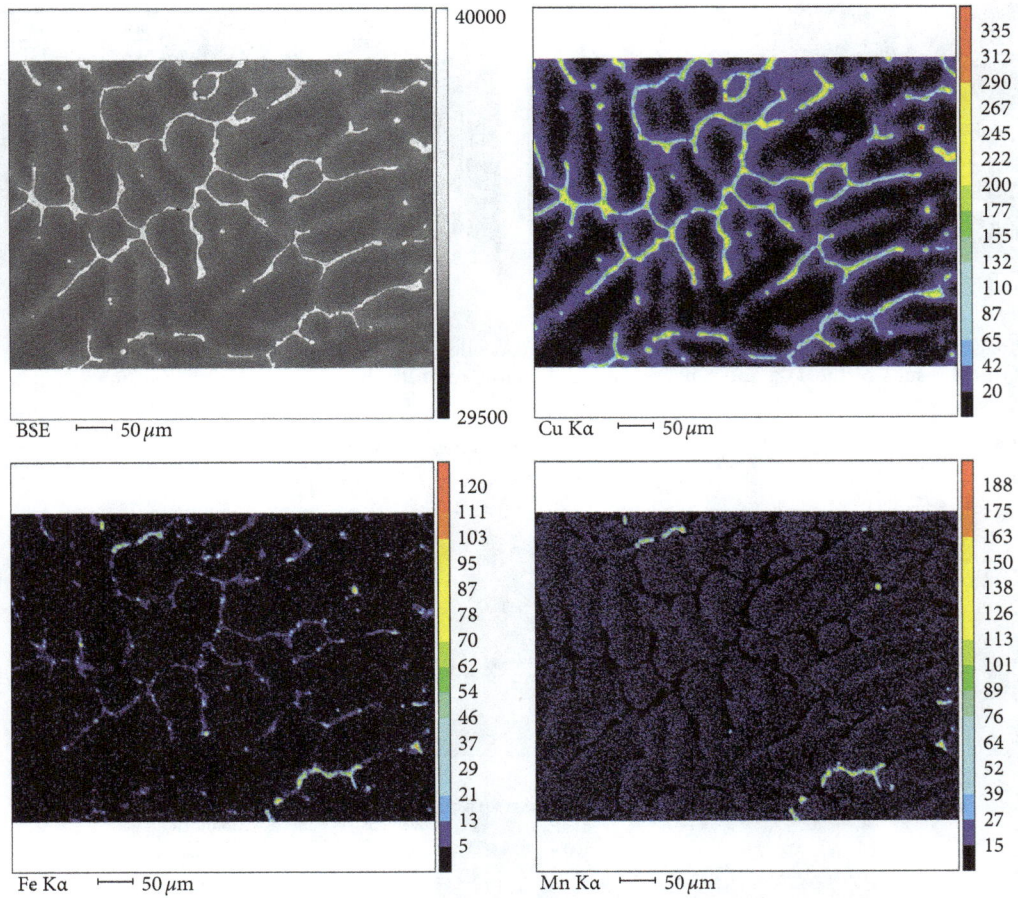

FIGURE 7: EPMA microstructure in the A$_2$ alloy as-cast, showing the distribution of solute elements of Cu, Fe, and Mn (for interpretation of the references to color in this figure legend, the reader is referred to the web version of this article).

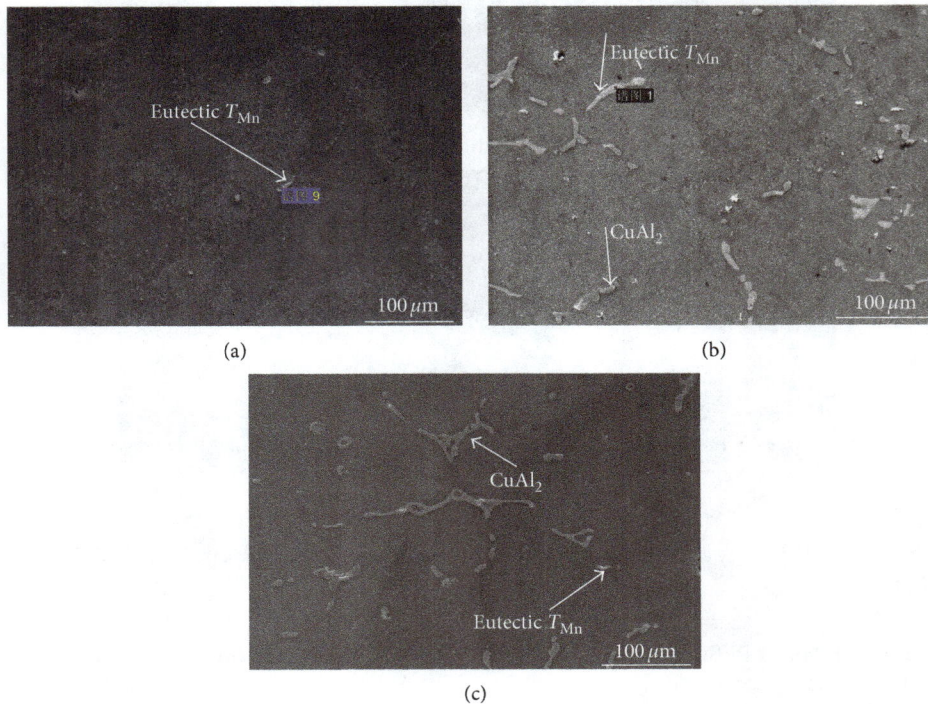

FIGURE 8: SEM images showing the variation of eutectic phases after solution treatment: (a) A$_1$ alloy; (b) A$_2$ alloy; (c) A$_3$ alloy.

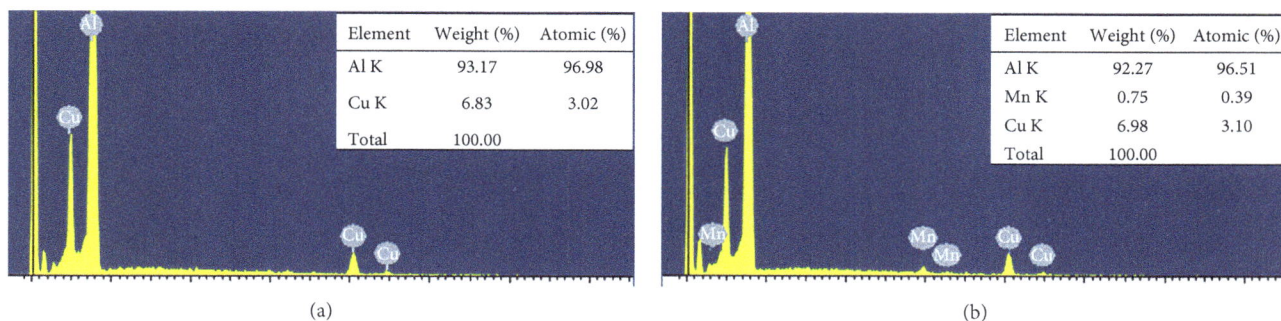

Element	Weight (%)	Atomic (%)
Al K	93.17	96.98
Cu K	6.83	3.02
Total	100.00	

(a)

Element	Weight (%)	Atomic (%)
Al K	92.27	96.51
Mn K	0.75	0.39
Cu K	6.98	3.10
Total	100.00	

(b)

FIGURE 9: EDS results of the regions marked by Arrow 2 (a) and Arrow 3 (b) in Figure 2(a).

precipitated around it, but in the PFB, no Mn trace is detected. Figure 7 illustrates that the regions where the $CuAl_2$ phase exists alone are highly Mn-poor. With Cu content in the alloy, the amount of the eutectic $CuAl_2$ phase formed in form of nonequilibrium is increased (as seen in Figures 4 and 5). During the course of solutionizing at 525°C for 6 hrs + 535°C for 6 hrs, first, this eutectic $CuAl_2$ phase in form of non-equilibrium has to be redissolved into matrix, and then, the remaining $CuAl_2$ phase becomes granulated. This evolution of redissolution and granulation of the eutectic $CuAl_2$ phase leaves the band area where it is highly Mn-poor. It is due to Mn-poor that no precipitation of T_{Mn} and $AlCu_3Mn_2$ particles takes place there, and thus, PFBs are formed among the PZs. In the alloy with higher Cu content, the amount of the eutectic $CuAl_2$ phase that would be dissolved is increased, and thus, the formed PFBs become more apparent. It is concluded that it is the redissolution and granulation of the eutectic $CuAl_2$ phase that lead to the formation of PFBs between PZs.

3.2. Contribution to Tension Properties. For the A_2 alloy, the tension mechanical properties at different states are as follows: as-cast, as-solutionized, and T6 that are labeled in Table 2. In case of the tension test at room temperature (25°C), solutionizing treatment results in a remarkable increase in both the UTS (ultimate tension strength) and YS (yield strength) by 74.1% and 74.4%, respectively, without sacrificing the elongation, and further aging treatment also considerably increases the UTS (from 256 MPa to 369 MPa), YS (from 157 MPa to 248 MPa), and elongation (from 4.8% to 6.3%) again. But, under the tension test at 300°C, the case is different. Solutionizing treatment produces an appreciable increase in UTS and YS by 29.2% and 36.8%, respectively; however, further aging treatment does not lead to obvious improvement of strength both in YS and UTS. Engineering stress verses strain curves shown in Figure 10 vividly exhibit the contribution of microstructure evolution to mechanical properties. The curves of tension at room temperature in Figure 10(a) are common as expected. Both the microstructure evolutions occurring during the solutionizing and aging courses contribute to their respective strengthening effects. Curves in Figure 10(b) reveal the physic process in strengthening at high temperature. In case of tension at 300°C, for the sample as-cast, hardening occurs with further

TABLE 2: Tension properties at 25°C and 300°C of the A_2 alloy at different states.

State		UTS (MPa)	YS (MPa)	Elongation (%)
As-cast	25°C	147 ± 9.0	90 ± 4.0	4.5 ± 0.7
	300°C	89 ± 8.0	68 ± 5.0	6 ± 0.5
As-solutionized	25°C	256 ± 18.0	157 ± 1.6	4.8 ± 0.7
	300°C	115 ± 6.0	93 ± 2.6	6.04 ± 2.1
T6	25°C	369 ± 22.0	248 ± 4.5	6.26 ± 0.8
	300°C	116 ± 11.0	96 ± 8.0	8.3 ± 4.5

plastic deformation beyond yielding, but it merely sustains for only about 4% strain in total, and then a collapse failure arrives suddenly. For samples as-solutionized and T6 tempered, the curves of the two samples are almost the same at the first half stage, with almost equal values both in UTS and YS. Surely, the values are much higher than that as-cast. And the hardening behavior also occurs with further plastic deformation beyond yielding. However, after the hardening reaches a maximum at about 4 w.% strain in total, slightly softening occurs. It continues to a much larger strain of about 8% and 10% in total, respectively, until the final collapse failure. This softening is from dynamic recovery or recrystallization during plastic deformation with a slow strain rate (about $0.0033 \, s^{-1}$) at high temperature, which needs more attention. Compared with the sample as-cast, what support does the samples get (as-solutionized and T6) to tolerate much larger plastic deformation? In the A_2 alloy as-cast, the microstructure is constituted with the dendritic α-Al phase and small amount of the eutectic T_{Mn} and θ phases. After solutionizing, a great number of fine T_{Mn} and $AlCu_3Mn_2$ dispersoid particles are precipitated in the A_2 alloy except for the redissolving of part of the eutectic θ phase (Figures 1–3 and 8). Once these T_{Mn} and $AlCu_3Mn_2$ particles are formed during the solutionizing course, they will almost not change whether in morphology or in size even when they undergo another thermal history with a temperature less than the solutionizing temperature. It is due to its complicated structure and much low diffusion coefficient of Mn in the Al solution [12, 18, 23]. In the sample after aging treatment, the needle-like θ'' and slender rod-like θ' phases in the conventional Al-Cu alloy have not been observed. Conversely, a number of very fine particles (dozens of nanometers in size) are found to precipitate on the rod-like T_{Mn} particles, as shown in

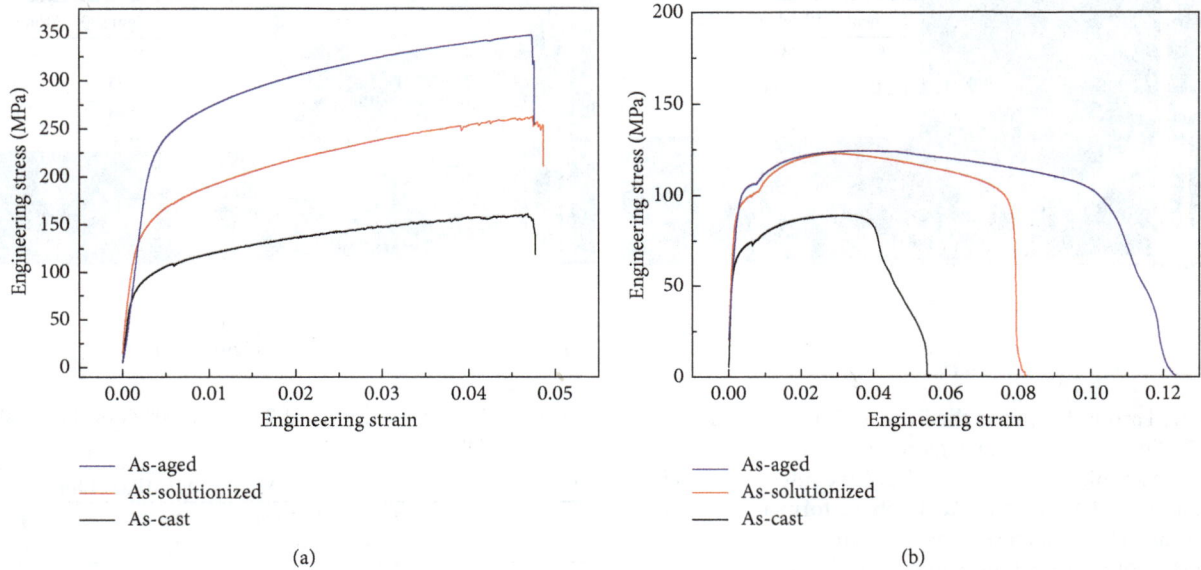

FIGURE 10: Typical engineering stress-strain curves of the A_2 alloy at different states: (a) tension at room temperature; (b) tension at 300°C.

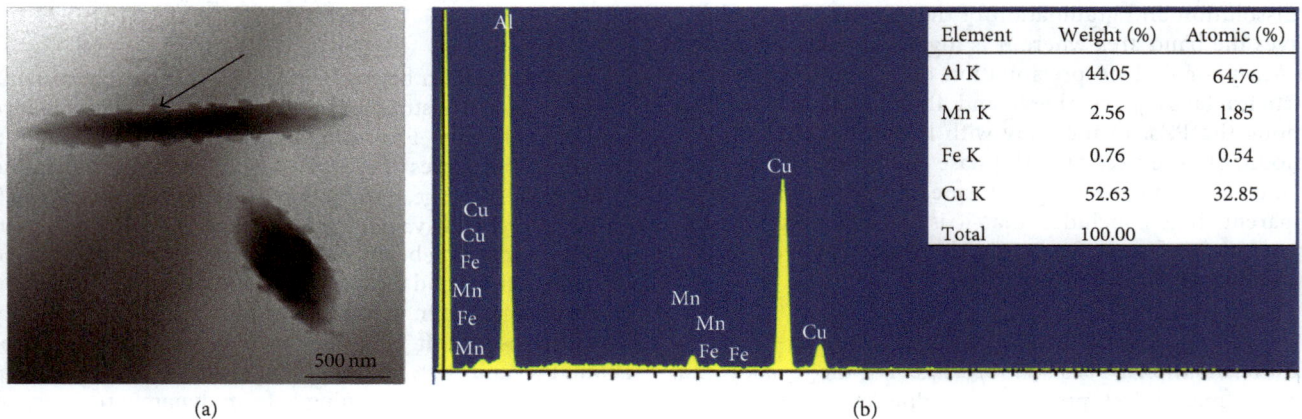

Element	Weight (%)	Atomic (%)
Al K	44.05	64.76
Mn K	2.56	1.85
Fe K	0.76	0.54
Cu K	52.63	32.85
Total	100.00	

FIGURE 11: TEM image showing very fine precipitation particles around/on the rod-like T_{Mn} particles in the A_2 alloy T6 tempered (a) and its EDS results (b).

Figure 11(a). Its EDS results (Figure 11(b)) indicate that it contains a large amount of Cu and Al and a small amount of Mn and Fe. However, due to its Cu/Al atom ratio close to 1 : 2, these fine precipitates may be the metastable θ phase. The precipitation behavior of these fine particles during aging may be responsible for the considerable increment in room temperature strength of the T6 samples compared with the as-solutionized samples. These particles precipitated during the aging course can strengthen the matrix in case of tension at room temperature. But usually, these particles formed during aging possess simple structure, and the diffusion coefficient of the constituents is high, such as Cu, Si, and Mg in the Al solution, and thus, these particles will be coarsened quickly even redissolved when they undergo another thermal history [24–27]. So, the particles of the Al-Cu-Mn phases and metastable θ phase precipitated during the solutionizing and aging courses, respectively, remarkably increase the YS and UTS at room temperature (Figure 10(a) and Table 2). But the microstructure evolution occurring in the aging course does not lead to an improvement in YS and UTS at 300°C (Table 2 and Figure 10(b)). It is a great number of T_{Mn} and AlCu$_3$Mn$_2$ particles precipitated during solutionizing that support the samples (as-solutionized and T6) to tolerate much larger plastic deformation. That is, the considerable increments in YS and UTS at 300°C of the A_2 alloy as-solutionized and T6 tempered are almost completely contributed to the dispersoid precipitation of a great number of T_{Mn} and AlCu$_3$Mn$_2$ particles. Meanwhile, it also indicates that the precipitation behavior in the aging course has no or little contribution to high temperature strength. On other words, if a component is working at a low temperature or room temperature, aging treatment is necessary. And if it is at high temperature, such as at/above 300°C, solutionizing treatment is enough, and aging treatment is completely unnecessary.

TABLE 3: Tension properties at 25°C and 300°C of the studied alloys T6 tempered.

Alloy		UTS (MPa)	YS (MPa)	Elongation (%)
A_1	25°C	153 ± 4.0	66 ± 2.0	8.4 ± 1.7
	300°C	53 ± 3.3	44 ± 2.9	11 ± 6.0
A_2	25°C	369 ± 22.0	248 ± 4.5	6.26 ± 0.8
	300°C	116 ± 11.0	96 ± 11.0	8.3 ± 4.5
A_3	25°C	323 ± 8.0	303 ± 3.0	1.08 ± 0.28
	300°C	134 ± 5.4	101 ± 4.5	10 ± 3.6

The tension mechanical properties at 25°C and 300°C of the studied alloys after T6 treatment are listed in Table 3. In case of tension at 25°C, the YS is increased significantly from 66 MPa to 303 MPa with Cu content in the alloy changing from 2.0 wt.% to 7.5 wt.%. But for UTS, the A_2 alloy has the maximum of 369 MPa, with about 45 MPa higher than that of the A_3 alloy. The A_3 alloy contains 7.5 wt.% Cu, with a higher Cu content than the A_2 alloy (4.5 wt. % Cu). In as-solutionized microstructure of the A_3 alloy, the amount of the survived eutectic θ phase at grain boundaries is much higher than that of the A_2 alloy (Figure 8). During tension at 25°C, because the amount of the T_{Mn} and $AlCu_3Mn_2$ particles in the A_3 alloy precipitated during solutionizing is larger than that in the A_2 alloy and its size is also finer, the YS of the A_3 alloy is higher than that of the A_2 alloy (Table 3), but more severe grain boundary brittleness in the A_3 alloy that is induced by the survived θ phase at grain boundaries makes the matrix lose the ability to tolerate large plastic deformation at room temperature; thus, only with a smaller strain, the sample is fractured. As a result, the elongation of the A_3 alloy (about 1.1%) is much less than that of the A_2 alloy (about 6.3%), and hence, the UTS is also decreased. In case of tension at 300°C, both the YS and UTS are increased with Cu content due to the increased precipitation hardening with the Cu content, consistent with the results in [28]. As concluded above, high temperature strength overridingly depends on the precipitation behavior during solutionizing. The number of particles precipitated during solutionizing is increased with Cu content in the alloy, and simultaneously, the particle size is decreased (Figure 3); the survived eutectic θ and T_{Mn} particles at the grain boundaries can also enhance the ability of grain boundaries to opposite plastic deformation at high temperature, and thus, the high temperature strength is considerably increased with Cu content in the alloy.

4. Conclusions

(1) A shell-core structure is found to form on primary α-Al dendrites in Al-Cu-Mn alloys due to non-equilibrium solidification, in which the Cu content in the shell is much higher than that in the core. The area of the Cu-rich regions (denoting the shell zones) is increased with Cu content in the alloy.

(2) A great amount of fine dispersoid Al-Cu-Mn particles are found to precipitate during the solutionizing course of the studied Al-Cu-Mn alloys. In the alloy with a low Cu content, only T_{Mn} ($Al_{20}Cu_2Mn_3$)

particles are precipitated. However, in the alloy with a high Cu content, $AlCu_3Mn_2$ particles are first demonstrated to precipitate besides T_{Mn}. The $AlCu_3Mn_2$ phase has a simple cubic structure with $a = 0.6904$ nm.

(3) This precipitation behavior is uneven. Besides the precipitation zones (PZs), there are no-particles zones (NPZs) and particle-free bands (PFBs) between PZs. The PZs in the as-solutionized microstructure are highly consistent with the Cu-rich regions in the as-cast microstructure. It is the supersaturation of the Cu solute in the Cu-rich regions and very low diffusion coefficient of manganese in the Al solution that drive the dispersoid precipitation of T_{Mn} and $AlCu_3Mn_2$ to occur in the Cu-rich regions during the solutionizing course.

(4) The evolution of redissolution and granulation of the eutectic $CuAl_2$ phase during solutionizing leaves band areas where they are highly Mn-poor. It is due to Mn-poor that no precipitation of T_{Mn} particles takes place there, and thus, PFBs are formed between the PZs.

(5) Results of the tension test strongly demonstrate that, compared with the samples as-cast, the considerable increments in YS and UTS at 300°C of the samples as-solutionized and T6 tempered are almost completely contributed to the dispersoid precipitation of a great number of T_{Mn} and $AlCu_3Mn_2$ particles during solutionizing, and the precipitation behavior in the aging course has no or little contribution to it.

(6) High temperature strength of Al-$x$wt.%Cu-1.0 wt.% Mn alloy is considerably increased with Cu content, which is closely tied with the increased number and simultaneously decreased size of T_{Mn} and $AlCu_3Mn_2$ particles precipitated during solutionizing.

Conflicts of Interest

The authors declare that they have no conflicts of interest.

Acknowledgments

This work was supported by the Jiangsu Key Laboratory Metallic Materials (Grant no. BM2007204) and the Fundamental Research Funds for the Central Universities (Grant no. 2242016k40011). The authors are also thankful to Sha-steel Iron and Steel Research Institute of Jiangsu Province for their useful assistance.

References

[1] G. Sha, R. K. W. Marceau, X. Gao, B. C. Muddle, and S. P. Ringer, "Nanostructure of aluminum alloy 2024: segregation, clustering and precipitation processes," *Acta Materialia*, vol. 59, no. 4, pp. 1659–1670, 2011.

[2] D. M. Yao, W. G. Zhao, H. L. Zhao, F. Qiu, and Q. C. Jiang, "High creep resistance behavior of the casting Al-Cu alloy modified by La," *Scripta Materialia*, vol. 61, no. 12, pp. 1153–1155, 2009.

[3] X. Li, K. Lei, P. Song et al., "Strengthening of aluminum alloy 2219 by thermo-mechanical treatment," *Journal of Materials Engineering & Performance*, vol. 24, no. 10, pp. 3905–3911, 2015.

[4] A. K. Jha, S. V. S. N. Murty, K. Sreekumar, and P. P. Sinha, "High strain rate deformation and cracking of AA 2219 aluminum alloy welded propellant tank," *Engineering Failure Analysis*, vol. 16, no. 7, pp. 2209–2216, 2009.

[5] S. C. Wang and M. J. Starink, "Precipitates and intermetallic phases in precipitation hardening Al–Cu–Mg–(Li) based alloys," *International Materials Reviews*, vol. 50, no. 4, pp. 193–215, 2005.

[6] Y. L. Zhao, Z. Q. Yang, Z. Zhang, G. Y. Su, and X. L. Ma, "Double-peak age strengthening of cold-worked 2024 aluminum alloy," *Acta Materialia*, vol. 61, no. 5, pp. 1624–1638, 2013.

[7] X. X. Feng, A. M. Kumar, and J. P. Hirth, "Mixed mode I/III fracture toughness of 2034 aluminum alloys," *Acta Metallurgica et Materialia*, vol. 41, no. 9, pp. 2755–2764, 1993.

[8] D. Tsivoulas, J. D. Robson, C. Sigli, and P. B. Prangnell, "Interactions between zirconium and manganese dispersoid-forming elements on their combined addition in Al-Cu-Li alloys," *Acta Materialia*, vol. 60, no. 13-14, pp. 5245–5259, 2012.

[9] Z. Q. Feng, Y. Q. Yang, B. Huang, M. H. Li, Y. X. Chen, and J. G. Ru, "Crystal substructures of the rotation-twinned T ($Al_{20}Cu_2Mn_3$) phase in 2024 aluminum alloy," *Journal of Alloys and Compouds*, vol. 583, pp. 445–451, 2014.

[10] Z. Shen, C. Liu, Q. Ding et al., "The structure determination of $Al_{20}Cu_2Mn_3$ by near atomic resolution chemical mapping," *Journal of Alloys and Compounds*, vol. 601, no. 9, pp. 25–30, 2014.

[11] Z. Chen, P. Chen, and S. Li, "Effect of Ce addition on microstructure of $Al_{20}Cu_2Mn_3$ twin phase in an Al-Cu-Mn casting alloy," *Materials Science and Engineering: A*, vol. 532, no. 3, pp. 606–609, 2012.

[12] Z. W. Chen, Q. Y. Fan, and K. Zhao, "Microstructure and microhardness of nanostructured Al-4.6Cu-Mn alloy ribbons," *International Journal of Minerals, Metallurgy, and Materials*, vol. 22, no. 8, pp. 860–867, 2015.

[13] S. Wang, C. Li, and M. Yan, "Determination of structure of $Al_{20}Cu_2Mn_3$ phase in Al-Cu-Mn alloys," *Materials Research Bulletin*, vol. 24, no. 10, pp. 1267–1270, 1989.

[14] W. J. Park and N. J. Kim, "Microstructural characterization of 2124 Al-SiCW composite," *Scripta Materialia*, vol. 36, no. 9, pp. 1045–1051, 1997.

[15] G. B. Johnston and E. O. Hall, "Studies on the Heusler alloys-II. The structure of Cu_3Mn_2Al," *Journal of Physics and Chemistry of Solids*, vol. 29, no. 2, pp. 201–207, 1968.

[16] B. Li, Q. L. Pan, C. P. Chen, and Z. M. Yin, "Effect of aging time on precipitation behavior, mechanical and corrosion properties of a novel Al-Zn-Mg-Sc-Zr alloy," *Transactions of Nonferrous Metals Society of China*, vol. 26, no. 9, pp. 2263–2275, 2016.

[17] A. Munitz and C. Cotler, "Aging impact on mechanical properties and microstructure of Al-6063," *Journal of Materials Science*, vol. 35, no. 10, pp. 2529–2538, 2000.

[18] H. C. Liao, Y. Y. Tang, X. J. Suo et al., "Dispersoid particles precipitated during the solutionizing course of Al-12 wt%Si-4 wt%Cu-1.2 wt%Mn alloy and its contribution to high temperature strength," *Materials Science and Engineering:A*, vol. 699, pp. 201–209, 2017.

[19] S. Maâmar and M. Harmelin, "On the transitions of the icosahedral and decagonal phases towards equilibrium phases in Al-Cu-Mn alloys," *Philosophical Magazine Letters*, vol. 64, no. 6, pp. 343–348, 1991.

[20] S. M. Skolianos, T. Z. Kattamis, and O. F. Devereux, "Microstructure and corrosion behavior of as-cast and heat-treated Al-4.5 Wt pct Cu-2.0 wt pct Mn alloys," *Metallurgical Transactions A*, vol. 20, no. 11, pp. 2499–2516, 1989.

[21] A.T Chen, L. Zhang, G. H. Wu, M. Sun, and W. C. Liu, "Influences of Mn content on the microstructures and mechanical properties of cast Al-3Li-2Cu-0.2Zr alloy," *Journal of Alloys and Compounds*, vol. 715, pp. 421–431, 2017.

[22] M. Chen and T. Z. Kattamis, "Dendrite coarsening during directional solidification of Al-Cu-Mn alloys," *Materials Science and Engineering: A*, vol. 247, no. 1-2, pp. 239–247, 1998.

[23] I. Häusler, C. Schwarze, M. U. Bilal et al., "Precipitation of T_1 and θ' phase in Al-4Cu-1Li-0.25Mn during age hardening: microstructural investigation and phase-field simulation," *Materials*, vol. 10, no. 2, p. 117, 2017.

[24] S. P. Ringer and K. Hono, "Microstructural evolution and age hardening in aluminium alloys: atom probe field-ion microscopy and transmission electron microscopy studies," *Materials Characterization*, vol. 44, no. 1, pp. 101–131, 2000.

[25] E. Cerri, E. Evangelista, and N. Ryum, "The relationship between microstructural and plastic instability in Al-4.0 wt pct Cu alloy," *Metallurgical and Materials Transactions A*, vol. 27, no. 10, pp. 2916–2922, 1996.

[26] E. Balducci, L. Ceschini, S. Messieri, S. Wenner, and R. Holmestad, "Thermal stability of the lightweight 2099 Al-Cu-Li alloy: tensile tests and microstructural investigations after overaging," *Materials & Design*, vol. 119, pp. 54–64, 2017.

[27] Z. Gao, J. H. Chen, S. Y. Duan, X. B. Yang, and C. L. Wu, "Complex precipitation sequences of Al-Cu-Li-(Mg) alloys characterized in relation to thermal ageing processes," *Acta Metallurgica Sinica*, vol. 29, no. 1, pp. 94–103, 2016.

[28] D. H. Xiao, J. N. Wang, D. Y. Ding, and S. P. Chen, "Effect of Cu content on the mechanical properties of an Al-Cu-Mg-Ag alloy," *Journal of Alloys and Compounds*, vol. 343, no. 1, pp. 77–81, 2002.

10

Effect of Powder Size on Microstructure and Mechanical Properties of 2A12Al Compacts Fabricated by Hot Isostatic Pressing

Xina Huang ⓘ,[1] Lihui Lang,[1] Gang Wang,[1] and Sergei Alexandrov ⓘ[1,2]

[1]*School of Mechanical Engineering and Automation, Beihang University, Beijing 100191, China*
[2]*Institute for Problems in Mechanics, Russian Academy of Sciences, Moscow 119526, Russia*

Correspondence should be addressed to Xina Huang; huangxina@126.com

Academic Editor: Hongchao Kou

This paper studied the effects of powder size on densification, microstructure, and mechanical properties of the hot isostatic-pressed 2A12 aluminum alloy powder compact. The results show that the near-fully dense powder compact can be successfully achieved and the smaller the powder is, the higher the relative density is. In addition, as the powder size decreases, the precipitated phases in the powder compact change from continuously point-like distribution at the junctions among powder particles to the concentrated distribution at the three-way intersections. Compared with the large powder, the tensile strength, yield strength, and elongation of the compact with the small powder were improved by 14%, 30.8%, and 48.6%, respectively.

1. Introduction

2A12 aluminum alloy (2A12Al) has low density, high specific strength, and good corrosion resistance, which makes it widely used for the structural components in aeronautics and astronautics [1].

The near-net shaping (NNS) technology has attracted widespread interests. It is considered as a reliable and low-cost material forming process. It has great potential to form the parts with complicated internal cavities, which are difficult to be fabricated by traditional methods such as forging, machining, and so on. The NNS technology is composed of several methods: powder injection molding (PIM), hot isostatic pressing (HIP), powder forging (PF), and cold isostatic pressing (CIP)/sintering [2, 3]. Usually, the products fabricated by PIM have inferior mechanical properties than those using other NNS technologies [4]. Higher mechanical strength can be obtained through PF, but this method is restricted to the large-scale components. HIP is an appropriate forming process for components with large and complex shapes. In addition, the HIPed parts have a finer and more homogeneous microstructure than the parts formed by other NNS technologies, which leads to an enhancement of the mechanical properties [5, 6].

The mechanical properties of HIPed parts are influenced by the densification of the metal powder. The densification process includes three main mechanisms: powder compaction and rearrangement, plastic deformation, and diffusion creep [7–9]. These densification mechanisms are significantly affected by the size and shape of the powder [10]. Longrasso and Koss investigated the influence of Ti powder shape on densification during HIP [8]. In addition, many papers studied the effects of matrix powder size on densification, hardness, microstructure, and wear resistance of composites [7, 11, 12]. However, to the authors' best knowledge, few studies have focused on the effect of monolithic aluminum alloy powder size on densification. Therefore, the aim of this work is to investigate the influence of 2A12Al powder size on densification behavior, microstructure, and mechanical properties of powder compacts prepared using HIP.

2. Experimental Procedures

2.1. Original Material. Nitrogen gas-atomized 2A12Al powder with different average sizes of 195 μm, 124 μm, and 35 μm was used as the original material and defined as powders 1, 2, and 3, respectively. They were supplied by ARI

TABLE 1: Basic characteristics of the powder.

Alloy powders	Powder size (μm)			Apparent density (%)	W (O) (%)
	D_{10}	D_{50}	D_{90}		
Powder 1	131	195	292	62.74	0.12
Powder 2	85	124	174	64.30	0.15
Powder 3	14	35	67	69.96	0.26

Forster (Beijing) Technology Development Co., Ltd. in China. The powder size, apparent density, and oxygen content are listed in Table 1. Among them, D_{10} indicates that the powder size smaller than the D_{10} corresponding size accounts for 10%. D_{50} usually represents the average size of powders. D_{90} shows that the powder size smaller than the D_{90} corresponding size accounts for 90%. The small powders with high surface energy are prone to be contaminated, resulting in a high oxygen content [13]. The chemical composition of the 2A12Al powder is given in Table 2.

Figure 1 shows the morphologies and the particle size distribution of powders. The powders are nearly spherical in shape, and the size of the powders 1, 2, and 3 is between 131 and 292 μm, 85 and 174 μm, and 14 and 67 μm, respectively. It can be found that there are plenty of satellite particles in the small powders (no. 3; Figure 1(c)), which can be attributed that metal liquid was broken into small droplets under the impact force. Compared with the large droplets, small droplets were formed more per unit time. The greater surface tension and under-cooling gave rise to the earlier solidification of small droplets than large droplets. The firstly solidified small droplets moved with a high speed and then collided with large droplets, resulting in the cold welding. The small droplets attached to the surface of large ones form the tiny satellite particles. These satellite particles can be filled into the gap between large powders, giving rise to a higher apparent density of powder 3.

2.2. HIP Process. Firstly, 2A12Al powders were filled into a 1060 pure Al (1060Al) circular cylinder capsule with 60 mm internal diameter, 120 mm height, and 1 mm wall thickness. The 1060Al capsule was used because it does not react with the original material. Subsequently, the capsule was degassed at a temperature of 400°C by an FJ-620 molecular pump until the vacuum degree was 1.0×10^{-4} Pa. And then, hot isostatic pressing was carried out in QIH-15 HIP equipment under the process parameters of 470°C, 130 MPa, and 3 h dwell time. Three samples of each powder size were prepared at least. After HIP, the capsule was removed by machining.

2.3. Microstructure and Mechanical Property. The relative density of powder compacts was measured based on Archimedes' principle, and the high-purity water was used in Archimedes' experiment. In addition, three HIPed specimens for each powder size were taken to carry out the tensile tests on a QJ210 electronic universal testing machine based on GB/T 228-2002 metallic material tensile testing in the ambient temperature. The microstructure and fracture morphology of HIPed 2A12Al powder compacts were observed by a JSM 6010 scanning election microscope (SEM). The element distribution

TABLE 2: Chemical composition of the 2A12Al powder (wt.%).

Element	Al	Cu	Mg	Ti	Si	Fe	Mn	Zn	Others
Content	Bal.	4.1	1.5	0.15	0.5	0.5	0.4	0.3	0.15

was characterized by an energy dispersive spectrometer (EDS, CamScan-3400). Before SEM and EDS, the specimens were sectioned, grounded, and polished. Then, the specimens were etched with the Keller reagent (1.5 ml HCl, 2.5 ml HNO$_3$, 1 ml HF, and 95 ml distilled H$_2$O).

3. Results and Discussion

3.1. The Variation of Dimension and Relative Density. Three samples were prepared using HIP for measuring the density and dimensional variation. Figure 2 shows the averaged relative density and the radial and axial dimensional variation rate of powder compacts with different powder sizes and measurement positions.

It can be found in Figure 2(a) that the near-fully dense powder compact can be successfully obtained by HIP. The smallest relative density is 96.2% obtained from powders with 131–292 μm size, which is higher than that prepared by traditional pressing or sintering (95%) [14]. The relative density increases with the decrease of powder size, which is similar to the results reported by Barringer and Bowen [15], and the maximum relative density of 97.6% is obtained when the powder size is between 14 and 67 μm. At the beginning of HIP, there is a large amount of voids among loose powders with a contact of point-to-point. With the increase of pressure, the loose powders translate or rotate and become close to each other. Besides that, the bridges among the powders collapse. Therefore, the powder particle rearrangement is the main densification mechanism [16, 17]. With the further increase of pressure, the normal stress on the powder surface increases and the yield stress of powder particles decreases with the increase of temperature. When the normal stress exceeds the yield stress, the plastic deformation of powder particles begins to occur. At this time, the plastic deformation of powders becomes the dominant densification mechanism. With the HIP process being carried on, the plastic deformation of powders and the friction among them are increased. Even if the temperature and pressure increase, the movement of powders is still restricted [18, 19]. The smaller the powder is, the easier the plastic deformation is and the better the densification is. Following rearrangement and plastic deformation, the previous interconnected voids are independent of each other and dispersed uniformly in the powder compact. In addition, the powder compact already has a certain density. The

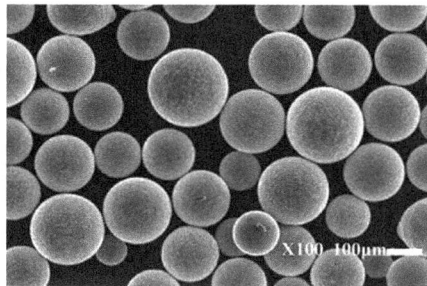

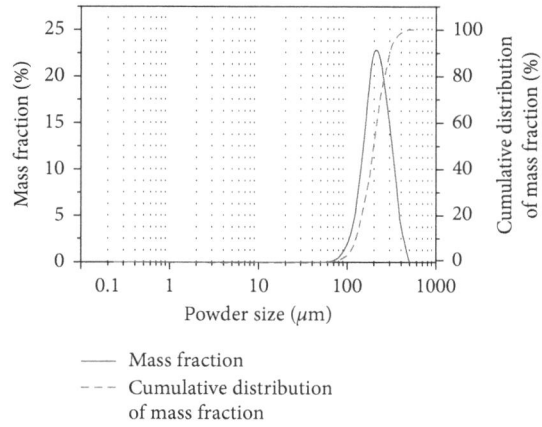

(a)

(b)

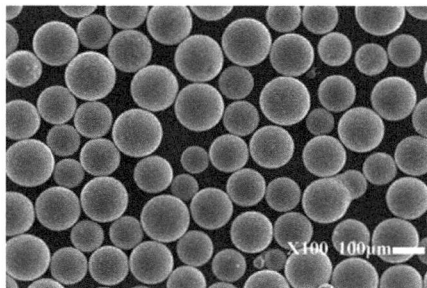

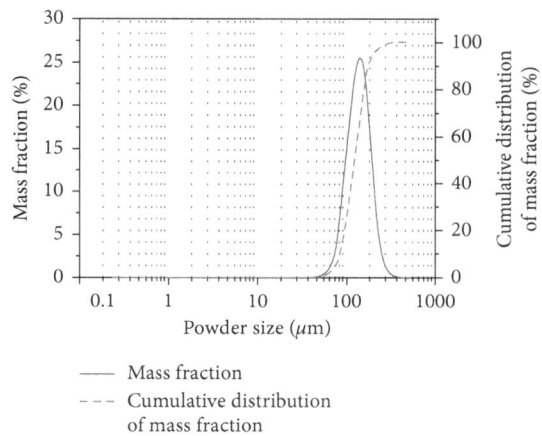

(c)

(d)

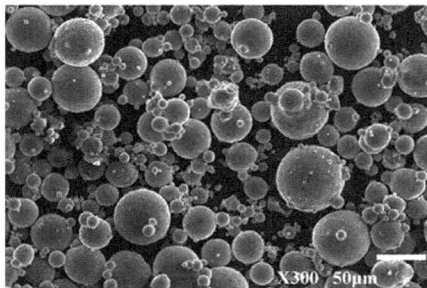

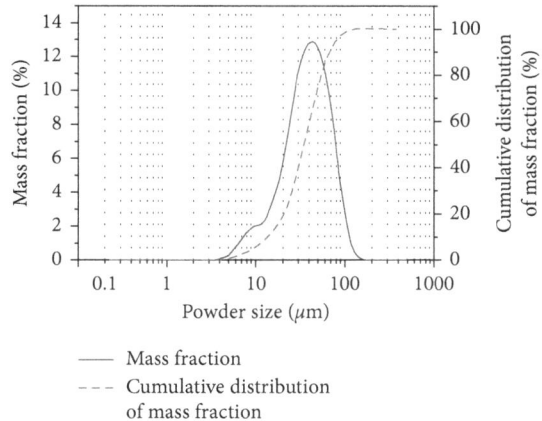

(e)

(f)

FIGURE 1: The morphologies of the powders 1 (a), 2 (b), and 3 (c).

voids are continuously spheroidized under the driven of the surface energy. Therefore, the void size decreases and the contact area of powders increases. With the occurrence of the spheroidization process, the number and size of the voids decrease and the relative density increases. The densification is mainly achieved by the diffusion of individual atoms or voids and the creep of the powder particles. The smaller the

powder is, the easier the spheroidization is and the better the densification is. Therefore, the smaller powder has the better densification and the higher relative density.

The change in relative density gives rise to the dimensional variation. As shown in Figure 2(b), the negative value indicates that the powder compact shrinks in both radial and axial directions during HIP. The trends of the dimensional shrinkage

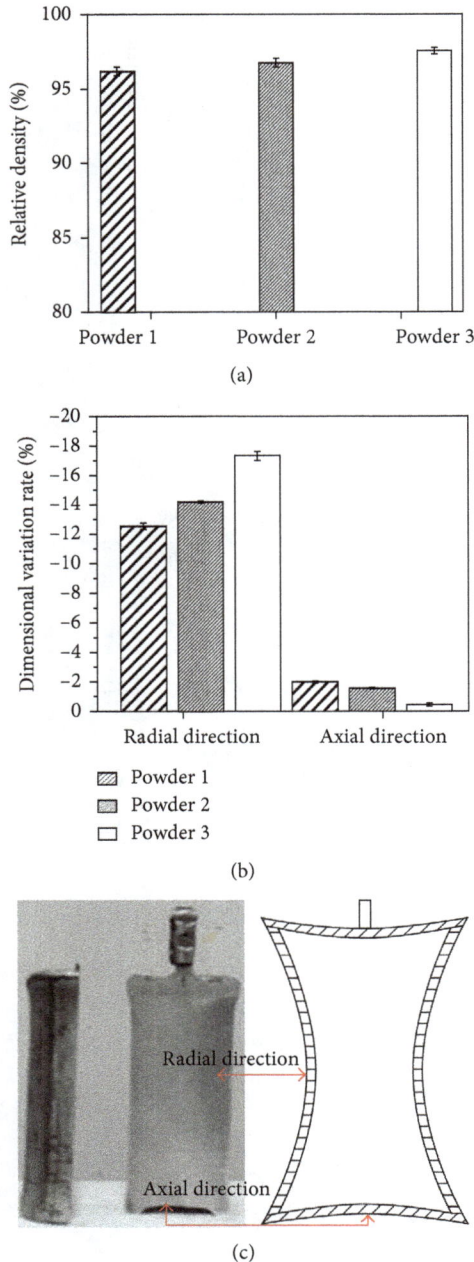

FIGURE 2: The relative density (a), the radial and axial dimensional variation rate (b), and the radial and axial direction dimensional measurement position (c).

and relative density are consistent. The radial dimensional shrinkage rate increased distinctly from 12.6% to 17.4% with the decrease of powder size, while the axial dimensional variation rate decreased slightly from 2.0% to 0.5% (Figure 2(b)). The increase in the radial dimensional shrinkage and the decrease in the axial dimensional variation rate are more obvious with the decrease of powder size, which is because the apparent density is higher and the particle gaps are smaller in the powder compact with small powders (powder 3) than that with large powders (powders 1 and 2). With the radial dimensional

shrinkage, the powder compact reaches the dense state quickly, which might prevent the axial dimensional variation.

Previous research indicated that radial and axial pressure of the cylinder part is different during HIP, as follows [20]:

$$P_r = \frac{P(2rL)}{2tL} = \frac{Pr}{t}, \tag{1}$$

$$P_a = \frac{P(\pi r^2)}{2\pi rt} = \frac{Pr}{2t}, \tag{2}$$

where P_r and P_a are the radial and axial pressure; P is the applied external pressure; and L, r, and t are the height, radius, and thickness of the capsule, respectively. According to (1) and (2), the radial pressure is twice the axial pressure; therefore, the axial dimensional variation rate is smaller than the radial. In addition, the upper cover was welded with the cylinder, the rigidity of the welding seam was large, and the solder was accumulated at the welding, leading to a small axial dimensional variation rate.

3.2. Microstructure. Figure 3 shows the micrographs of powder compacts with three different powder sizes after HIP. No obvious internal voids are observed, indicating that the near-fully dense powder compacts are achieved. Prior particle boundary (PPB) of powders was formed because of a large amount of second-phase particles existing at the powder boundaries and can be clearly seen in all compacts. In powders 1 and 2, the powder size is large and uniform; as a consequence, the PPB of each powder was retained in the final compacts, and the precipitated phases (as seen in Figures 3(a) and 3(b)) were distributed at the powder particle boundaries in the continuous point-like form. However, in powder 3 (Figure 3(c)), the PPB of satellite particles almost disappeared, whereas the PPB of larger particles was apparent, which shows that the precipitated phases were heavily distributed at the three-way intersections among powder particles, instead of the continuously point-like form at the junctions. In addition, coarse precipitated phases can be found at the three-way intersections, which are composed of Al, Cu, Mg, and O elements (Figure 3(d)). The Cu content of the original material is much lower than precipitated phases at the three-way intersections. Therefore, it can be deducted that Cu in the original powder diffused and aggregated at the intersections, resulting in the formation of Al-Cu-Mg precipitated phases, which is similar to the results in [15].

It is noted that the PPB tended to be straight and the angle between boundaries was near 120° after HIP (vertex of Figure 3(b)) In light of the observation above, it is suggested that the deformation law of the powder under high temperature and pressure is similar to the grain growth law, that is, the grain boundary tends to be straight to reduce the surface area and surface energy. Moreover, the small particles were annexed by adjacent large particles during boundary migration. The angle between boundaries tended to migrate closer to 120°, which is similar to the results in [21]. Nevertheless, the polygonal particles would gradually shrink or grow until the grain size is six. At this time, the interface is in equilibrium, the particles no longer move, and

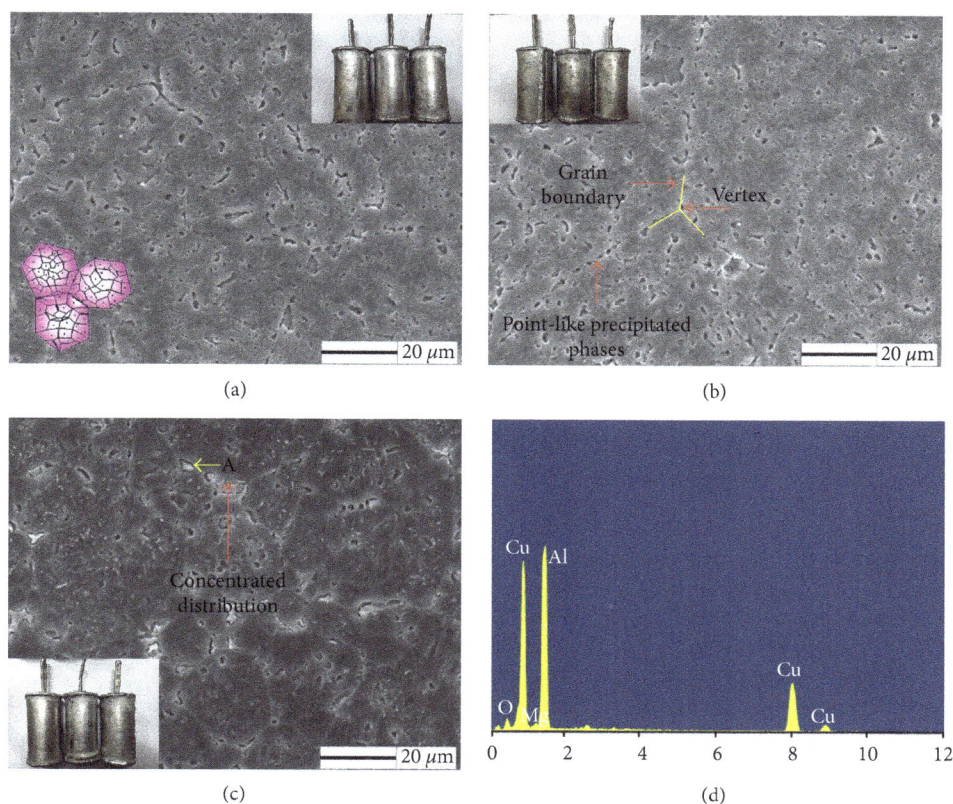

FIGURE 3: The micrographs of the powder compacts (a) 1, (b) 2, and (c) 3 and (d) the EDS map of point A in (c).

TABLE 3: Mechanical properties of the 2A12Al powder.

Alloy powders	Tensile strength R_m (MPa)	Yield strength $R_{p0.2}$ (MPa)	Elongation A (%)
1	265	182	1.8
2	284	191	2
3	308	263	3.5

the edges are straight with an angle of 120° [21]. In powder 3, the large particles contact with each other firstly and then small particles. In addition, the overall temperature of small particles was higher than that of large ones [22], and the plasticity was better. In the further forming process, a majority of the stress among large particles was undertaken by small particles. The deformation of large particles became slow and even stopped, which is similar to the rigid movement of the rigid body. The deformation of small particles promoted the large-scale movement of particles [23]. Therefore, the particle boundary deformation of powder 3 was severe, and part of the small particles even fused together. For the powders 1 and 2, the shape is more uniform and the size is larger than powder 3, resulting in a more difficult particle rearrangement [24]. Hence, the relative density of powder 3 is higher than that of the powders 1 and 2, which is consistent with the results of relative density measurement.

3.3. Mechanical Properties of 2A12 Compacts after HIP.
Table 3 shows the variation of mechanical properties of the

samples fabricated with the three powder sizes. With the decrease of powder size, the ultimate tensile strength, proof strength of plastic extension, and elongation of powder compacts after HIP increased from 265 MPa to 308 MPa, from 182 MPa to 263 MPa, and from 1.8% to 3.5%, respectively. The precipitated phases were distributed at the junctions in the continuous uniform point-like form in the powders 1 and 2. The effective diffusion connection could not be formed, which would have a deleterious effect on the mechanical properties. However, the higher densification and the finer PPB of small powders are favorable for forming a sound bonded interface, which improves the mechanical properties. The tensile strength, yield strength, and elongation increased by 14%, 30.8%, and 48.6%, respectively. Besides, the oxygen content of powder 3 was higher than that of the powders 1 and 2. Such interstitial atoms may act as the obstacles to the dislocation motions; as a result, the strength would be improved with the decrease of powder size [21].

Figure 4 is the fracture morphology of powder compact 1 (Figure 4(a)) and powder compact 3 (Figure 4(b)). The fracture morphology of powder compact 2 is similar to powder

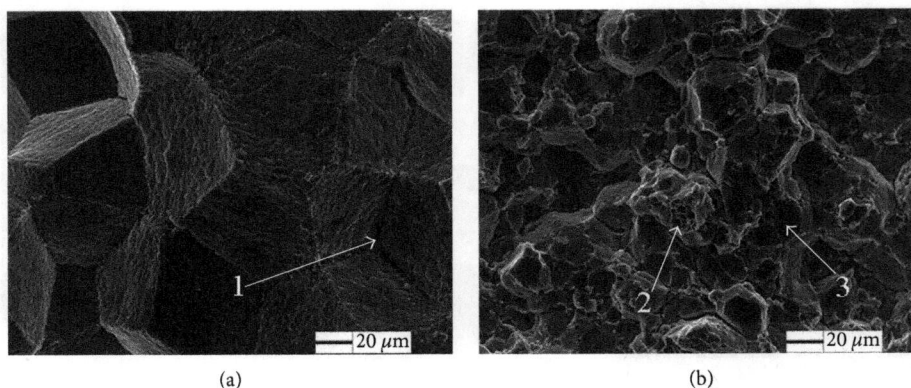

(a) (b)

FIGURE 4: The fracture morphologies of the powder compacts 1 and 2 (a) and 3 (b).

compact 1. In the powders 1 and 2, the fracture only occurred at the interfaces among particles (arrow 1 in Figure 4(a)); therefore, the fracture feature can be deduced as brittle. In powder 3, dimples appeared in some areas (arrow 2 in Figure 4(b)), indicating the ductile fracture characteristic, which is beneficial for improving the plasticity of the powder compact. In addition, the alloy element precipitation of powder 3 was embedded in the interior of particles (arrow 3 in Figure 4(b)), enhancing the pinning effect and preventing dislocation movement, which would help to increase the strength.

4. Conclusions

The following conclusions can be drawn from this study:

(1) Near-fully dense 2A12Al compacts by HIP could be successfully achieved. The maximum relative density was 97.6%, which was obtained with the finest powder. Moreover, higher relative density was accompanied with larger radial and smaller axial dimensional variation rates.

(2) The PPB of large powders tended to be straight, and the angle between boundaries was near 120° after HIP. However, the PPB of small powders underwent larger deformation, some small particles even fused together, and the PPB of satellite particles almost disappeared.

(3) With the decrease of powder size, the precipitated phase changed from continuous point-like distribution at the junctions to the concentrated distribution in three-way intersection of particles, which made the diffusion bonding of interface tighter and was beneficial to improve the tensile strength, yield strength, and elongation.

(4) The fracture mechanism of the powder compact with the large powder is brittle fracture, while the plastic fracture characteristic can be found in the powder compact with the small powder.

Conflicts of Interest

The authors declare that they have no conflicts of interest.

References

[1] J. Yan, X. Y. Zeng, M. Gao, J. Lai, and T. X. Lin, "Effect of welding wires on microstructure and mechanical properties of 2A12 aluminum alloy in CO_2 laser-MIG hybrid welding," *Applied Surface Science*, vol. 255, no. 16, pp. 7307–7373, 2009.

[2] D. L. Zhang, S. Raynova, V. Nadakuduru, P. Cao, B. Gabbitas, and B. Robinson, "Consolidation of titanium, and Ti-6Al-4V alloy powders by powder compact forging," *Materials Science Forum*, vol. 618-619, pp. 513–516, 2009.

[3] L. Wang, Z. B. Lang, and H. P. Shi, "Properties and forming process of prealloyed powder metallurgy Ti-6Al-4V alloy," *Transactions of Nonferrous Metals Society of China*, vol. 17, pp. s639–s643, 2007.

[4] R. M. German, "Progress in titanium metal powder injection molding," *Materials*, vol. 6, no. 12, pp. 3641–3662, 2013.

[5] Y. M. Kim, E. P. Kim, Y. B. Song, S. H. Lee, and Y. S. Kwon, "Microstructure and mechanical properties of hot isostatically pressed Ti-6Al-4V alloy," *Journal of Alloys and Compounds*, vol. 603, pp. 207–212, 2014.

[6] K. T. Kim and H. C. Yang, "Densification behavior of titanium alloy powder under hot isostatic pressing," *Powder Metallurgy*, vol. 44, no. 1, pp. 41–46, 2001.

[7] E. A. Diler, A. Ghiami, and R. Ipek, "Effect of high ratio of reinforcement particle size to matrix powder size and volume fraction on microstructure, densification and tribological properties of SiC_p reinforced metal matrix composites manufactured via hot pressing method," *International Journal of Refractory Metals and Hard Materials*, vol. 52, pp. 183–194, 2015.

[8] B. K. Lograsso and D. A. Koss, "Densification of titanium powder during hot isostatic pressing," *Metallurgical Transactions A*, vol. 19, no. 7, pp. 1767–1773, 1998.

[9] E. Arzt, "The influence of an increasing particle coordination on the densification of spherical powders," *Acta Metallurgica*, vol. 30, no. 10, pp. 1883–1890, 1982.

[10] D. Bouvard, "Densification behavior of mixtures of hard and soft powders under pressure," *Powder Technology*, vol. 111, no. 3, pp. 231–239, 2000.

[11] T. Fan, C. L. Xiao, Y. R. Sun, and H. B. Li, "Microstructure and properties of SiC particle reinforced aluminum matrix composites by powder metallurgy method," *Applied Mechanics and Materials*, vol. 457-458, pp. 131–134, 2014.

[12] E. K. Omyma and A. Fathy, "Effect of SiC particle size on the physical and mechanical properties of extruded Al matrix nanocomposites," *Materials & Design*, vol. 54, pp. 348–353, 2014.

[13] Y. M. Kim, E. P. Kim, J. W. Noh, S. H. Lee, Y. S. Kwon, and I. S. Oh, "Fabrication and mechanical properties of powder metallurgy tantalum prepared by hot isostatic pressing," *International Journal of Refractory Metals and Hard Materials*, vol. 48, pp. 211–216, 2015.

[14] C. Padmavathi and A. Upadhyaya, "Sintering behavior and mechanical properties of Al–Cu–Mg–Si–Sn aluminum alloy," *Transactions of the Indian Institute of Metals*, vol. 64, no. 4-5, pp. 345–357, 2011.

[15] E. A. Barringer and H. K. Bowen, "Effects of particle packing on the sintered microstructure," *Applied Physics A Solids and Surfaces*, vol. 45, no. 4, pp. 271–275, 1988.

[16] M. M. Goudarzi and F. Akhlaghi, "Effect of nanosized SiC particles addition to CP Al and Al–Mg powders on their compaction behavior," *Powder Technology*, vol. 245, pp. 126–133, 2013.

[17] H. R. Hafizpour, A. Simchi, and S. Parvizi, "Analysis of the compaction behavior of Al–SiC nanocomposites using linear and non-linear compaction equations," *Advanced Powder Technology*, vol. 21, no. 3, pp. 273–278, 2010.

[18] P. J. Denny, "Compaction equations: a comparison of the Heckel and Kawakita equations," *Powder Technology*, vol. 127, no. 2, pp. 162–172, 2002.

[19] S. Mahdavi and F. Akhlaghi, "Effect of SiC content on the processing, compaction behavior, and properties of Al6061/SiC/Gr hybrid composites," *Journal of Materials Science*, vol. 46, no. 5, pp. 1502–1511, 2011.

[20] D. P. Delo and H. R. Piehler, "Early stage consolidation mechanisms during hot isostatic pressing of Ti-6Al-4V powder compacts," *Acta Materialia*, vol. 47, no. 9, pp. 2841–2852, 1999.

[21] N. K. Li, G. Ling, and B. Nie, *Aluminum Alloy Material and Heat Treatment Technology*, Vol. 287, Metallurgical Industry Press, Beijing, China, 2012.

[22] G. H. Xu, X. H. Zhang, C. M. Zhao, L. Wang, and Z. W. Yin, "Microstructure of PM TC11 alloy and its effect on mechanical behavior," *Aerospace Materials & Technology*, vol. 3, pp. 110–113, 2013.

[23] Y. E. Bing, R. M. Matsen, and C. D. David, "Finite-element modeling of titanium powder densification," *Metallurgical and Materials Transactions A*, vol. 43, no. 1, pp. 381–390, 2012.

[24] K. Kondah, A. Kimura, and R. Watanabe, "Effect of Mg on sintering phenomenon of aluminum alloy powder particle," *Powder Metallurgy*, vol. 44, no. 2, pp. 161–164, 2001.

11

Statistical Model for the Mechanical Properties of Al-Cu-Mg-Ag Alloys at High Temperatures

A. M. Al-Obaisi,[1,2] **E. A. El-Danaf,**[1] **A. E. Ragab,**[3] **M. S. Soliman,**[1] **and A. N. Alhazaa**[4,5]

[1]*Mechanical Engineering Department, College of Engineering, King Saud University, P.O. Box 800, Riyadh 11421, Saudi Arabia*
[2]*Mechanical Engineering Department, King Abdulaziz University, Jeddah, Saudi Arabia*
[3]*Industrial Engineering Department, College of Engineering, King Saud University, P.O. Box 800, Riyadh 11421, Saudi Arabia*
[4]*Physics & Astronomy Department, Faculty of Science, King Saud University, P.O. Box 2455, Riyadh 11451, Saudi Arabia*
[5]*King Abdullah Institute for Nanotechnology (KAIN), King Saud University, Riyadh, Saudi Arabia*

Correspondence should be addressed to E. A. El-Danaf; edanaf@ksu.edu.sa

Academic Editor: Markus Bambach

Aluminum alloys for high-temperature applications have been the focus of many investigations lately. The main concern in such alloys is to maintain mechanical properties during operation at high temperatures. Grain coarsening and instability of precipitates could be the main reasons behind mechanical strength deterioration in these applications. Therefore, Al-Cu-Mg-Ag alloys were proposed for such conditions due to the high stability of Ω precipitates. Four different compositions of Al-Cu-Mg-Ag alloys, designed based on half-factorial design, were cast, homogenized, hot-rolled, and isothermally aged for different durations. The four alloys were tensile-tested at room temperature as well as at 190 and 250°C at a constant initial strain rate of $0.001\,s^{-1}$, in two aging conditions, namely, underaged and peak-aged. The alloys demonstrated good mechanical properties at both aging times. However, underaged conditions displayed better thermal stability. Statistical models, based on fractional factorial design of experiments, were constructed to relate the experiments output (yield strength and ultimate tensile strength) with the studied process parameters, namely, tensile testing temperature, aging time, and copper, magnesium, and silver contents. It was shown that the copper content had a great effect on mechanical properties. Also, more than 80% of the variation of the high-temperature data was explained through the generated statistical models.

1. Introduction

For decades, the market of lightweight materials has been growing and enlarging year by year, due to the increasing demand for energy saving. Aluminum alloys are still one of the main lightweight materials under investigation. Development and design of high strength aluminum alloys to operate at elevated temperatures is getting great attention. Applications for high-temperature aluminum alloys include supersonic aviation and automotive components. However, stability of precipitates at high temperature is still a major concern. Strength at elevated temperature starts to deteriorate after a specific time because of precipitates coarsening. Thus, the objective of many researches, recently, is to develop and design new aluminum alloys with precipitates that are stable at high temperatures. The use of Al-Cu-Mg-Ag aluminum

alloys (AA2139 and AA2519) has increased substantially in aircraft and military applications due to their low density, exceptional toughness, and moderately high-temperature stability [1–3].

The Concorde was the most famous supersonic civil aircraft. It was adopted by French and British Airways. The Concorde alloy was 2618A (Al-2.2%Cu-1.5%Mg-1%Fe-1%Ni-0.2%Si). The precipitates of 2618A are stable at elevated temperature and this is the reason behind choosing this alloy; even so, the mechanical properties are not in the same level when compared with conventional aerospace aluminum alloys like 2024-T6 and 7075-T6 [4]. The temperature on the skin of the airplane body is about 127°C, due to air friction at a speed of Mach 2.05 [5]. Despite its speed, the Concorde failed economically, because it can only carry up to 100 passengers and was not able to fly for a long distance.

Therefore, Al-Cu-Mg-Ag system was proposed to replace 2618A and other conventional aerospace aluminum alloys, 2024-T6 and 7075-T6, in view of the fact that these new alloys give high thermal stability and good mechanical properties [5, 6]. Many recent publications [1, 7, 8] have focused on investigating the evolution of microstructure and mechanical properties in these alloys.

A superior combination of high thermal stability and good mechanical properties comes from special precipitates called Ω [5]. Bakavos et al. [9] used transmission electron microscopy (TEM) to explore the habit planes of Ω precipitates. It was found that the habit planes of Ω with the matrix are $\{111\}_\alpha$. Regarding the morphology of Ω precipitates, Lumley and Polmear [10] stated that Ω has an orthorhombic plate-like shape. For the composition of Ω phase, researchers are still uncertain about it. Lumley and Polmear [10] mentioned that the composition of Ω was close to that of Al_2Cu with Mg and Ag detected at α/Ω interfaces. Gable et al. [11] studied the stability of Ω phase at different aging temperatures of 200 and 250°C. It was shown that the density of Ω phase plates decreases intensely if the alloy is aged at a temperature at 250°C or higher for 30 minutes or longer. Also, Gable et al. [11] showed that the thickness of Ω phase increases significantly if the alloy is aged at 250°C. Xiao et al. [12] confirmed this observation where the Ω phases are thermally stable at temperatures below 200°C.

Bakavos et al. [9] investigated the precipitates of two alloys of Al-Cu-Mg with and without silver (Ag) addition. It was stated that the Ω phase was observed in both alloys. However, Ω phases in Al-Cu-Mg-Ag were finer compared with Al-Cu-Mg free of Ag. Gable et al. [11] showed that the content of Ω phase is related to the content of Mg. Ω phase works with the other well-known precipitates θ' and S to enhance the mechanical properties significantly. The habit planes of θ' phases are $\{001\}_\alpha$ and have a composition of Al_2Cu and tetragonal plate-like shape [9, 13]. For S phases, the habit planes are $\{001\}_\alpha$ and have a composition of Al_2CuMg and a cubic shape [10]. The dominant phases in the Al-Cu-Mg-Ag system which has a great effect on enhancing the mechanical properties are Ω and then θ'. S has minor significance [10]. It has been shown [14] that the precipitation sequence is as follows:

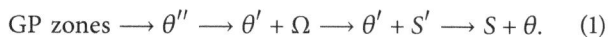

$$\text{GP zones} \longrightarrow \theta'' \longrightarrow \theta' + \Omega \longrightarrow \theta' + S' \longrightarrow S + \theta. \quad (1)$$

Song et al. [15] studied the mechanical properties of three alloys, A2618 (Al-2.2%Cu-1.5%Mg-1%Fe-1%Ni-0.2%Si), Al-8Cu-0.5Mg free of Ag, and Al-8Cu-0.5Mg-0.6Ag, at a wide range of temperatures from 20 to 300°C. It was shown that the alloy with the addition of low content of Ag has better mechanical properties at low and high temperatures compared with other alloys. Xia et al. [16] investigated the effect of heat exposure on the mechanical properties of the aged alloy (Al-4.72Cu-0.45Mg-0.54Ag-0.17Zr). The aging process was conducted at a temperature of 165°C for 2 h (underaged condition). The heat exposure was implemented

at 200°C for different times of exposure starting from zero to 100 h. It was shown that there was an initial increase in strength with increasing duration of heat exposure, whereas after 10 h the strength started to decrease. The ultimate tensile stress (UTS) after 100 h of exposure was about 400 MPa, whereas it was 430 MPa before exposure [16].

Liu et al. [17] studied the creep behavior of Al-5.33Cu-0.79Mg-0.48Ag-0.30Mn-0.14Zr. The study on the three alloys was conducted at a temperature of 150°C as well as at stress of 150 to 300 MPa for underaged conditions. It was shown that the steady creep rates were 0.12, 0.06, 0.03, and 0.01% per hour at 150, 200, 250, and 300 MPa, respectively [17]. Lumley et al. [18] explored the creep behavior of two alloys, that is, traditional aviation aluminum alloy (Al 2024) and experimental alloy with composition of Al-5.6Cu-0.45Mg-0.45Ag-0.3Mn-0.18Zr for two different aging conditions of underaged and fully hardened (T6) conditions. The creep test parameters were temperature of 300°C and stress of 150 MPa. It was obvious that the underaged condition gave a lower creep rate for both alloys. Al 2024 displayed secondary creep after 200 h and 400 h for T6 and underaged conditions, respectively, while the experimental alloy with low content of Ag did not show a secondary creep behavior for both aging conditions. Based on this result, Lumley and Polmear [10] investigated the creep behavior of underaged conditions for the previous experimental alloy that contained a low amount of Ag extensively at different creep circumstances. The creep test condition was a temperature of 130°C and stress of 200 MPa for 20000 h. It was represented that the creep rate percentage was about 0.4 and there was no secondary creep observed along the duration of the test [10].

Al-Obaisi et al. [19] studied the aging characteristic of eight different compositions of Al-Cu-Mg-Ag based on full-factorial design at three different aging temperatures of 160, 190, and 220°C for a wide range of aging durations. Statistical modeling was constructed between hardness values and process inputs comprising aging temperatures and times, as well as weight percentages of alloying elements through Minitab software. It was presented that changing of weight percentages of alloying elements changed the hardness values significantly. Also, it was deduced that aging at 190°C gave good hardness values with reasonable aging duration that could be appealing to industry needs [19]. This temperature was used as an aging temperature for the current study.

Most researchers focused on studying the mechanical properties of a single composition of Al-Cu-Mg-Ag and mostly at one aging condition. A complete study of mechanical properties at high temperatures for different aging conditions and different compositions of Al-Cu-Mg-Ag is still required. This is the stimulation for the current work. Based on the fractional factorial design of the experiment, four alloys were prepared. They were tensile-tested at room temperature, 190°C, and 250°C and at two different aging conditions (underaged and peak-aged). Thus, the mechanical properties of the four alloys were related to the process parameters (tensile testing temperature, aging time, and alloying elements percentage) through statistical modeling.

TABLE 1: Chemical compositions of the used alloys.

Alloy number	wt.% Cu	wt.% Mg	wt.% Ag	wt.% Al
1	5.0 (+)	0.5 (−)	0.3 (−)	Balance
2	3.0 (−)	1.0 (+)	0.3 (−)	Balance
3	3.0 (−)	0.5 (−)	0.6 (+)	Balance
4	5.0 (+)	1.0 (+)	0.6 (+)	Balance

TABLE 2: Chemical analysis of the investigated alloys (wt.%)[*].

Alloy number	Cu	Mg	Ag	Al	Cu/Mg ratio
1	5.29	0.46	0.30	Balance	11.5
2	3.24	0.95	0.31	Balance	3.4
3	3.15	0.47	0.62	Balance	6.8
4	5.11	0.96	0.61	Balance	5.32

[*]The rest of the alloying elements had the percentages Si ≤ 0.05, Fe ≤ 0.2, Ni ≤ 0.03, Cr ≤ 0.056, Zn ≤ 0.031, Ti ≤ 0.016, and Mn ≤ 0.006.

TABLE 3: Aging time for each aging condition.

Alloy number	Underaged		Peak-aged	
	Aging time	Hardness (HV)	Aging time	Hardness (HV)
1	30 min	124	2 h	167
2	30 min	103	2 h	132
3	30 min	97	8 h	127
4	10 min	165	1 h	172

2. Methodology

2.1. Materials Preparation.
Three alloying elements, namely, Cu, Mg, and Ag, were added to Al, with two levels for each, (−) and (+), based on fractional factorial design. Since it is a fractional factorial design, four alloys with compositions presented in Table 1 were cast in a steel mold. If a full factorial was employed, eight alloys would have been required [19]. The dimensions of the cast ingots were $100 \times 40 \times 15$ mm. The homogenizing process was conducted at 540°C for 24 h. The four alloys were elementally analyzed through arc and spark excitation and the chemical analysis is displayed in Table 2. Then, the four alloys were hot-rolled at 450°C. During the rolling process, 80% of the total thickness was reduced.

Tensile samples were wire-cut from the rolled alloys such that the tensile axis was parallel to the rolling direction. The tensile samples had a gage length of 10 mm and a cross-sectional area of 4×1.5 mm^2. Solution treatment was carried out at a temperature of 540°C for Alloys 1 and 4 that have a higher content of Cu (5 wt.%) and at 500°C for Alloys 2 and 3 that have a low content of Cu (3 wt.%) to make sure that all four alloys were taken to a single phase region. Then, the four alloys were water-quenched. The aging process was implemented in a salt bath constituted of 50% potassium nitrate (KNO$_3$) and sodium nitrite (NaNO$_2$) at a temperature of 190°C with different aging times, and then the samples were water-quenched. The corresponding hardness values and aging times are displayed in Table 3. Tensile testing was performed for each sample at room temperature, 190°C, and 250°C for each aging condition. The tensile test was carried out on an Instron machine model 3388 equipped with a data

monitoring system. The testing temperature is controlled to be within ±2°C. The tensile data is processed using an Excel sheet and corrected for machine compliance.

The microstructure study was performed using SEM model JEOL 6610 LV and TEM JEOL model JEM-2100F-HR, operated at 200 kV. Thin foil, ~300 nm thickness, for TEM investigation was prepared using focused ion beam system (JEOL JEM9320 FIB).

2.2. Design of Experiments (DOE).
Fractional factorial is a well-known technique to be used in material and manufacturing experimentation. Several researchers made use of the technique in the process of investigating the significant factors in experiments [20–22]. A fractional factorial design reduces significantly the number of runs required, particularly in screening experiments where many factors are studied in order to decide the relative importance amongst them. The price of this reduction in the number of runs is the sacrifice of some higher order interactions.

In this research, a 2^{k-1} fractional factorial design was used with five factors, two levels each. The number of runs needed is 16 runs compared to 32 runs in a full-factorial (2^k) design. The factors and levels evaluated in the research are listed in Table 4. Four numerical factors are used, namely, testing temperature and Cu, Mg, and Ag wt.%. The fifth factor (aging time) was treated as a categorical parameter rather than numerical since the aging time for each alloy was different. Its levels are given as U (−1) and P (+1). It is worth noting that the model was built to study, solely, the high-temperature mechanical properties.

TABLE 4: Factors and their levels in the experiment fractional factorial design for the high-temperature study.

Factors	Index	Levels	
		Low (−1)	High (+1)
Aging time	A	Underaged (U)	Peak-aged (P)
Testing temperature (°C)	B	190	250
Cu (wt.%)	C	3.0	5.0
Mg (wt.%)	D	0.5	1.0
Ag (wt.%)	E	0.3	0.6

TABLE 5: Mechanical properties at room and high temperatures for each aging condition of all four alloys.

Alloy number	Testing temp. (°C)	Yield strength (MPa)		Ultimate tensile strength (UTS) (MPa)	
		Underaged	Peak-aged	Underaged	Peak-aged
1	20°C	305	420	412	480
	190°C	315	350	354	372
	250°C	250	247	260	247
2	20°C	260	275	366	385
	190°C	250	210	310	237
	250°C	200	190	214	195
3	20°C	265	310	352	359
	190°C	190	250	220	271
	250°C	190	185	201	195
4	20°C	300	400	405	447
	190°C	295	320	320	341
	250°C	200	210	220	245

3. Results and Discussion

3.1. Mechanical Properties. Figures 1(a)–1(c) represent examples of the engineering stress-strain curves for Alloy 4 tested at room temperature (a), 190°C (b), and 250°C (c), respectively, for the underaged and peak-aged conditions at 190°C.

Table 5 summarizes the yield strength (YS) and ultimate tensile strength (UTS) values for each aging condition of all four alloys at room and high temperatures. As expected, the stress values decrease with temperature. For ultimate tensile strength (UTS) values, Alloy 1 gave the highest values for both underaged and peak-aged conditions. However, Alloy 3 gave the lowest values. The difference between Alloy 1 and Alloy 3 was around 120 MPa at room temperature and about 50 MPa at high temperature for peak-aged conditions, while for underaged conditions the difference was about 60 MPa at both room and high temperatures. For yield strength (YS) values, Alloy 1 gave the highest values while Alloy 2 gave the lowest values at room temperature, but Alloy 3 exhibited the lowest values at high temperature. The difference between Alloy 1 and Alloy 2 was around 145 and 45 MPa at room temperature for peak-aged and underaged conditions, respectively. For high temperature, the difference between Alloys 1 and 3 was about 60 MPa for both aging conditions. Mostly, the stress values of peak-aged conditions were higher than of underaged conditions except at a temperature of 250°C at which underaged conditions were higher, while in

Alloy 2 the underaged condition is more superior compared to the peak-aged condition at both 190 and 250°C. In Alloy 4, stress values of the peak-aged condition were higher than of the underaged condition at all testing temperatures.

Figures 2–5 show the sensitivity of mechanical properties (UTS and YS) with temperatures of both underaged and peak-aged conditions of all four alloys. The data was linearly fitted and the negative slope was taken as an indicator of the thermal stability, in a sense that smaller negative slope can be interpreted as higher thermal stability. The negative slopes of yield strength for underaged conditions were between −0.17 and −0.37. Alloys 3 and 4 had the highest negative slopes, while Alloys 1 and 2 had the lowest negative slopes. However, the negative slopes of ultimate tensile strength (UTS) for underaged conditions were between −0.59 and −0.74. Alloys 3 and 4 had the highest negative slopes, while Alloys 1 and 2 had the lowest negative slopes.

The sensitivity of mechanical properties with temperatures of the peak-aged condition for all four alloys is as follows. For yield strength, the negative slopes were between −0.37 and −0.75. Alloys 1 and 4 had the highest negative slopes, while Alloys 2 and 3 had the lowest negative slopes. However, the negative slopes of ultimate tensile strength (UTS) were between −0.67 and −0.93. Alloys 1, 2, and 4 had higher negative slopes, while Alloy 3 had a lower negative slope. It is obvious that the sensitivity of peak-aged conditions is larger than of the underaged condition, which indicates that

(a)

(b)

(c)

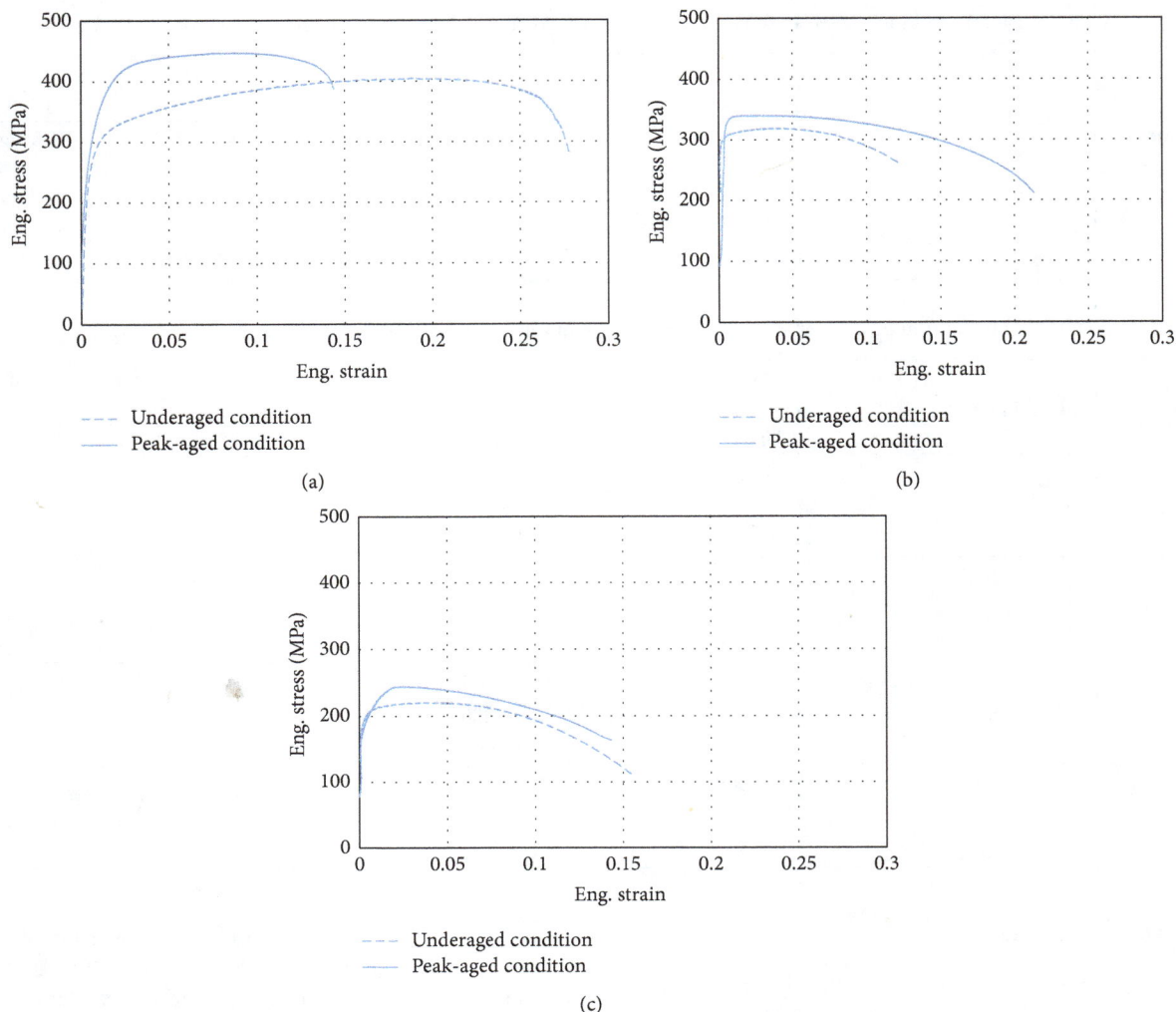

FIGURE 1: (a) Eng. stress-strain curves of Alloy 4 tested at room temperature for both aging conditions. (b) Eng. stress-strain curves of Alloy 4 tested at 190°C for both aging conditions. (c) Eng. stress-strain curves of Alloy 4 tested at 250°C for both aging conditions.

the underaged condition exhibits higher thermal stability. These observations are consistent with the results of Bai et al. [13].

It would be beneficial to estimate the behavior of the Al-Cu-Mg-Ag system by comparing it with the Concorde alloy (A2618). The following data is extracted from [10]. It was shown that the yield strength values at 20, 150, 200, and 250°C were 372, 303, 179, and 62 MPa, respectively. For ultimate tensile strength, the values were 441, 345, 221, and 90 MPa. The negative slope of the relation between yield strength and temperatures is −1.3, while for ultimate tensile strength it is −1.46. Therefore, Alloys 1 and 4 of peak-aged conditions, from the current study, gave better mechanical properties and lower sensitivity to temperatures. Also, it will be of interest to compare the present results with the behavior of Al 7075, which is one of the high strength aerospace aluminum alloys, at high temperatures. Polmear and Couper [23] showed that the yield strength of this alloy at room temperature is about 500 MPa, while at 190 and 250°C the yield strength is 200 and

50 MPa, respectively. Thus, the performance of this alloy is catastrophic at high temperatures.

The superior thermal stability for some of the current alloys is attributed to a special new phase named Ω that forms as thin platelet precipitates on $\{111\}_\alpha$ planes (α is an Al-based solid solution) and has either a hexagonal or an orthorhombic shape [24–26]. Since $\{111\}_\alpha$ are slip planes in α Al-based solid solution alloys, precipitation of Ω in these alloys tends to improve resistance to dislocation slip and improve mechanical properties [27].

3.2. Microstructure. A TEM study was conducted on Alloy 4 in the peak-aged condition as presented in Figure 6. Figure 6(a) represents a STEM dark field image of the alloy which shows the precipitated phases in bright color exhibiting different morphologies (rod and spherical shaped). The size of the precipitates ranged from 200 to 300 nm. Figure 6(b) presents an EDS spectrum taken at one of these particles, showing the elemental composition in wt.% to be Al 94.2, Cu

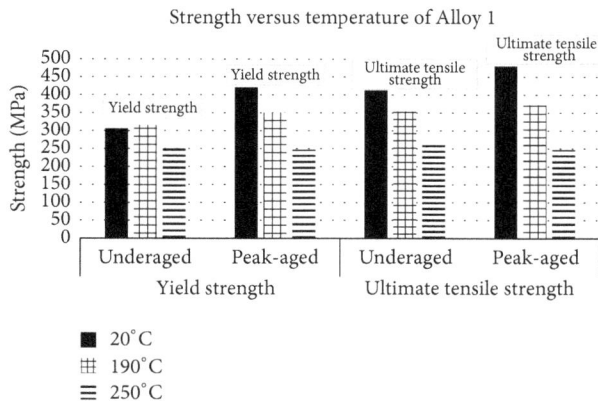

FIGURE 2: Yield strength (YS) and ultimate tensile strength (UTS) at various temperatures for both aging conditions of Alloy 1.

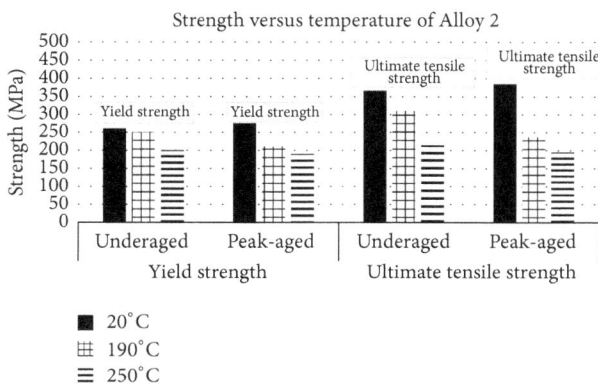

FIGURE 4: Yield strength (YS) and ultimate tensile strength (UTS) versus temperature for both aging conditions of Alloy 3.

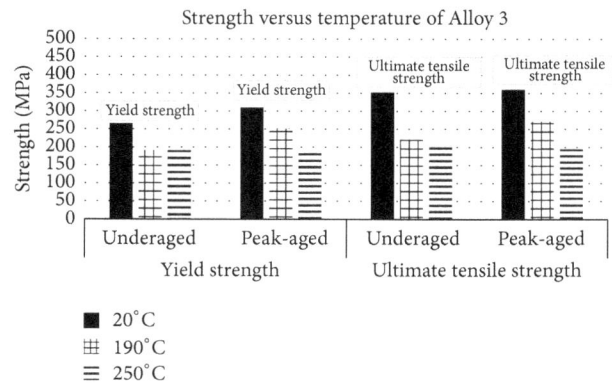

FIGURE 3: Yield strength (YS) and ultimate tensile strength (UTS) at various temperatures for both aging conditions of Alloy 2.

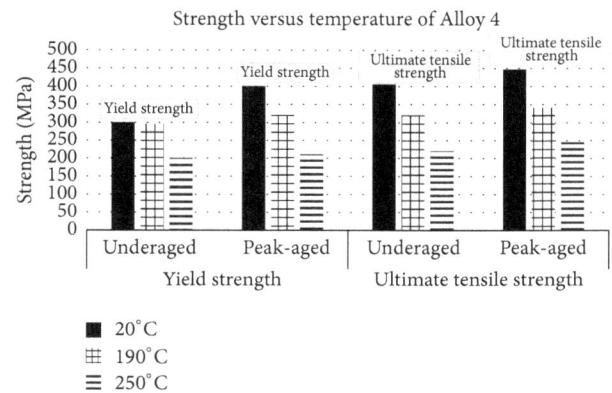

FIGURE 5: Yield strength (YS) and ultimate tensile strength (UTS) versus temperature for both aging conditions of Alloy 4.

4.3, and Ag 1.5 wt.%. The amount of Mg was not detectable as it could be traces. This particle could be an Ω phase, according to the definition previously mentioned.

A detailed fractography investigation was carried out for Alloy 2, since its mechanical properties were more thermally stable for the two aging conditions. Figure 7 shows SEM images of the fractured surfaces of Alloy 2 tested at room temperature, 190°C, and 250°C, respectively, for different aging conditions. The images of the underaged conditions for all testing temperatures show that the dominant fracture mode is transgranular fracture regarding the observed dimples along the fracture surfaces. These dimples become shallow at higher testing temperatures. For peak-aged conditions, the fractured surfaces show a combined fracture mode including transgranular and intergranular fracture modes. This observation is consistent with the deduction that the underaged conditions are more thermally stable compared with peak-aged conditions. Some of the particles distributed on the fractured surfaces were chemically analyzed through energy dispersion spectroscopy (EDS) technique. Figure 8 shows the EDS spectrum of one of the particles distributed on the fractured surface of Alloy 2 tested at room temperature for the peak-aged condition. The particle size is about 1 μm.

The EDS spectrum displays clustering of Cu and Al atoms and to a much lesser extent the Mg and Ag atoms which could imply a coarsened phase of Ω precipitates. Figure 9 presents the EDS spectrum of one of the particles distributed on the fractured surface of Alloy 2 tested at 250°C for the peak-aged condition. The particle size is about 1 μm. The EDS spectrum exhibits clustering of Cu, Mg, and Al atoms which suggests a coarsened phase of S precipitates.

3.3. Statistical Analysis. To prepare the fractional factorial design matrix, a full-factorial design was built for the basic factors (A, B, C, and D) and a generator ($E = CD$) was used to define the levels of the remaining factor (E) in the matrix. Generators are, mainly, interactions of the basic factors that determine how a subset of experiments is selected from full set runs. The alias structure, given in Table 6, illustrates the confounding between factors and interactions due to the reduction in total runs in a fractional factorial design.

The set of experiments with measured responses is illustrated in Table 7. Responses yield strength and ultimate tensile strength are indexed as Y and U, respectively. Since the design did not include replicates, the third-, fourth-, and fifth-level interactions were removed to free some

(a)

(b)

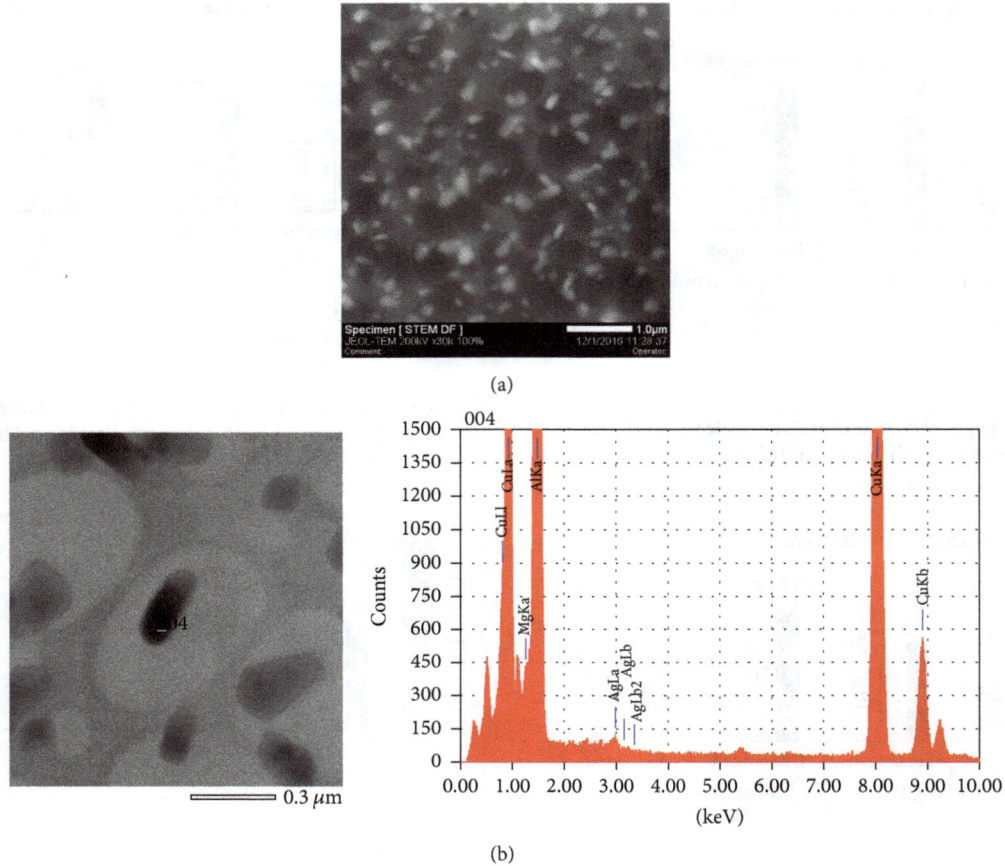

FIGURE 6: (a) STEM dark field image for Alloy 4 in the peak-aged condition. (b) EDS spectrum for the elemental composition of one selected precipitate.

TABLE 6: Alias structure for the high-temperature study.

Contrast	Estimates
1	$A + ACDE$
2	$B + BCDE$
3	$C + DE$
4	$D + CE$
5	$E + CD$
6	AB
7	$AC + ADE$
8	$AD + ACE$
9	$AE + ACD$
10	$BC + BDE$
11	$BD + BCE$
12	$BE + BCD$
13	$ABC + ABDE$
14	$ABD + ABCE$
15	$ABE + ABCD$

degrees of freedom for error estimation in order to test the significance of the effects of more important factors and second-order interactions. Analysis of variance was used to estimate the significance of each factor and interaction. Regression analysis was conducted to correlate each response to its significant parameters.

Analysis of variance (ANOVA) was used to decide which model terms (representing the studied process parameters and their interactions) affect significantly the experimental outputs. In ANOVA, the role of each term in the variability of experimental outputs is calculated as its adjusted sum of squares (Adj. SS). The value of Adj. SS of each term with respect to the total Adj. SS represents the contribution of this term to the total variability. Adj. MS for each term represents an estimate of population variance and is calculated by dividing its Adj. SS by its degrees of freedom. The F-value is then calculated for each term by dividing its Adj. MS by the error Adj. MS. A higher F-value indicates that the data contradicts more the test null hypothesis (which assumes nonsignificance of the considered term.) Another item to be calculated is the P value. A lower P value corresponds to a higher F-value. P value less than the test confidence level (generally taken as 0.05) indicates significance of the considered term.

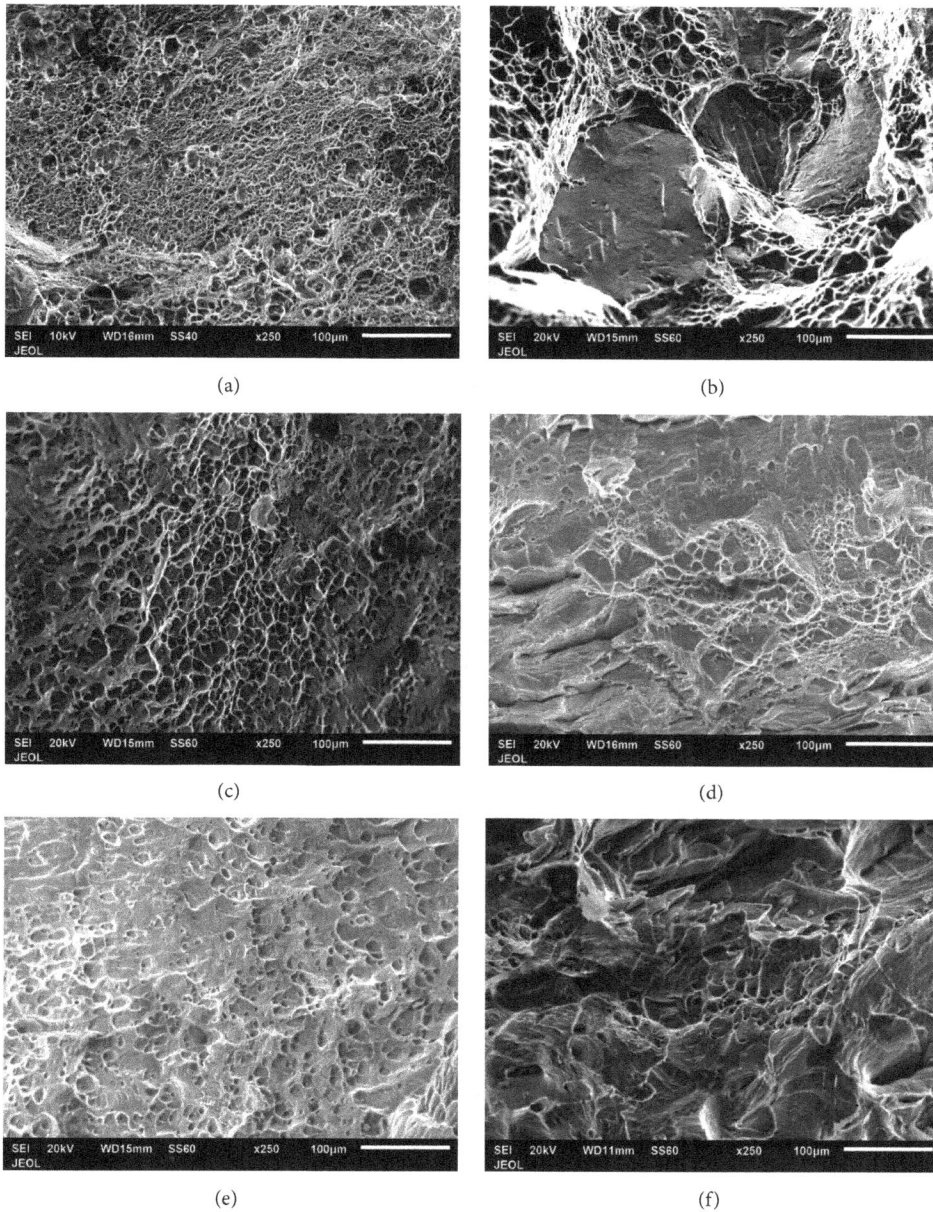

FIGURE 7: SEM images of the fractured surfaces of Alloy 2 produced by tensile testing: (a) underaged condition tested at room temperature, (b) peak-aged condition tested at room temperature, (c) underaged condition tested at 190°C, (d) peak-aged condition tested at 190°C, (e) underaged condition tested at 250°C, and (f) peak-aged condition tested at 250°C.

The ANOVA results for yield strength are given in Table 8. Model terms with P value > 0.05 are not significant and hence were removed from the model unless they are a part of a higher order interaction or their removal has a significant negative effect on the coefficient of determination (R-squared). The model has an F-value of 23.7 with a P value of about 0.000 implying that the model is significant relative to noise. Significant terms are B (temperature), C (copper content), and the interaction BC. Recall that the effect of C includes the interaction DE (Mg and Ag) as given by the alias structure. Assuming that the third- and fourth-level interactions are negligible, the effects of factor B and interaction BC are calculated with no interference from other terms. The results suggest that aging time and magnesium content do not affect the alloy yield strength. The values of the adjusted sum of squares (Adj. SS) show that the variability in the measured yield strength comes mainly from the copper content and testing temperature.

The model analysis summary is illustrated in Table 9. The adjusted R-squared equal to 0.86 implies that the model represents 86% of the variation in the data. The predicted R-squared was calculated as 0.78, within an acceptable difference from the adjusted R-squared (<0.2), proving that the model is not overfit and has a good predictability.

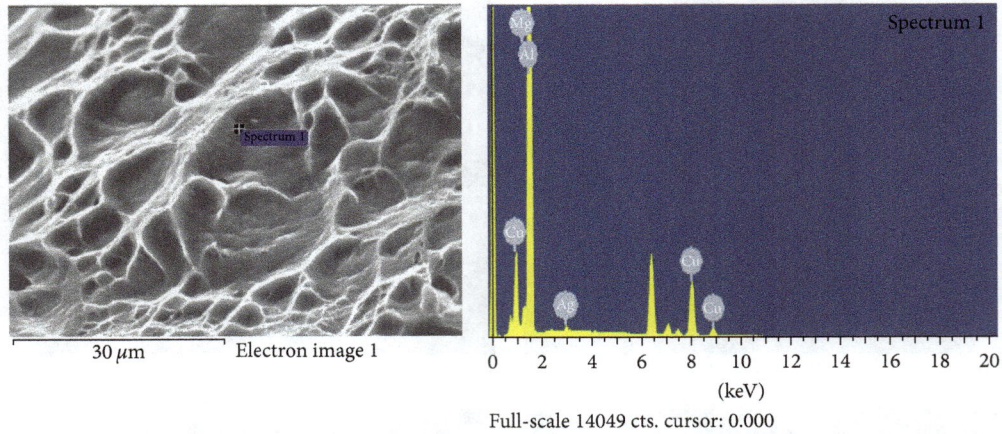

Full-scale 14049 cts. cursor: 0.000

FIGURE 8: EDS spectrum at the particle taken from fractured surfaces of Alloy 2 peak-aged condition produced by tensile testing at room temperature (Ag: 0.74 wt.%, Mg: 0.76 wt.%, Cu: 20.35 wt.%, and Al: 78.15 wt.%).

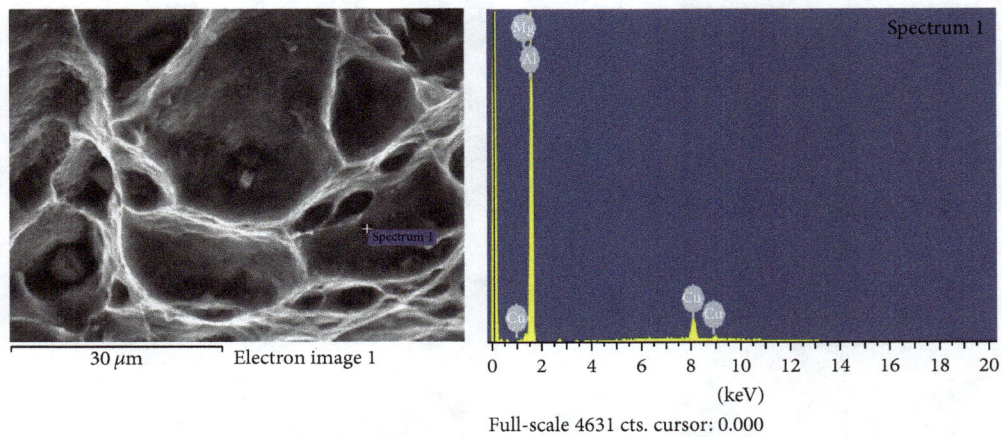

Full-scale 4631 cts. cursor: 0.000

FIGURE 9: EDS spectrum at the particle taken from fractured surfaces of Alloy 2 peak-aged condition produced by tensile testing at 250°C (Mg: 1.95 wt.%, Cu: 28.96 wt.%, and Al: 69.09 wt.%).

TABLE 7: The DOE and experimental results for the high-temperature study.

Time	Temp.	Cu	Mg	Ag	Yield strength (MPa)	Ultimate tensile strength (MPa)
A	B	C	D	E	Y	U
−1	−1	−1	−1	1	190	219.7
1	−1	−1	−1	1	250	270.5
−1	1	−1	−1	1	190	201.0
1	1	−1	−1	1	185	194.5
−1	−1	1	−1	−1	315	353.8
1	−1	1	−1	−1	350	372.0
−1	1	1	−1	−1	250	259.5
1	1	1	−1	−1	247	247.1
−1	−1	−1	1	−1	250	310.0
1	−1	−1	1	−1	210	236.6
−1	1	−1	1	−1	200	213.8
1	1	−1	1	−1	190	195.0
−1	−1	1	1	1	295	320.0
1	−1	1	1	1	320	341.0
−1	1	1	1	1	200	220.0
1	1	1	1	1	210	245.0

TABLE 8: Analysis of variance (ANOVA) results of yield strength for the high-temperature study.

Source	DF	Adj. SS	Adj. MS	F-value	P value
Model	4	38548	9637.1	23.73	0
Linear	3	35008	11669.4	28.74	0
B	1	16129	16129	39.72	0
C	1	17030	17030.2	41.94	0
E	1	1849	1849	4.55	0.056
2-way interactions	1	3540	3540.2	8.72	0.013
B ∗ C	1	3540	3540.2	8.72	0.013
Error	11	4467	406		
Total	15	43015			

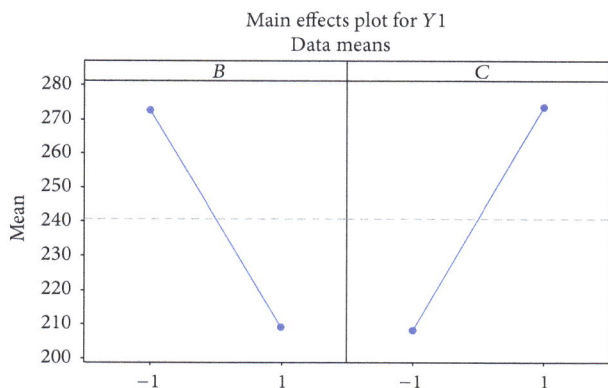

FIGURE 10: Main effect of significant factors on yield strength for the high-temperature study.

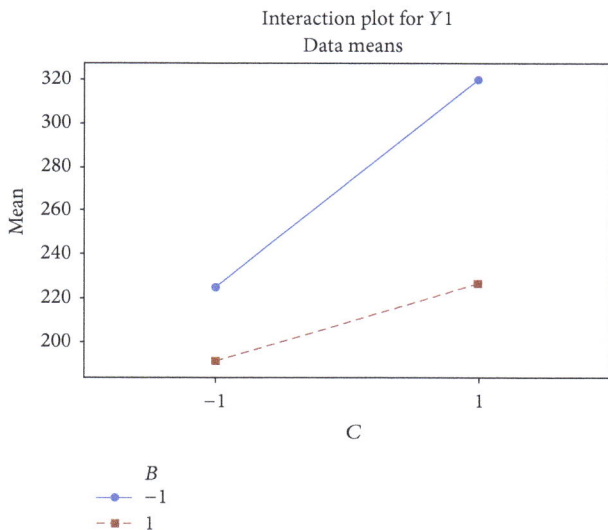

FIGURE 11: Interaction plot for yield strength for the high-temperature study.

Figures 10 and 11 show the main effect plot and interaction plot for yield strength, respectively. Figure 10 illustrates that increasing the temperature reduces the yield strength, while increasing the copper content increases the yield strength.

TABLE 9: Model summary of yield strength for the high-temperature study.

S	R-sq.	R-sq. (adj.)	R-sq. (pred.)
20.15	89.62%	85.84%	78.03%

In Figure 11, the interaction is visible as the two lines are not parallel. It is clear that the effect of copper content on increasing the yield strength is reduced as the temperature increases.

Equation (2) gives the regression model for yield strength:

$$Y = -61 + 0.925B + 141.7C - 71.7E - 0.496B * C. \quad (2)$$

The ANOVA results for ultimate tensile strength are given in Table 10. The model has an F-value of 21.7 with a P value of about 0.000 implying that the model is significant relative to noise. Significant terms are B (temperature), C (copper content), E (silver content), and the interactions AE and BC. Recall that the effect of C includes the interaction DE (Mg and Ag) and the effect of E includes the interaction CD as given by the alias structure. Assuming that the third- and fourth-level interactions are negligible, the effects of factor B and interactions BC and AE are calculated with no interference from other terms. The results suggest that aging time and magnesium content do not affect the tensile strength. However, the aging time has an interactive effect with the silver content. The values of the adjusted sum of squares (Adj. SS) show that the variability in the measured ultimate tensile strength comes mainly from the testing temperature followed by the copper content. Previous work [7] reported that the tensile strength did not increase directly with increasing Mg content in Al-Cu-Mg-Ag alloy, which agrees with the current results. On the other hand, though the aging time plays a major role in varying the hardness values, it did not impart a major effect on the tensile properties (yield and ultimate) especially for the testing temperature of 250°C. However, at the 190°C testing temperature, the variation in yield and ultimate strength is present but not with a systematic trend, which could have led to the present statistical prediction of the model. It is worth noting that aging was conducted at 190°C, for all samples; thus, for the testing temperature of 190°C, the effect of under- and peak-aged precipitate conditions was obvious. This effect is

TABLE 10: Analysis of variance (ANOVA) results of ultimate tensile strength for the high-temperature study.

Source	DF	Adj. SS	Adj. MS	F-value	P value
Model	6	48921.4	8153.6	21.65	0
Linear	4	44883.8	11221	29.8	0
A	1	1	1	0	0.961
B	1	26219.7	26219.7	69.64	0
C	1	16725	16725	44.42	0
E	1	1938.2	1938.2	5.15	0.049
2-way interactions	2	4037.6	2018.8	5.36	0.029
$A * E$	1	1951.4	1951.4	5.18	0.049
$B * C$	1	2086.2	2086.2	5.54	0.043
Error	9	3388.7	376.5		
Total	15	52310.2			

TABLE 11: Model summary of ultimate tensile strength for the high-temperature study.

S	R-sq.	R-sq. (adj.)	R-sq. (pred.)
19.4043	93.52%	89.20%	79.53%

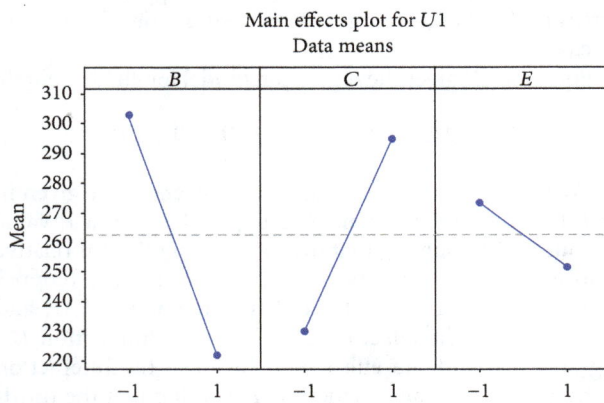

FIGURE 12: Main effect of significant factors on ultimate tensile strength for the high-temperature study.

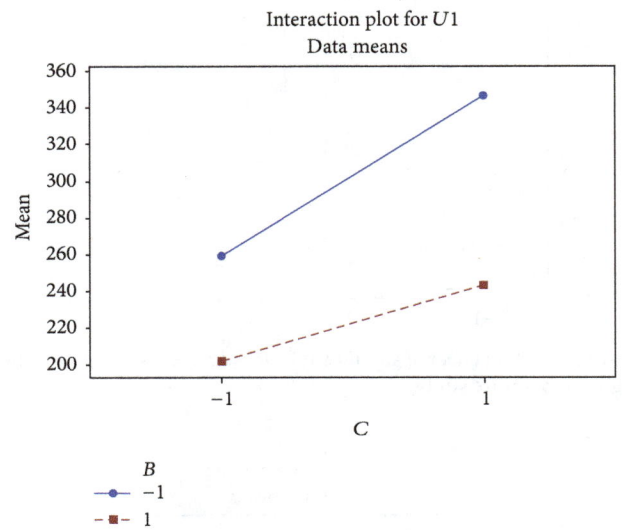

FIGURE 13: Interaction plot of copper content and temperature BC for ultimate tensile strength for the high-temperature study.

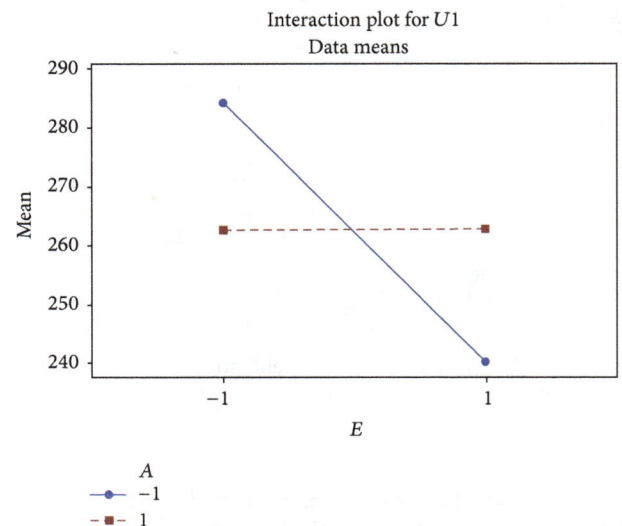

FIGURE 14: Interaction plot of silver content and time AE for ultimate tensile strength for the high-temperature study.

expected to be less pronounced at higher testing temperature due to the precipitates coarsening that limits the strain hardening processes.

The model analysis summary is illustrated in Table 11. The adjusted R-squared equal to 0.89 implies that the model represents 89% of the variation in the data while the predicted R-squared was calculated as 0.79.

Figures 12 to 14 show the main effect plot and interaction plots, respectively, for ultimate tensile strength. Figure 12 illustrates that increasing the temperature reduces the tensile strength. Increasing the copper content increases the tensile strength while increasing the silver content reduces it. In Figure 13, the interaction is visible as the two lines are not parallel. It is clear that the effect of copper content on increasing the tensile strength is reduced with increasing temperature. Figure 14 shows that although factor A (aging time) is not significant by itself, it seems that increasing

TABLE 12: Model predicted values versus measured values for validating alloy (Al-3Cu-0.5Mg-0.3Ag) for the high-temperature study.

	Yield strength (MPa)			Ultimate tensile strength (MPa)		
	Measured	Predicted	Relative % error	Measured	Predicted	Relative % error
Underaged (190°C)	200	236	15.3%	239	248	3.6%
Peak-aged (190°C)	240	236	−1.7%	245	292	16.1%
Underaged (250°C)	165	202	18.3%	173	190	9%
Peak-aged (250°C)	160	202	20.8%	173	234	26%

the time at high silver content reduces the tensile strength significantly.

Equation (3) gives the regression model for tensile strength:

$$U1 = 128 + 0.173B + 116.1C - 73.4E + 73.6A * E$$
$$- 0.381B * C. \tag{3}$$

3.3.1. Validation. To validate the model, an extra alloy was prepared and tested under the same testing conditions as the main designed alloys. The validating alloy had Cu (3 wt.%), Mg (0.5 wt.%), and Ag (0.3 wt.%). The alloy was tested under two temperatures (190 and 250°C) and under two conditions of aging (underaged and peak-aged). The measured yield strength and tensile strength of the alloy are given in Table 12 in comparison to model predicted values. Note that the value of factor *A* (time) is substituted in the equations as −1 for the underaged condition and +1 for the peak-aged condition, while the other factors are substituted by their actual values.

Paired *t*-test was used to estimate the significance of the difference between the measured and the predicted values. The test proved no significant difference with P value = 0.081 for the yield strength and 0.074 for ultimate tensile strength. However, as the model was built on a fractional factorial design, it might lack some significant interactions that were omitted in the current study. This may be the reason for some large errors (above 20%) shown in Table 12. A more compressive model should be considered as an extension to this study.

4. Conclusions

Four alloys with different compositions of Al-Cu-Mg-Ag were cast. The alloys were homogenized and hot-rolled. Tensile samples were cut from the rolled sheet. These samples were solution-treated and then aged at 190°C for different aging times comprising underaged and peak-aged conditions. Tensile testing was conducted at room and high temperatures. Changing weight percentages of alloying elements had significant effects on the mechanical properties. The sensitivity of mechanical properties for temperatures was measured through calculating the negative slopes of the yield and ultimate tensile strength variation with temperature.

A mathematical model for the variation of yield strength and ultimate tensile strength was built to relate them with alloying elements content, aging time, and tensile testing temperatures. Both models represent more than 80% of the variation in the data, which represents the reliability of the models. Copper content was most significant for increasing the yield strength and ultimate tensile strength. As expected, increasing temperature reduces the values of YS and UTS.

The promising mechanical properties and the lower sensitivity to high temperatures of Al-Cu-Mg-Ag make them a potential replacement for A2618 and other aluminum alloys used in high-temperature applications such as supersonic aviation and automobile industry.

Conflicts of Interest

The authors declare that they have no conflicts of interest.

Acknowledgments

The authors are thankful to the financial and logistic support of King Abdullah Institute for Nanotechnology and the Deanship of Scientific Research, King Saud University, Riyadh, Saudi Arabia.

References

[1] M. Gazizov and R. Kaibyshev, "Low-cyclic fatigue behaviour of an Al–Cu–Mg–Ag alloy under T6 and T840 conditions," *Materials Science and Technology (United Kingdom)*, pp. 1–11, 2016.

[2] P. Lequeu, "Advances in aerospace aluminum," *Advanced Materials & Processes*, pp. 47–49, 2008.

[3] M. Macar, *Investigation of Dynamics of Behavior of Aluminum Alloy Armor Materials [Ph.D. thesis]*, Middle East technical University, 2014.

[4] J. C. Williams and E. A. Starke Jr., "Progress in structural materials for aerospace systems," *Acta Materialia*, vol. 51, no. 19, pp. 5775–5799, 2003.

[5] J. S. Robinson, R. L. Cudd, and J. T. Evans, "Creep resistant aluminium alloys and their applications," *Materials Science and Technology*, vol. 19, no. 2, pp. 143–155, 2003.

[6] C. L. Lach and M. S. Domack, "Characterization of Al-Cu-Mg-Ag Alloy RX226-T8 Plate," NASA/TM-2003-212639, 2003.

[7] S. Bai, P. Ying, Z. Liu, J. Wang, and J. Li, "Quantitative transmission electron microscopy and atom probe tomography study of Ag-dependent precipitation of Ω phase in Al-Cu-Mg alloys," *Materials Science and Engineering: A*, vol. 687, pp. 8–16, 2017.

[8] S. Bai, X. Zhou, Z. Liu, P. Xia, M. Liu, and S. Zeng, "Effects of Ag variations on the microstructures and mechanical properties of Al-Cu-Mg alloys at elevated temperatures," *Materials Science and Engineering A*, vol. 611, pp. 69–76, 2014.

[9] D. Bakavos, P. B. Prangnell, B. Bes, and F. Eberl, "The effect of silver on microstructural evolution in two 2xxx series Al-alloys

with a high Cu:Mg ratio during ageing to a T8 temper," *Materials Science and Engineering A*, vol. 491, no. 1-2, pp. 214–223, 2008.

[10] R. N. Lumley and I. J. Polmear, "The effect of long term creep exposure on the microstructure and properties of an underaged Al-Cu-Mg-Ag alloy," *Scripta Materialia*, vol. 50, no. 9, pp. 1227–1231, 2004.

[11] B. M. Gable, G. J. Shiflet, and J. Starke, "Alloy development for the enhanced stability of Ω precipitates in Al-Cu-Mg-Ag alloys," *Metallurgical and Materials Transactions A: Physical Metallurgy and Materials Science*, vol. 37, no. 4, pp. 1091–1105, 2006.

[12] D. H. Xiao, J. N. Wang, D. Y. Ding, and H. L. Yang, "Effect of rare earth Ce addition on the microstructure and mechanical properties of an Al-Cu-Mg-Ag alloy," *Journal of Alloys and Compounds*, vol. 352, no. 1-2, pp. 84–88, 2003.

[13] S. Bai, Z. Liu, Y. Li, Y. Hou, and X. Chen, "Microstructures and fatigue fracture behavior of an Al-Cu-Mg-Ag alloy with addition of rare earth Er," *Materials Science and Engineering A*, vol. 527, no. 7-8, pp. 1806–1814, 2010.

[14] A. Cho and B. Bes, "Damage tolerance capability of an Al-Cu-Mg-Ag alloy(2139)," *Materials Science Forum*, vol. 519-521, no. 1, pp. 603–608, 2006.

[15] M. Song, K.-H. Chen, and L.-P. Huang, "Effects of Ag addition on mechanical properties and microstructures of Al-8Cu-0.5Mg alloy," *Transactions of Nonferrous Metals Society of China (English Edition)*, vol. 16, no. 4, pp. 766–771, 2006.

[16] Q. K. Xia, Z. Y. Liu, and Y. T. Li, "Microstructure and properties of Al-Cu-Mg-Ag alloy exposed at 200°C with and without stress," *Transactions of Nonferrous Metals Society of China (English Edition)*, vol. 18, no. 4, pp. 789–794, 2008.

[17] X. Y. Liu, Q. L. Pan, X. L. Zhang et al., "Creep behavior and microstructural evolution of deformed Al-Cu-Mg-Ag heat resistant alloy," *Materials Science and Engineering A*, vol. 599, pp. 160–165, 2014.

[18] R. N. Lumley, A. J. Morton, and I. J. Polmear, "Enhanced creep performance in an Al-Cu-Mg-Ag alloy through underageing," *Acta Materialia*, vol. 50, no. 14, pp. 3597–3608, 2002.

[19] A. M. Al-Obaisi, E. A. El-Danaf, A. E. Ragab, and M. S. Soliman, "Precipitation Hardening and Statistical Modeling of the Aging Parameters and Alloy Compositions in Al-Cu-Mg-Ag Alloys," *Journal of Materials Engineering and Performance*, pp. 1–13, 2016.

[20] J. C. Lourenço, M. I. S. T. Faria, A. Robin, L. P. Prisco, and M. C. Puccini, "Influence of process parameters on localized corrosion of AA7075 alloy during the production of aeronautic components," *Materials and Corrosion*, vol. 66, no. 12, pp. 1498–1503, 2015.

[21] T. A. El-Taweel and S. Haridy, "An application of fractional factorial design in wire electrochemical turning process," *International Journal of Advanced Manufacturing Technology*, vol. 75, no. 5-8, pp. 1207–1218, 2014.

[22] E. M. Salleh, H. Zuhailawati, S. Ramakrishnan, and M. A.-H. Gepreel, "A statistical prediction of density and hardness of biodegradable mechanically alloyed Mg-Zn alloy using fractional factorial design," *Journal of Alloys and Compounds*, vol. 644, pp. 476–484, 2015.

[23] I. J. Polmear and M. J. Couper, "Design and development of an experimental wrought aluminum alloy for use at elevated temperatures," *Metallurgical Transactions A*, vol. 19, no. 4, pp. 1027–1035, 1988.

[24] K. M. Knowles and W. M. Stobbs, "The structure of 111 age-hardening precipitates in Al–Cu–Mg–Ag alloys," *Acta Crystallographica Section B*, vol. 44, no. 3, pp. 207–227, 1988.

[25] B. C. Muddle and I. J. Polmear, "The precipitate Ω phase in Al-Cu-Mg-Ag alloys," *Acta Metallurgica*, vol. 37, no. 3, pp. 777–789, 1989.

[26] A. Garg, Y. C. Chang, and J. M. Howe, "Precipitation of the Ω phase in an Al-4.0Cu-0.5Mg alloy," *Scripta Metallurgica et Materiala*, vol. 24, no. 4, pp. 677–680, 1990.

[27] M. Grujicic, G. Arakere, C.-F. Yen, and B. A. Cheeseman, "Computational investigation of hardness evolution during friction-stir welding of AA5083 and AA2139 aluminum alloys," *Journal of Materials Engineering and Performance*, vol. 20, no. 7, pp. 1097–1108, 2011.

12

Study on the Gap Flow Simulation in EDM Small Hole Machining with Ti Alloy

Shengfang Zhang, Wenchao Zhang, Yu Liu, Fujian Ma, Chong Su, and Zhihua Sha

School of Mechanical Engineering, Dalian Jiaotong University, Dalian 116028, China

Correspondence should be addressed to Wenchao Zhang; traum525@gmail.com

Academic Editor: Gianfranco Palumbo

In electrical discharge machining (EDM) process, the debris removed from electrode material strongly affects the machining efficiency and accuracy, especially for the deep small hole machining process. In case of Ti alloy, the debris movement and removal process in gap flow between electrodes for small hole EDM process is studied in this paper. Based on the solid-liquid two-phase flow equation, the mathematical model on the gap flow field with flushing and self-adaptive disturbation is developed. In our 3D simulation process, the count of debris increases with number of EDM discharge cycles, and the disturbation generated by the movement of self-adaptive tool in the gap flow is considered. The methods of smoothing and remeshing are also applied in the modeling process to enable a movable tool. Under different depth, flushing velocity, and tool diameter, the distribution of velocity field, pressure field of gap flow, and debris movement are analyzed. The statistical study of debris distribution under different machining conditions is also carried out. Finally, a series of experiments are conducted on a self-made machine to verify the 3D simulation model. The experiment results show the burn mark at hole bottom and the tapered wall, which corresponds well with the simulating conclusion.

1. Introduction

Titanium (Ti) alloy is an excellent candidate for aerospace, biomedical applications and ocean development owing to the high specific strength and excellent corrosion resistance. In a traditional hole drilling the high tensile strength and low thermal conductivity of Ti alloy can result in a large machining force, high machining temperature, high tool wear, and poor accuracy, especially for deep small hole machining.

EDM could remove the material by spark erosion, which produces the local high temperature to melt and vaporize the material at the workpiece surface. During EDM process [1], the tool electrode does not contact the workpiece, which results in the tiny machining force. Therefore, the deformation is small and the machining accuracy is excellent, which is suitable to manufacture the deep small hole on Ti alloy. In general, a lot of debris is generated during EDM process in small deep hole. The debris could affect the dielectric strength of gap and discharge stability during EDM process, which

results in the concentrated discharge and low machining efficiency [2]. The debris is also likely to form secondary discharge between the tool and wall, which results in the tapered side wall of hole. Even though the flushing is introduced to remove the debris from the machining gap, because of the nonthoroughness of debris removal, it is difficult to obtain satisfactory machining accuracy [3]. So the debris removal process is one of the most important challenges for EDM application in deep small hole machining.

There are many researchers studying the debris removal mechanism in EDM process for deep small hole. Koenig et al. built a mathematical model of bottom gap flow field with flushing and calculated the pressure and velocity field [4]. Masuzawa et al. simulated the flow field with flushing and the debris distribution in the gap [5]. Takeuchi and Kunieda simulated the velocity field of debris and took the dielectric movement and bubble expansion into consideration [6]. Allen and Chen investigated the material removal for micro-EDM on molybdenum by using a Matlab-based thermonumerical model, which simulated single spark

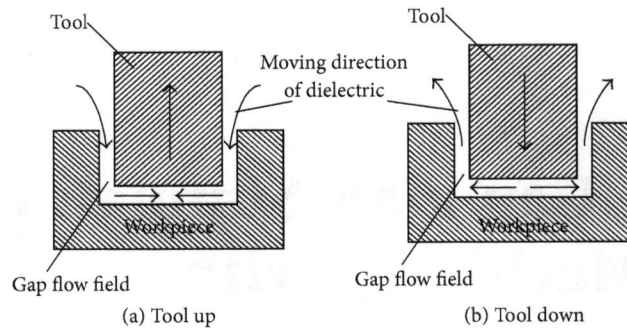

FIGURE 1: Schematics of tool movement model.

discharge process [7]. Wang et al. simulated the flow field in EDM machining, and debris aggregated at the corner increased the concentrated discharging, which destroyed the edge of tool [8]. Xie et al. simulated the holes array with ultrasonic assisted in EDM and concluded that the ultrasonic vibration of tool had influence on the debris exclusion from discharge gap [9]. Cetin et al. investigated the effects of electrode jump parameters on machining speed and depth experimentally in linear motor equipped electrical discharge machining [10]. Wang et al. made researches on the debris and bubble movement with Fluent and summarized that the bubble generated at the bottom was the main factor to expel the debris. But the expelling effect became weak when the bubble reached the side gap [11]. Kong et al. simulated the motion of debris with bubbles under with and without ultrasound [12]. Mastud et al. conducted simulations on computational modeling of the debris motion and its interaction with the dielectric fluid under low-amplitude vibrations imparted via a magnetorestrictive actuator [13]. Mullya and Karthikeyan analyzed the fluid flow along the narrow gap of microelectrodischarge-milling process for different machining conditions by computational fluid dynamics simulation [14].

For above available results, most researchers only developed the single discharge machining models. Actually, the debris is continuously generated in EDM process. Besides, the flow disturbation generated by the movement is not considered. So the debris movement process for multicycle discharge with tool movement in EDM is not fully understood.

This paper develops a mathematical model that considers tool movement in solid-liquid two-phase gap flow field and a 3D model to simulate the tool movement and debris generation when the tool electrode conducts self-adaptive movement. Besides, the debris is generated continuously in the gap between the tool and workpiece. Such simulation is much closer to the real machining process.

2. Debris Generation and Tool Movement in EDM

EDM as a nontraditional machining method uses the electrothermal effect of pulsed spark discharge between tool and workpiece to remove the material in dielectric fluid. When the distance between the electrodes is reduced to 10^2 μm, the electric field intensity between electrodes becomes greater than the strength of the dielectric. The voltage applied breaks down the dielectric and induces the plasma channel between two electrodes. Then the power supply instantaneously releases the energy through the channel, which generates the temperature of about $1 \times 10^{4\circ}$C and a very high pressure in a local minimal area [15]. Consequently, a small quantity of metal is directly molten or even vaporized due to the heat effect. Once the voltage stops, the spherical debris is removed, and microdischarging craters appear on both tool and workpiece surfaces. Simultaneously, the insulating property of dielectric is restored.

As the process going on, a certain amount of material will be removed and the debris will be generated continuously. The accumulative debris can influence the break down process and machining efficiency. During the small hole machining process, the debris is regularly swept away from the machining region by flow of flushing. However, in a deep small hole machining, the flushing effect on debris at the bottom of the hole becomes weak. In addition to the flushing, the tool motion could facilitate the debris removal. As the operation progresses, the servo mechanism controls tool to make self-adaptive movement and maintains a proper gap. Meanwhile the servo movement of tool generates the disturbation to the gap flow. When the tool moves up, a negative pressure zone forms at the bottom and fresh dielectric is drawn into the gap, which lowers the concentration of debris, as shown in Figure 1(a). On the contrary, a positive pressure zone forms when the tool moves down, just as shown in Figure 1(b), the debris spreads to the surroundings, and a little debris is carried away with the fluid into the fresh dielectric. During the debris removal process, the major factor to drive debris is the drag force, for the gap flow can be seen as a solid-liquid two-phase flow.

3. Gap Flow Mathematical Model of Solid-Liquid Two-Phase Flow

3.1. Drag Force Analysis. In solid-liquid two-phase flow field, the drag force of particles F_d is the most basic form of interaction between particles and fluid. In order to obtain the

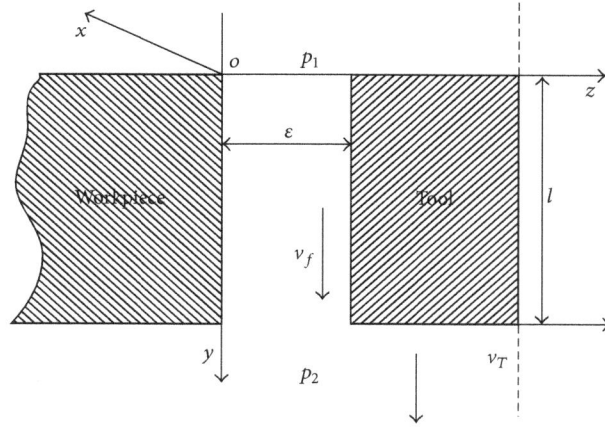

FIGURE 2: Schematics of interelectrode gap.

drag force equation of solid-liquid two-phase flow, this paper assumes the conditions as follows:

(1) The particle is considered to be spherical.

(2) The particle reflects back when colliding with the wall or other particles.

(3) The ambient temperature and machining region temperature are considered as a constant, no heat exchanges.

(4) The fluid field is considered to be infinite, inviscid, and uncompressible.

(5) There is no friction con the inner wall of hole.

Figure 2 shows a partial view of the side gap. The tool and workpiece are considered as two parallel plates. Only the tool moves during simulation. Therefore, the velocity difference is generated in the gap, which results in the pressure difference along the depth (y-axis) changes. v_T and v_f are the tool speed and liquid speed, respectively.

Naviers-Stokes equation can be simplified as follows:

$$-\frac{1}{\rho}\frac{dp}{dy} + \lambda\frac{d^2v_f}{dz^2} = 0, \tag{1}$$

where ρ is the fluid density, p is the fluid pressure, λ is coefficient, and v_f is the fluid velocity along y-axis. The pressure p along the y-axis declines uniformly, which can be presented as

$$\frac{dp}{dy} = -\frac{\Delta p}{l}, \tag{2}$$

where $\Delta p = p_1 - p_2$. When the tool moves down, v_T points to the positive y-axis and Δp is less than 0. When the tool moves up, v_T points to the negative y-axis and Δp is greater than 0. By combining (1) and (2), we get the equation

$$\frac{d^2v_f}{dz^2} = -\frac{\Delta p}{\mu l}, \tag{3}$$

where μ is dynamic viscous coefficient of fluid. By integrating (3), we get the fluid speed as

$$v_f = -\frac{\Delta p}{2\mu l}z^2 + C_1 z + C_2, \tag{4}$$

where C_1 and C_2 are coefficients; given that the tool has a speed of 0.01 m/s, finally the boundary conditions are listed as follows:

$$z = \varepsilon,$$
$$v_f = 0.01,$$
$$z = 0, \tag{5}$$
$$v_f = 0,$$

where ε is the z-axis position. C_1 and C_2 can be solved by combing (4) and (5).

Considering the impact of flushing on fluid velocity, the fluid velocity v'_f is shown as

$$v'_f = v_f + v_t, \tag{6}$$

where v_t is the speed at a certain position in the case of flushing and can be solved by unsteady flow equation (7) of solid-liquid two-phase flow.

$$\frac{\partial \rho}{\partial t} + \frac{\partial}{\partial y}(\rho v_t) = 0,$$
$$\rho\left(\frac{\partial v_t}{\partial t} + v_t\frac{\partial v_t}{\partial y}\right) = -\frac{\partial p}{\partial y} - \rho g\cos\theta - \frac{P}{A}\tau_w,$$
$$\frac{\partial}{\partial t}\left[\rho\left(e + \frac{v_t^2}{2}\right)\right] + \frac{\partial}{\partial y}\left[\rho v_t\left(h + \frac{v_t^2}{2}\right)\right]$$
$$= \frac{1}{A}\left(\frac{\partial q_e}{\partial y} - \frac{\partial \omega}{\partial y}\right) - \rho v_t g\cos\theta, \tag{7}$$

where g is acceleration of gravity, τ_w is the average shearing stress of the wall, e is the thermal energy of unit mass, h is

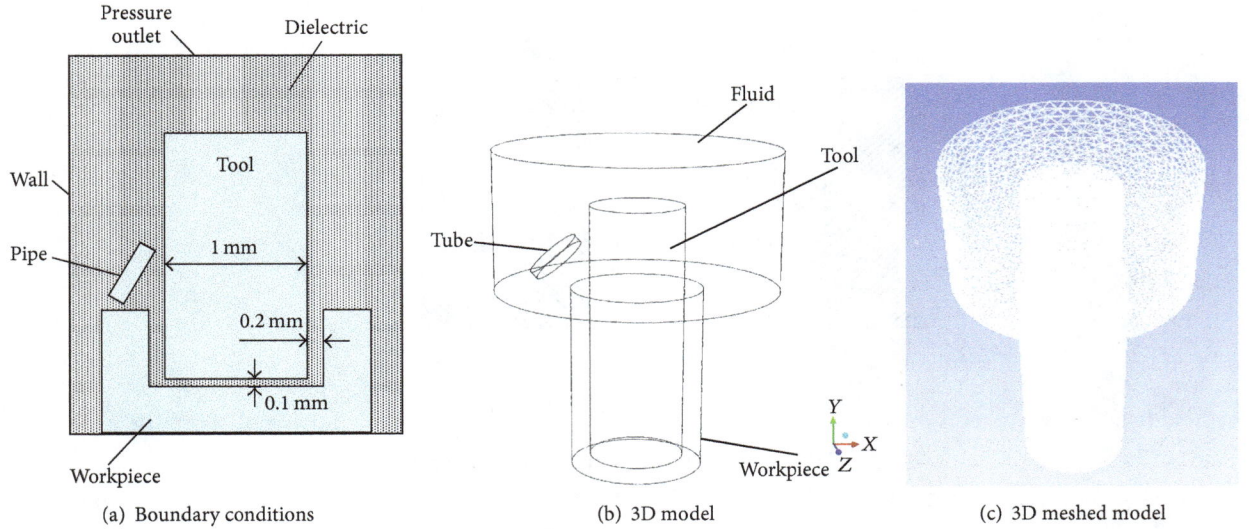

(a) Boundary conditions   (b) 3D model   (c) 3D meshed model

FIGURE 3: The schematics of 2D boundary conditions and 3D meshed model.

the enthalpy, q_e is the heat transfer rate, and ω is the power. Bruce et al. summarized an equation to get the drag force on the particles, which can be presented as [16]

$$F_d = 6\pi\mu r_p \left(v'_f - v_p\right)\sigma, \tag{8}$$

where r_p is the radius of particle, v_p is the particle velocity, and σ is the correction factor. Each step of v_p can be solved by iterating equations. Initially, the actual drag force equation of solid-liquid two-phase flow considering the effect of tool movement can be shown as follows:

$$F_{d0} = 6\pi\mu r_p \left[\frac{\Delta p_0}{2\mu l}\left(\varepsilon \cdot z - z^2\right) + v_{t0} - v_{p0}\right] \tag{9}$$

as $F = ma$, $m = (4/3)\pi r^3 \rho$, and the initial acceleration of particle a_{p0} can be presented as

$$a_{p0} = \frac{9\mu\left[\left(\Delta p_0/2\mu l\right)\left(\varepsilon \cdot z - z^2\right) + v_{t0} - v_{p0}\right]}{2r_p^2 \cdot \rho_p}, \tag{10}$$

where ρ_p is particle density. After several iteration, the nth time step of particle velocity $v_{p(n)}$ and moving distance $S_{p(n)}$ can be solved as follows:

$$v_{p(n)} = a_{p(n-1)} \cdot t,$$

$$S_{p(n)} = \sum_{i=1}^{n} S_{p(i)} = \sum_{i=1}^{n} v_{p(i)} t. \tag{11}$$

From the equations above, we deduce that when the depth of gap (y) increases, the fluid velocity v'_f and drag force F_d decline and the debris velocity v_p and moving distance S_p decrease. On the contrary, if flushing velocity v_t increases, the fluid velocity v'_f and drag force F_d increase and the debris velocity v_p and moving distance S_p increase.

3.2. Simulating Model for the Interelectrode Gap Field.

Figure 3 shows the simulation model for the interelectrode gap flow field in EDM. The side and bottom gap are set to 0.2 and 0.1 mm, respectively. The debris generated in single pulse is distributed in a random position at the bottom with the macrofunction provided by Fluent [17].

The boundary conditions are shown in Figure 3(a). The interface boundary is applied, because the flushing tube and working liquid need to exchange data. The upper boundary condition is the pressure outlet and rest boundaries are the walls as default. Figure 3(b) presents the 3D model, where the flushing is ejected from the tube outlet into the gap. In Figure 3(c), the machining zone is considered as a 3D cylinder model, which is established and meshed in the Gambit. The meshed model is imported into Fluent. The secondary development interface UDF module is used to simulate the debris removal process, because the shape of flow field is constantly changing in with the servo movement of the tool.

This paper assumes simulation cycle and discharge cycle at 0.02 s and 20 μs. 1000 discharges occur during the simulation process. Only half of the discharges are effective. The initial count of debris is 2550 at the bottom, which distributes into three layers and each layer uniformly contains 850 debris. During the simulating process, each discharge generates 30 debris in the random position, and the ratio of Ti alloy debris to Cu debris is 2 to 1, of which Ti alloy and Cu debris are in blue and red, respectively [18]. During the process, 17550 debris are generated. The simulating parameters are listed in Table 1. Flushing velocity is obtained from the previous experiment.

4. Simulation Results and Analysis

4.1. Impact of Tool Movement on Gap Flow Field without Flushing.
Without flushing, the simulation results of 1 mm

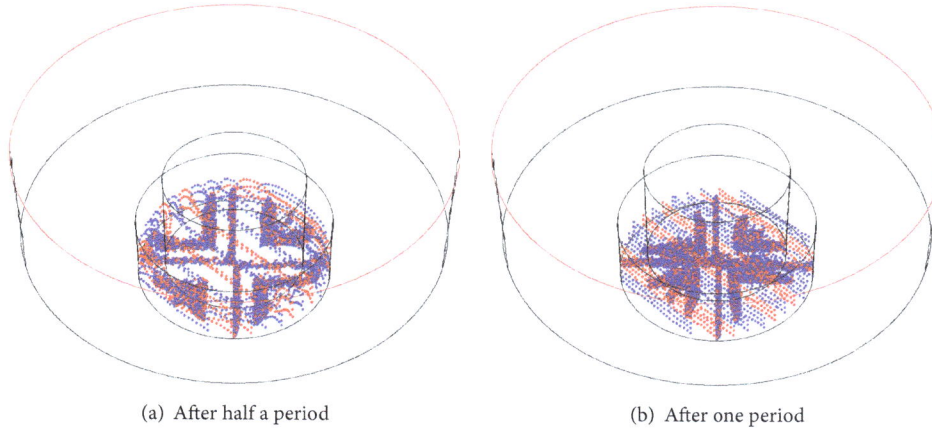

(a) After half a period

(b) After one period

FIGURE 4: Gap flow field with 0.5 mm deep hole.

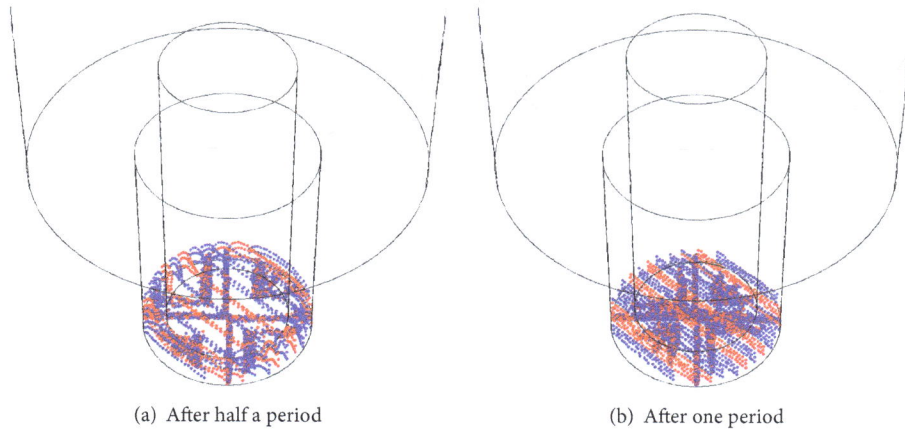

(a) After half a period

(b) After one period

FIGURE 5: Gap flow field with 2 mm deep hole.

TABLE 1: Continuous discharge simulating parameters.

Parameter	Description
Dielectric fluid	Deionized water
Tool	Copper
Workpiece	TC4
Tool velocity	0.01 m/s
Polarity	Positive
Diameter (mm)	1, 2
Flushing vel. (m/s)	0, 2, 5
Depth (mm)	0.5, 2.0, 3.5, 5.0

tool diameter are shown in Figures 4 and 5. Figures 4 and 5 show the gap flow field with 0.5 mm and 2.0 mm deep hole, respectively. In Figures 4 and 5, the tool initially moves down from the top to the bottom and finally returns to the original position. During the tool moving down process, a positive pressure is formed at the bottom, which drives the dielectric and debris at the bottom to the surrounding region, as shown in Figures 4(a) and 5(a). But no debris is carried out from bottom because the fluid velocity is low. In the next half period, the tool begins to move up, which results in the negative pressure at bottom. The negative pressure causes the fresh dielectric drawn into the gap, which pushes the debris to the center, as shown in Figures 4(b) and 5(b). From Figures 4 and 5, no debris is removed out of machining region. One conclusion could be drawn that, without flushing, the depth of the small hole has little influence on the debris distribution. It should be noticed that the accumulative debris stayed at the bottom increase the probability of concentrated discharges, which could generate the burn marks.

(a) Velocity field

(b) Pressure field

(c) Debris distribution

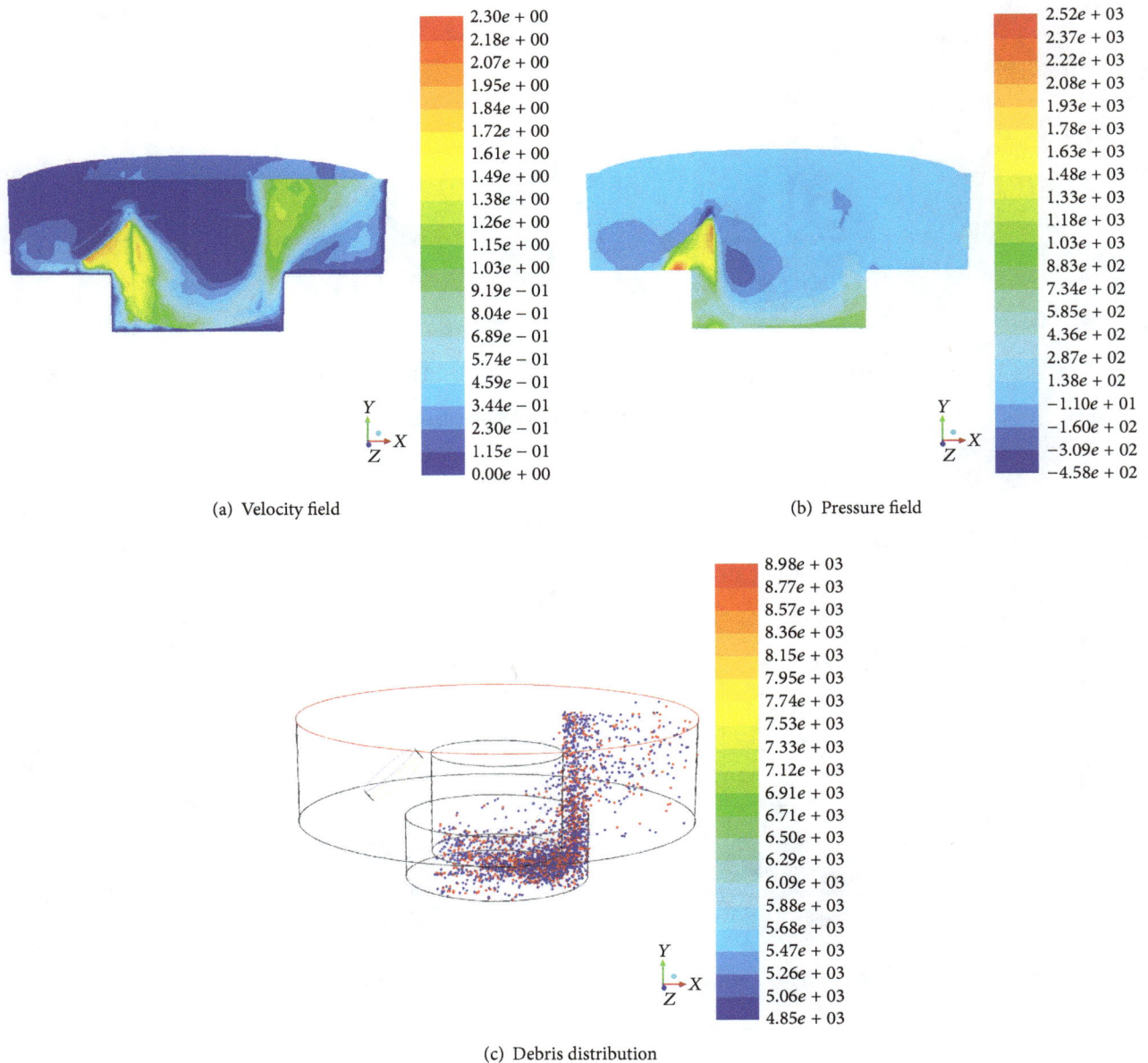

FIGURE 6: Schematics of gap flow field with flushing at 2 m/s and 0.5 mm deep hole.

4.2. Impact Analysis of Gap Flow Field with 1 mm Tool Diameter. In ordinary EDM machining, the flushing is introduced to bring fresh dielectric fluid into the gap and lower the debris concentration. In order to conduct the impact of flushing on the debris distribution, the flushing is applied in simulation model. In the model with flushing, a dielectric tube aimed at the gap between tool and workpiece induces the flow through the machining gap. Figures 6–13 show the simulation results in the case of 1 mm tool diameter after one period.

Figure 6 presents the gap flow field with flushing at 2 m/s and 0.5 mm deep hole. As shown in Figure 6(a), the highest fluid velocity of 2.3 m/s appears at the wall opposite to flushing direction. The fluid velocity at the bottom and right

side gap is large. The fluid velocity of right side outlet is lower than that of left side inlet. The fluid velocity declines with the increase of hole depth. The highest pressure 2520 Pa appears at the inlet and the negative pressure zone also appears near the inlet, as shown in Figure 6(b). In Figure 6(c), the debris moves along the gap to the right side. Besides, a part of debris escapes from the machining areas due to the effect of flushing, as well as the hole with small depth. There is more debris left at the right corner than those at the left, which corresponds well with the velocity distribution in Figure 6(a).

Figure 7 presents the gap flow field with flushing at 5 m/s and 0.5 mm deep hole. Compared with Figure 6, owing to the higher flushing velocity, the values of velocity field and

(a) Velocity field

(b) Pressure field

(c) Debris distribution

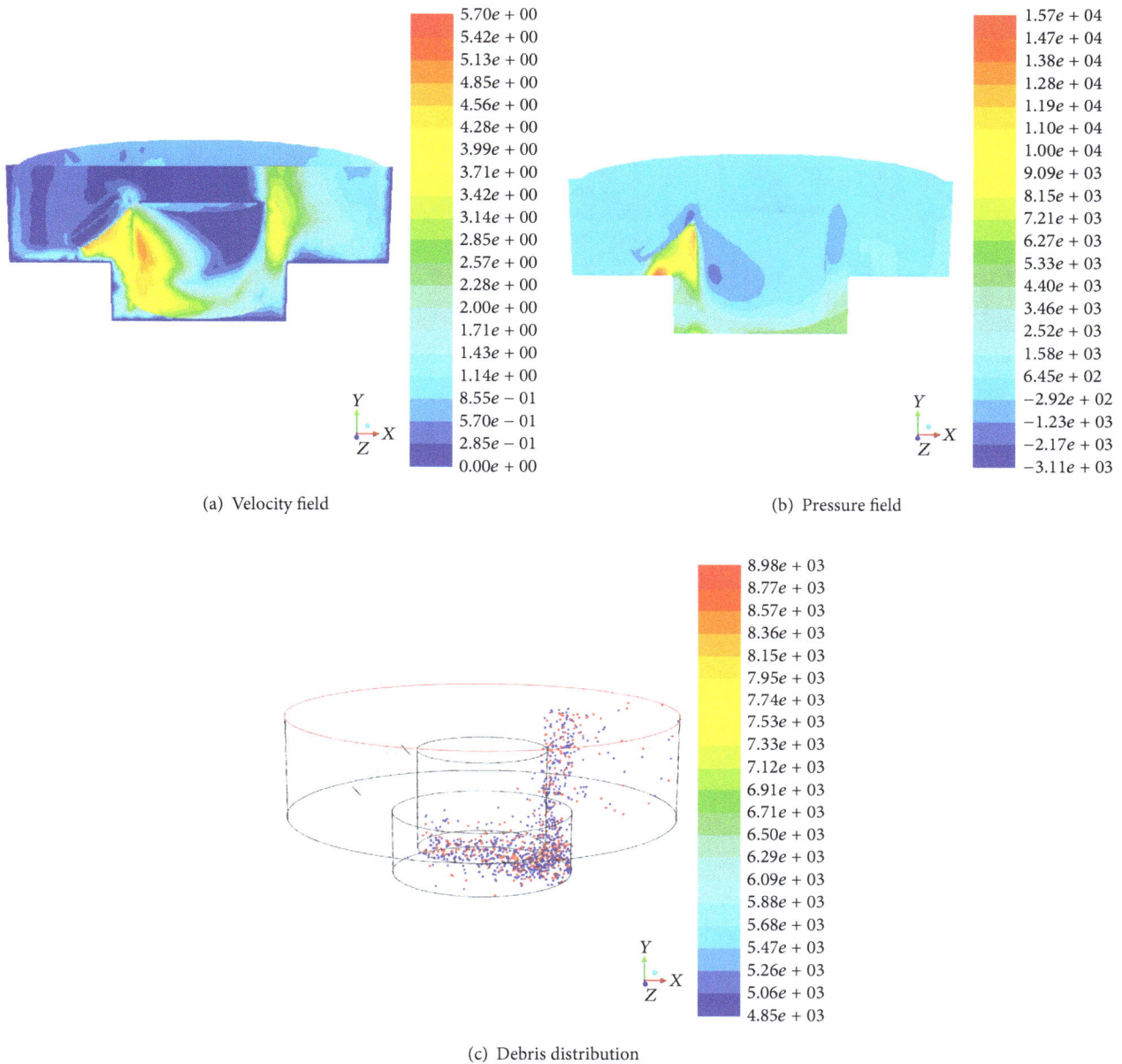

FIGURE 7: Schematics of gap flow field with flushing at 5 m/s and 0.5 mm deep hole.

pressure field increase abruptly, and the highest fluid velocity and highest pressure reach 5.7 m/s and 15700 Pa, respectively. Compared with Figure 6, more debris is carried out of the hole, which results in the fact that less debris remains at the machining gap. The simulation results correspond well with the mathematical model. For Figures 6 and 7, a low fluid velocity zone appears at the lower right corner of tool. More debris accumulates at the low fluid velocity zone, which increases the probability of concentrated discharges and blunts the edge of tool.

Figure 8 presents the gap flow field with flushing at 2 m/s and 2.0 mm deep hole. Compared with Figure 6, the flushing effect at bottom of hole becomes poor, which results in the poor debris removal at bottom. Only tiny of debris is flushed away, as shown in Figure 8(c). More debris is accumulated at the right corner of bottom.

Figure 9 presents the gap flow field with flushing at 5 m/s and 2.0 mm deep hole. With high velocity, more fluid is injected into the deep gap, even at the bottom gap. So compared with Figure 7, more debris is carried away by the fluid. But compared with Figure 7, as the hole becomes deeper, there are more debris left at the bottom. The simulation results correspond well with the mathematical model. In Figures 8 and 9, a lot of debris accumulates at the right side gap, which

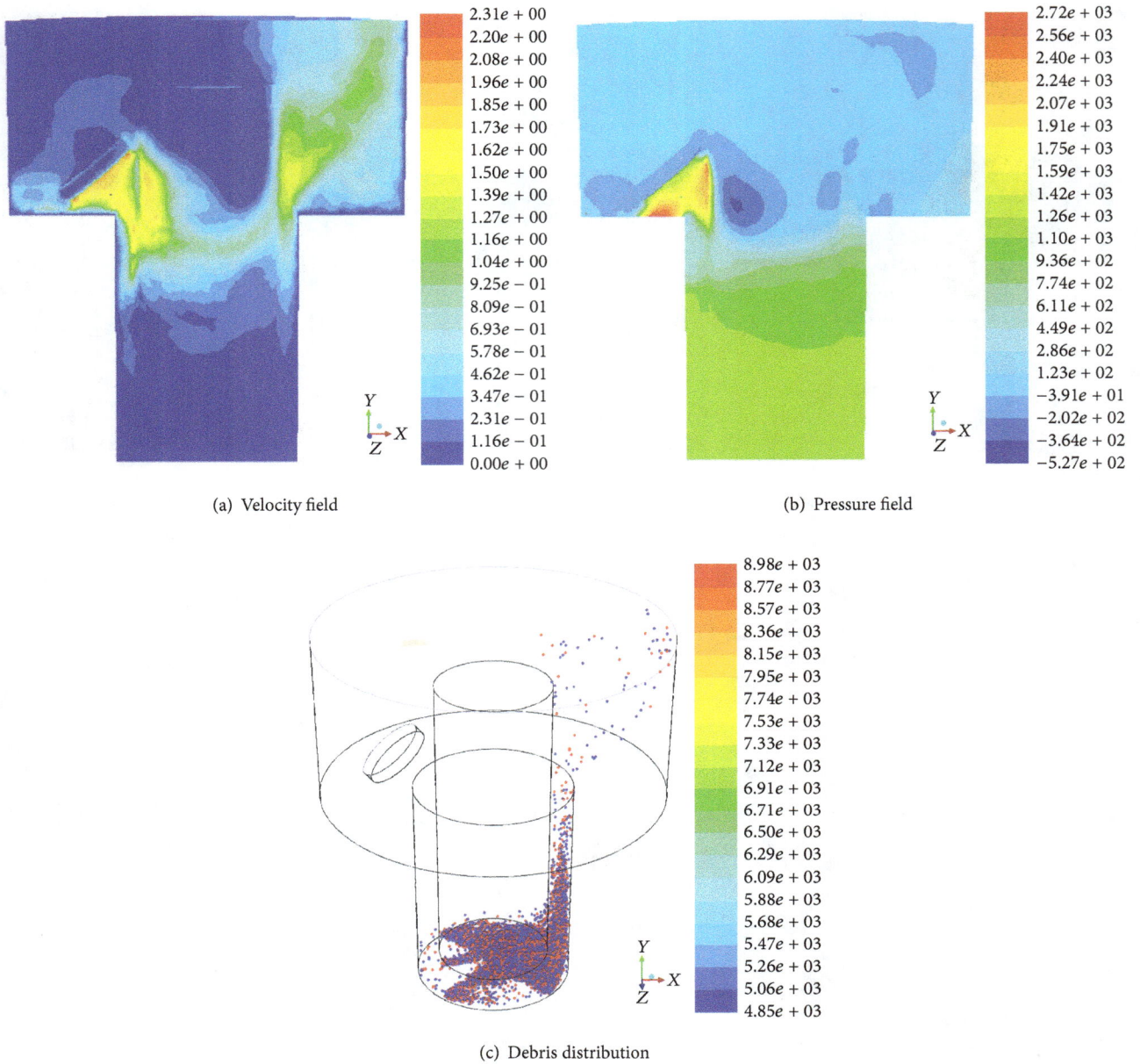

(a) Velocity field

(b) Pressure field

(c) Debris distribution

FIGURE 8: Schematics of gap flow field with flushing at 2 m/s and 2 mm deep hole.

increases the probability of secondary discharges between the sides of tool and right wall of hole and then increases the inclination of right wall of hole.

As shown from Figures 10 and 11, 3.5 mm deep hole is considered with different flushing velocity. From the simulation results, no flushing fluid is injected into the bottom gap of hole with depth over 3.5 mm. The pressure decreases sharply at the bottom, where the velocity is close to zero and the fluid velocity in this region is concerned with the tool movement. As the disturbance induced by the tool movement is small, all the debris stays at the bottom. In the case of 5 mm deep hole a similar result is obtained.

Compared with Figures 6–9, a conclusion can be made that when the depth to diameter ratio exceeds 3, no debris could be removed from the gap. The debris accumulates at the bottom and may create concentrated discharges between the workpiece and tool.

4.3. Impact of Tool Movement on Gap Flow Field with 2 mm Tool Diameter.
Figures 12–19 are the simulation results in the case of 2 mm tool diameter. Figure 12 shows the gap flow field with flushing at 2 m/s and 0.5 mm deep hole. The highest velocity 2.75 m/s appears at the wall opposite to flushing direction, where a low pressure zone forms. The debris is

(a) Velocity field

(b) Pressure field

(c) Debris distribution

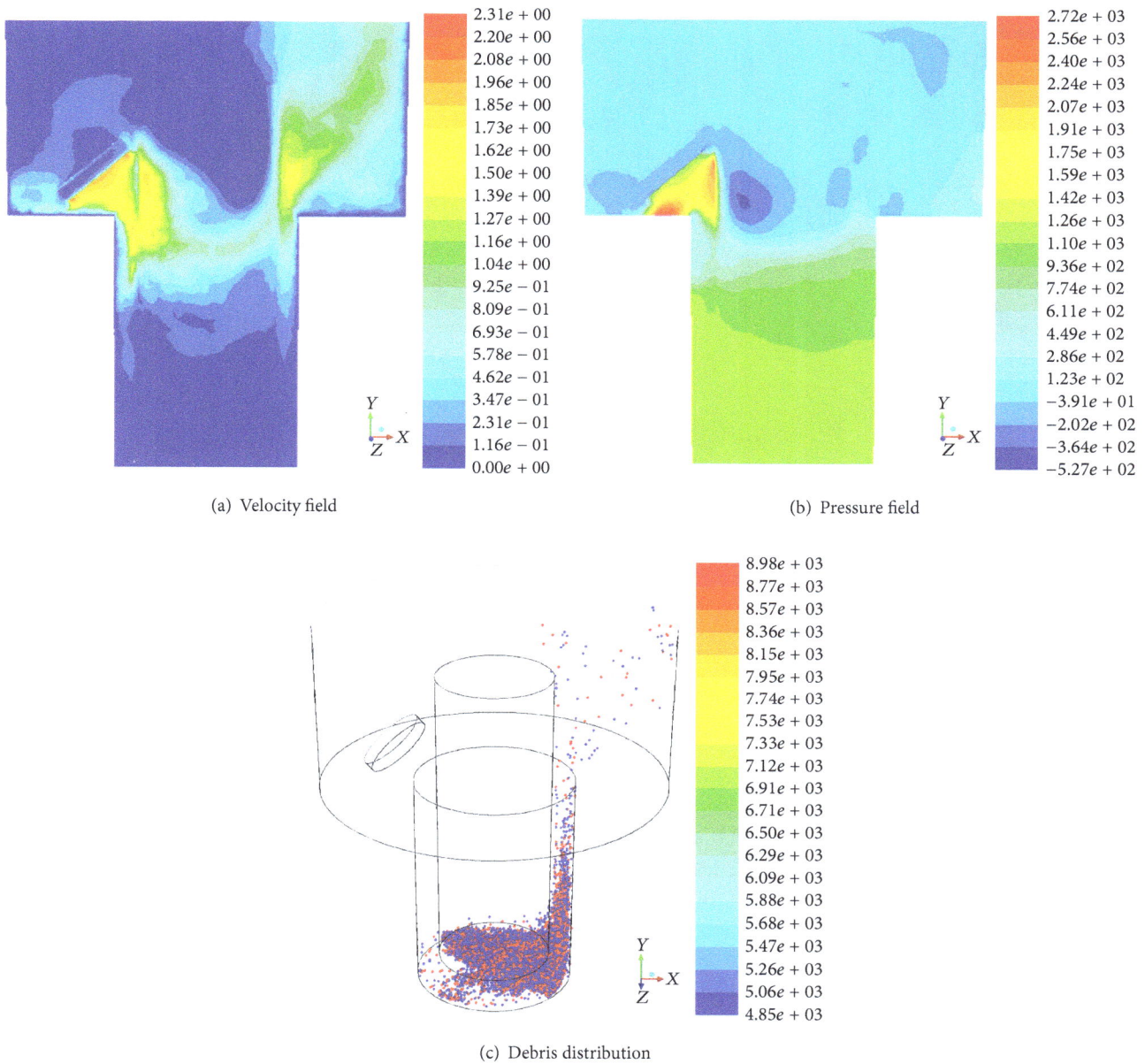

FIGURE 9: Schematics of gap flow field with flushing at 5 m/s and 2 mm deep hole.

flushed to the right corner of bottom, and the debris is flushed out from the right gap to unmachined area. Compared with Figure 6, both velocity and pressure are much higher and more debris is carried away. In case of Figure 12 the tool diameter is twice of that in Figure 6, which increases the curvature radius of gap. So the flushing fluid suffers the less resistance during the flowing, which is positive for the debris to be flushed out. However, with the increase of tool diameter, the diameter of the bottom gap increases. Large bottom gap increases the debris removal distance, which is negative for the debris to be flushed out.

Figure 13 shows the gap flow field with fluid flushing at 5 m/s and 0.5 mm deep hole. In Figure 13(a), the largest

velocity reaches 6.67 m/s and the highest pressure increases to 22800 Pa. There is sufficient pressure for the flushing injecting into the gap and driving more debris to the right side gap. Compared with Figure 12, less debris remains at the machined area.

Figures 14 and 15 are the gap flow fields of different flushing velocity with 2 mm deep hole. Compared with Figures 8 and 9, under the same machining condition, the fluid velocity at the bottom becomes larger in the case of a large tool diameter. More debris is flushed away and less debris remains at the bottom. Besides, in Figures 14 and 15, the debris is flushed out approximately in a line from the right gap and then dispersed near workpiece surface. It is confirmed

(a) Velocity field

(b) Pressure field

(c) Debris distribution

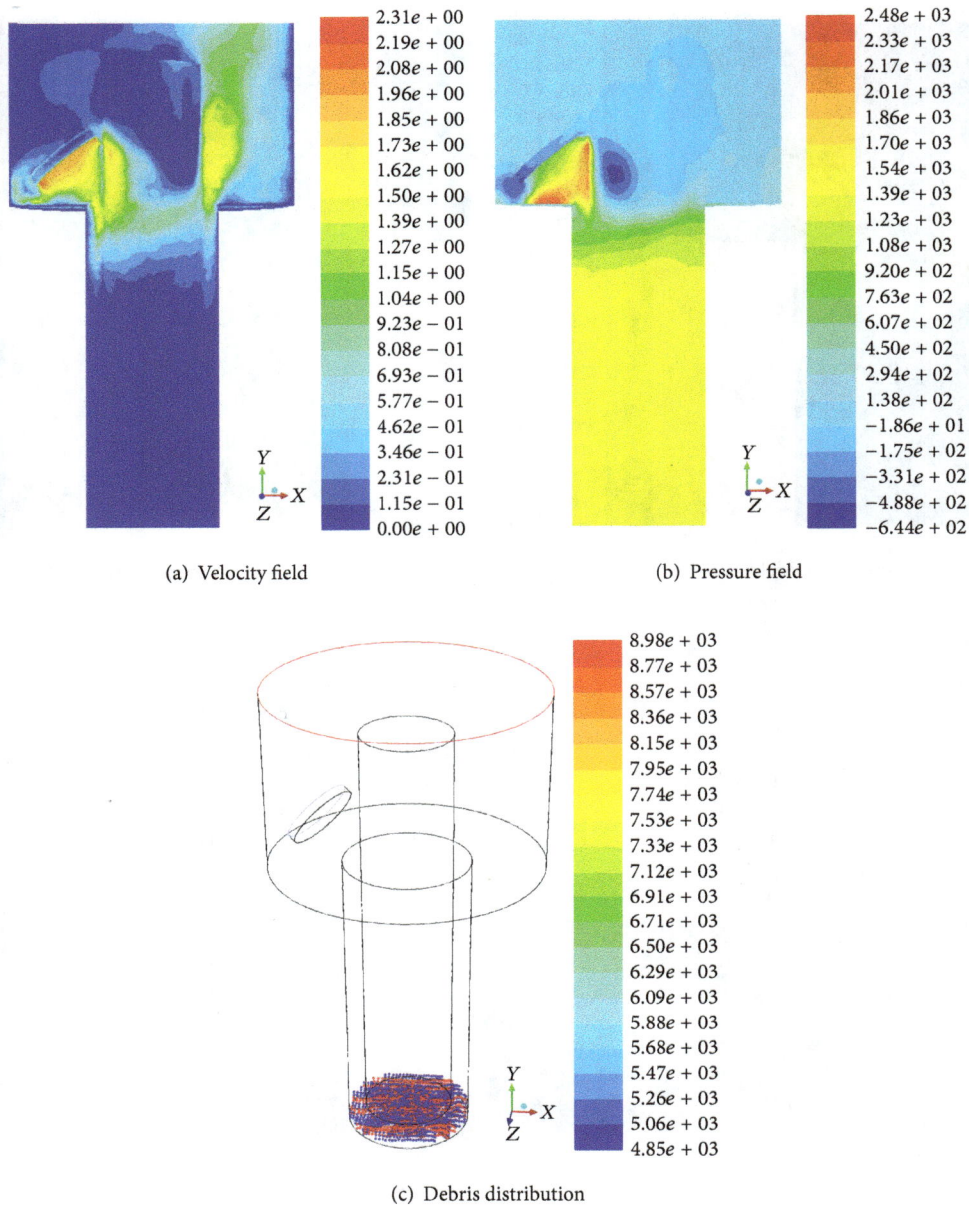

FIGURE 10: Schematics of gap flow field with flushing at 2 m/s and 3.5 mm deep hole.

that the fluid velocity at the bottom is low and a larger velocity appears in the rightmost gap, which corresponds well with the velocity field. When the fluid reaches the outlet, the velocity suddenly increases, which results in the dispersion of debris. A lot of debris accumulates at the right side gap, which increases the probability of secondary discharges between the tool and right wall of hole and then increases the inclination of right wall of hole. The low velocity region also appears at the right corner of bottom gap. The debris easily stays at the low velocity region, which could cause discharge between debris and workpiece. Then the burn mark could happen in practical operation.

Comparing 3.5 mm deep hole with 5 mm, as shown in Figures 16–19, the flushing liquid is injected into the gap and there is sufficient pressure to remove the debris. The debris can escape from 3.5 mm deep hole. On the contrary, in case of 5 mm deep hole, the flushing liquid is hardly injected into the bottom gap. Besides, the bottom velocity is only concerned with the tool movement and no debris could be flushed out from 5 mm deep hole. In this situation, the discharge stability decreases.

Combining the above analysis, one conclusion could be drawn that, for the 2 mm tool diameter, the flushing fluid contains less resistance than those of 1 mm tool diameter,

(a) Velocity field

(b) Pressure field

(c) Debris distribution

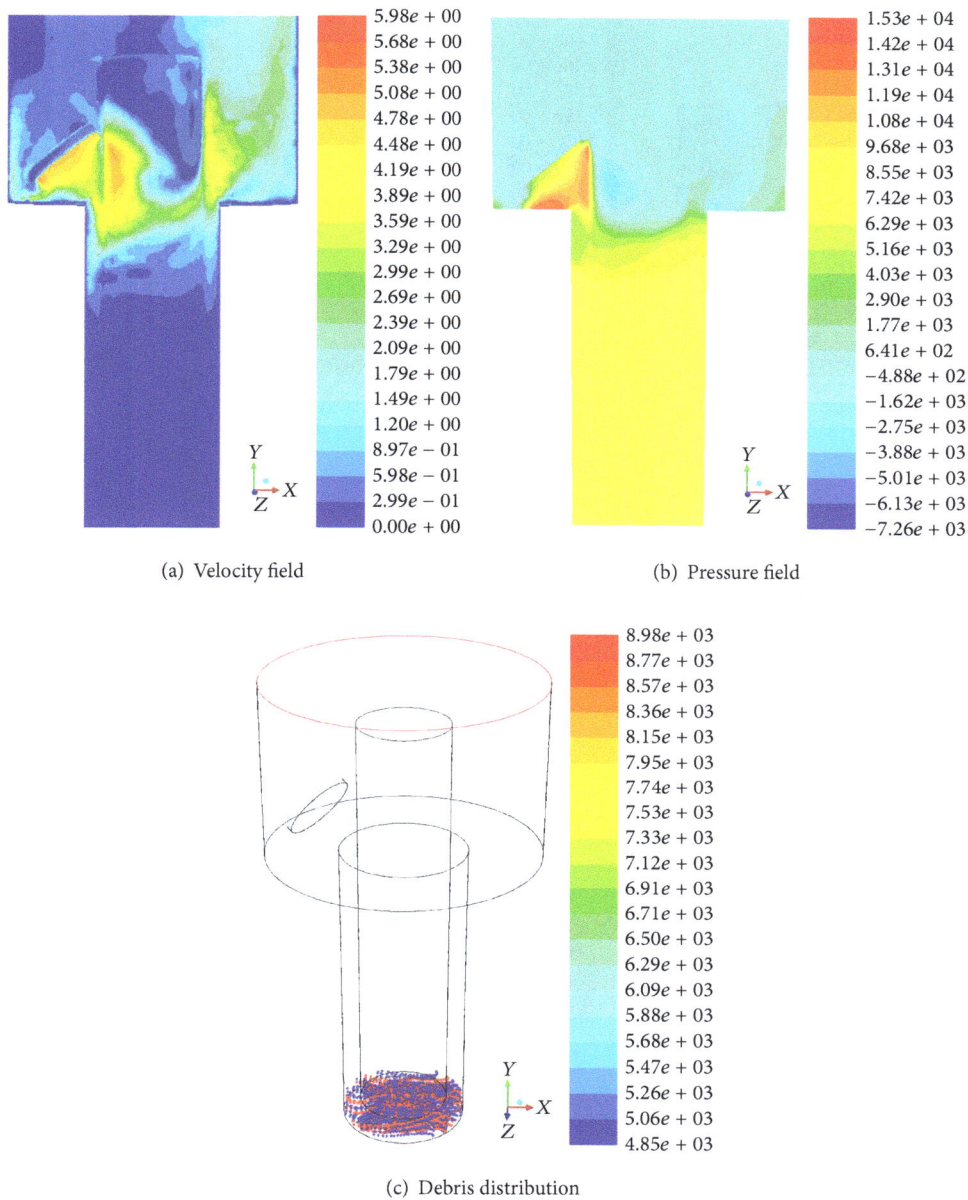

FIGURE 11: Schematics of gap flow field with flushing at 5 m/s and 3.5 mm deep hole.

which results in the fluid with higher velocity. Finally, more debris is taken away by the flushing fluid. On the contrary, in situation with lower fluid velocity, the flushing fluid with insufficient drag force could hardly bring the debris out of bottom gap. Such phenomenon could be deduced for deep hole with micro- or smaller diameter.

5. Impact of Machining Parameters on Debris Distribution

In order to obtain the debris distribution in machining region, the coordinate of each debris is exported for statistics. For the convenience of analysis, the machining region is divided into 5 zones, as shown in Figure 20(a). Zone S1 is the left bottom gap, zone S2 is the right bottom gap, zone S3 is the left side gap, zone S4 is the right side gap, and S5 is the dielectric zone above the workpiece. Figure 20(b) shows the bird view of the machining zone. In the following section, the debris distributions under different parameters are analyzed.

5.1. Impact of Hole Depths on the Debris Distribution. Figure 21 presents the debris distribution with different hole depths. As shown in Figure 21(a), for the depth of 0.5 mm, S1–S5 contain the debris in 1.1%, 15.4%, 0.0%, 3.4%, and 80.1%. The final inequation for the debris concentration is S5 > S2 > S4 > S1 > S3. Most debris gets out from the machining zone.

(a) Velocity field

(b) Pressure field

(c) Debris distribution

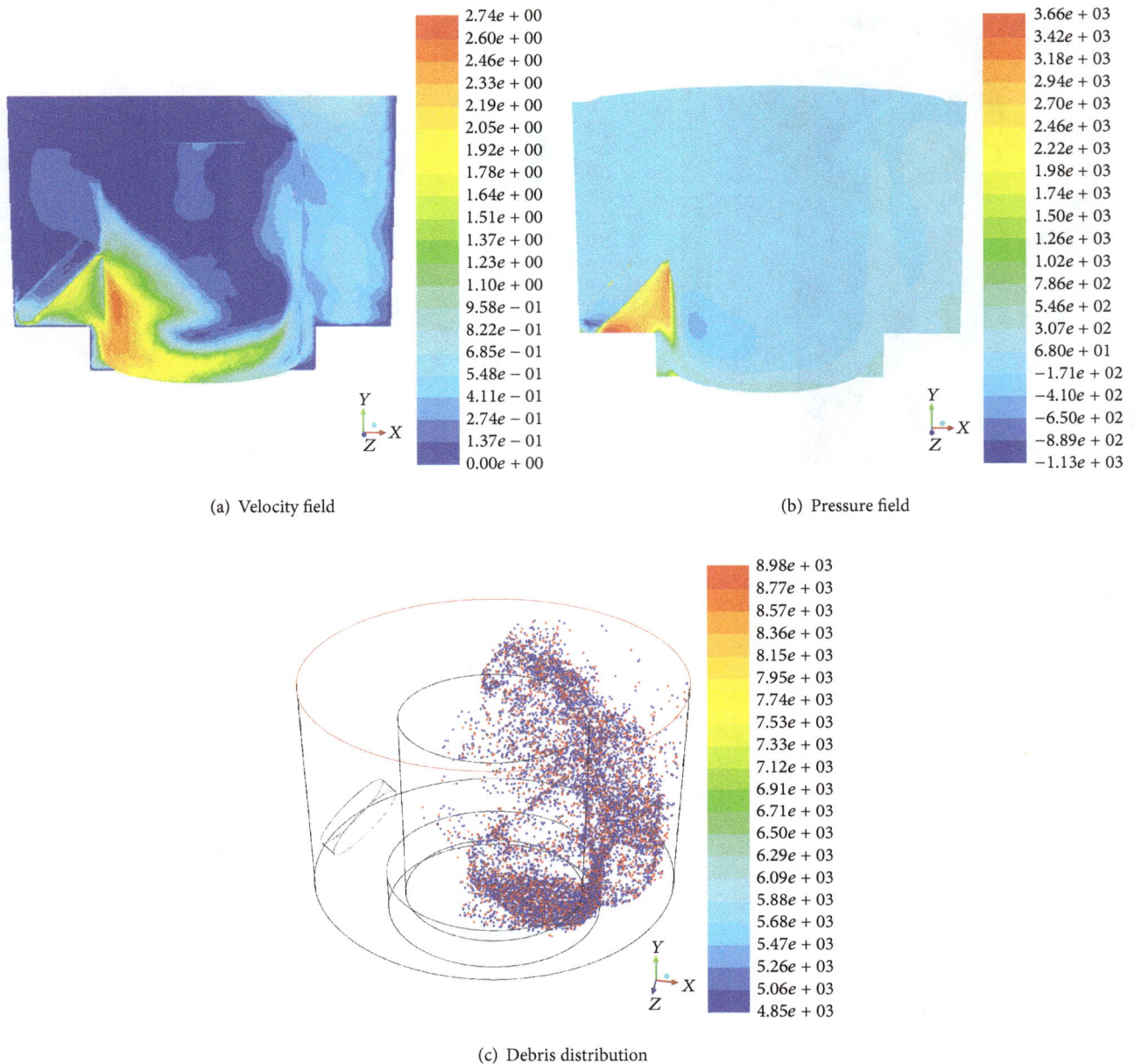

FIGURE 12: Schematics of gap flow field with flushing at 2 m/s and 0.5 mm deep hole.

For the debris remaining at machining zone, major part of debris is distributed at the right bottom gap (zone S2). For the 0.5 mm deep hole, the sufficient pressure at bottom gap could flush most debris out of hole. The debris distribution of hole with 2.0 mm deep hole is similar to that of 0.5 mm deep hole. In case of the 3.5 mm deep hole, most debris stays in zone S2 (65.3%); secondly in zone S1 (17.9%), only a little debris is distributed in zone S5 (4.8%); that is to say, a little debris is removed from the gap; a little debris is in the process of escaping (S4 (12.1%)). Most debris stays at the bottom (zones S1 and S2) and right side gap (zone S4) of the machining area. The reason of this is that as the hole is deep, the flush

of tube has a little impact on the gap near the bottom, and a little dielectric is injected into this region. Compared with 0.5 mm and 2 mm deep hole, only a little debris is carried away. However, in the case of 5 mm deep hole, almost no debris is carried away from the hole.

With the constant of tool diameter and fluid velocity in 5 m/s, the debris distribution is similar to that of fluid velocity in 2 m/s, except the debris distribution in S4 zone, just as shown in Figure 21(b). In the shallow hole (0.5 mm, 2 mm) machining, there is more debris in the condition of lower velocity than that of higher velocity, while, in the deep hole (3.5 mm, 5 mm) machining, there is less debris in the

(a) Velocity field

(b) Pressure field

(c) Debris distribution

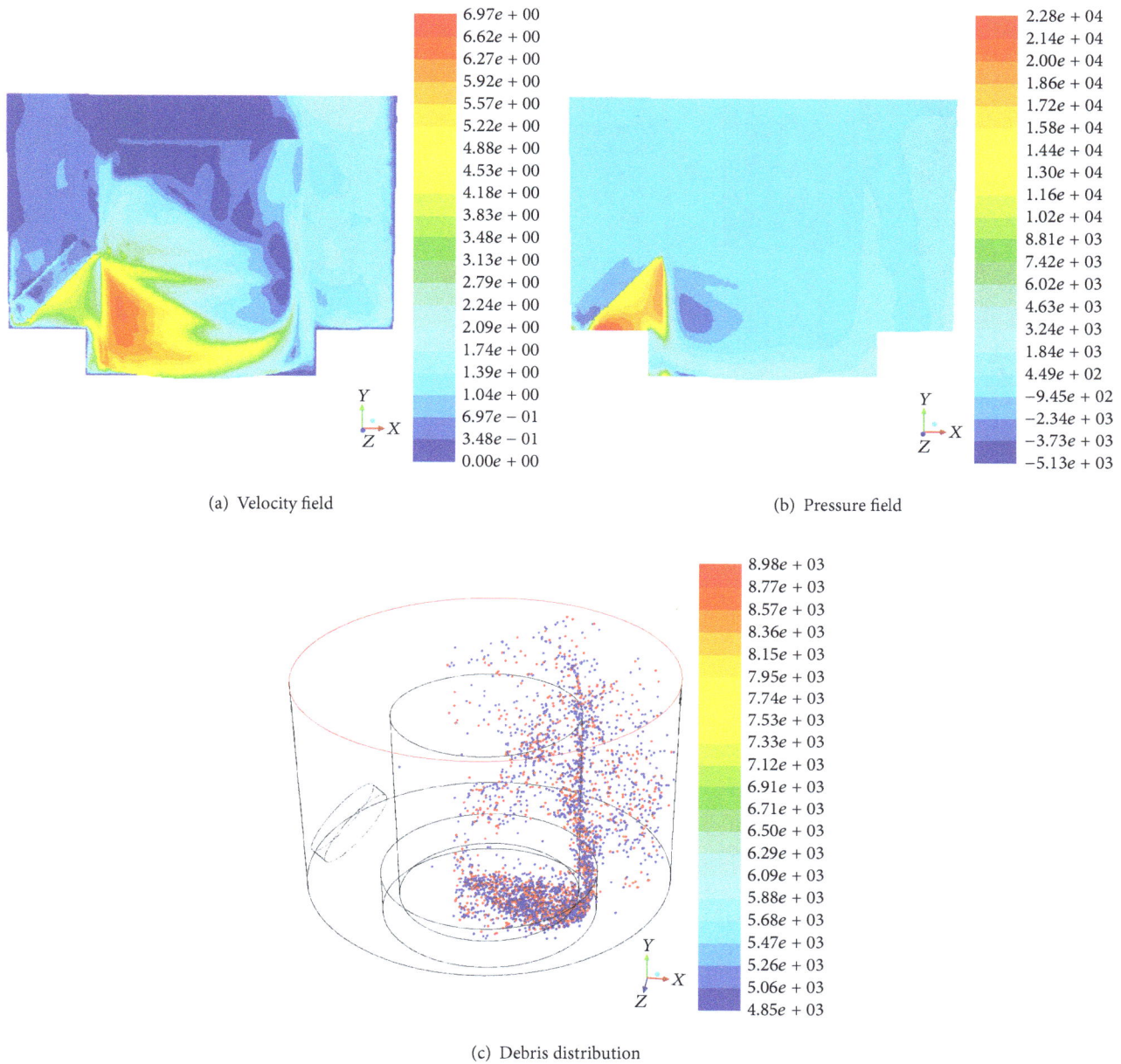

FIGURE 13: Schematics of gap flow field with flushing at 5 m/s and 0.5 mm deep hole.

condition of lower velocity than that of higher velocity. The reason is that, in a shallower hole, debris flushed from S4 to S5 is less than that from S2 to S4; thus, S4 has more debris, while, in a deeper hole, the situation is on the contrary. Comparing Figure 21(a) with Figure 21(b), a conclusion can be made that, with a high velocity and a shallow hole, it is much easier for the debris to be washed away.

5.2. Impact of Tool Diameters on the Debris Distribution. Figure 22 presents the impact of different tool diameters on the debris distribution with flushing velocity at 5 m/s.

As shown in Figure 22(a), there is more debris in zones S1 (23.1%) and S2 (60.4%) than other zones in the case of 1 mm tool diameter; however most debris is swept to S5 (9.3%) in the case of 2 mm tool diameter. A large diameter is more beneficial than small diameter for removing debris. It is because, with a large diameter, the curvature radius is large, the flushing can directly get into the gap with less resistance, and the debris is easily washed away. However, with a small diameter, the curvature radius is smaller, the flushing flows against the wall, the velocity declines sharply, and there is insufficient pressure to take away the debris. As a result, the debris could hardly be removed from the gap.

(a) Velocity field

(b) Pressure field

(c) Debris distribution

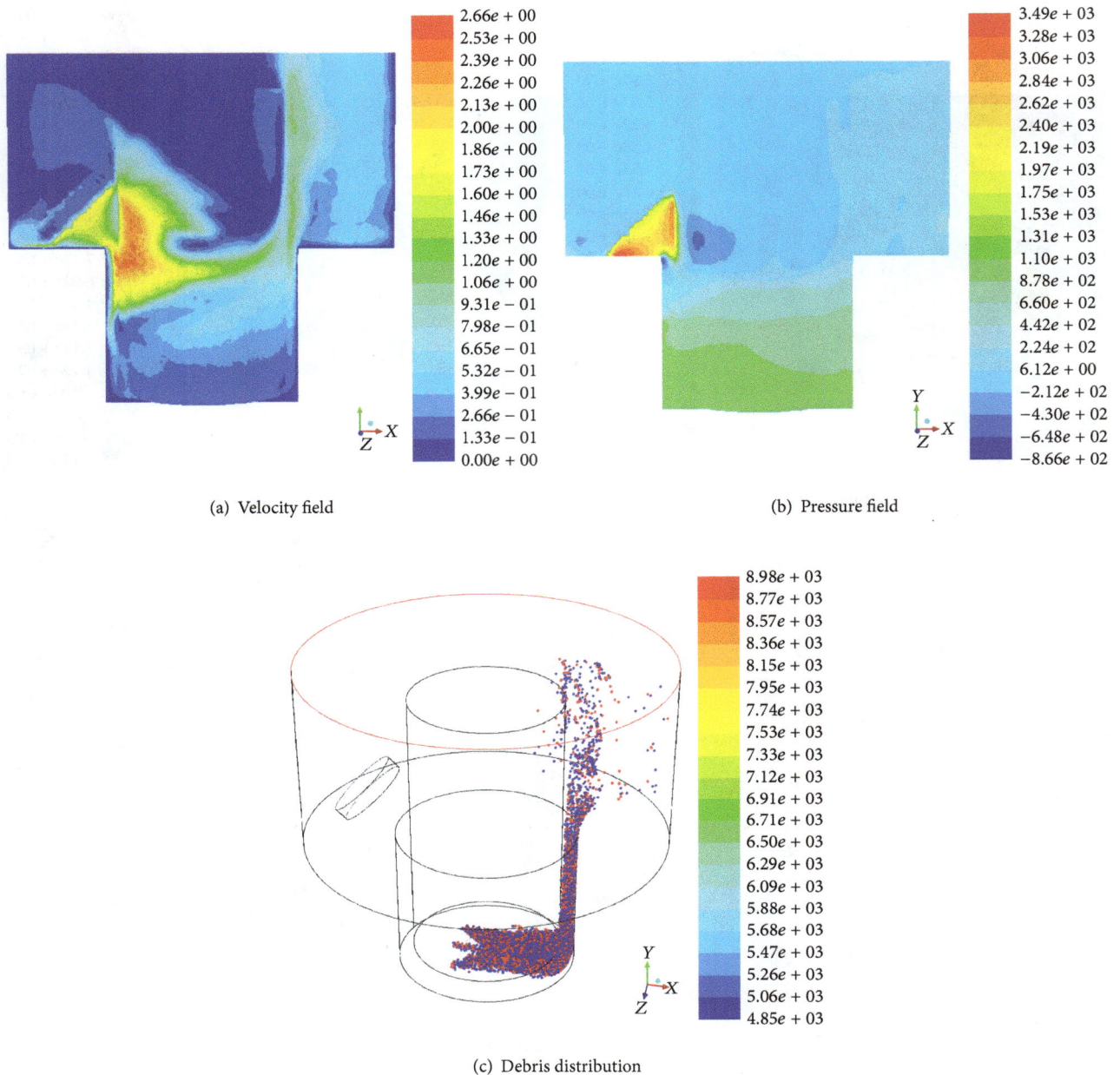

FIGURE 14: Schematics of gap flow field with flushing at 2 m/s and 2 mm deep hole.

Figure 22(b) presents a statistical result in a deeper hole. As shown in Figure 22(b), most debris stays in S1 (50.9%) and S2 (49.1%); almost no debris is carried away, as there is no enough pressure for the dielectric flowing into the gap; debris stays at the bottom. In the case of 1 mm tool diameter, the debris distributed in S1 and S2 is almost the same; however, in the case of 2 mm tool diameter, there is more debris in S2 than S1, which indicates that some debris in the left bottom gap is washed to the right. The impact of tool diameters on the debris distribution is proven again.

Comparing Figure 22(a) with Figure 22(b), in the case of 1 mm tool diameter, zone S1 has less debris than zone S2 (see Figure 22(a)), whereas for the deeper hole (see Figure 22(b)), the quantity of debris in zones S1 and S2 is similar. In the case of 2 mm tool diameter, most debris is swept away from the hole (see Figure 22(a)), whereas for a deeper hole, the diameter is of limited effect on the debris distribution.

A conclusion can also be made that, in the case of a shallow hole, a larger diameter is conducive to remove more debris, whereas in a deeper hole, the tool diameter has little

(a) Velocity field

(b) Pressure field

(c) Debris distribution

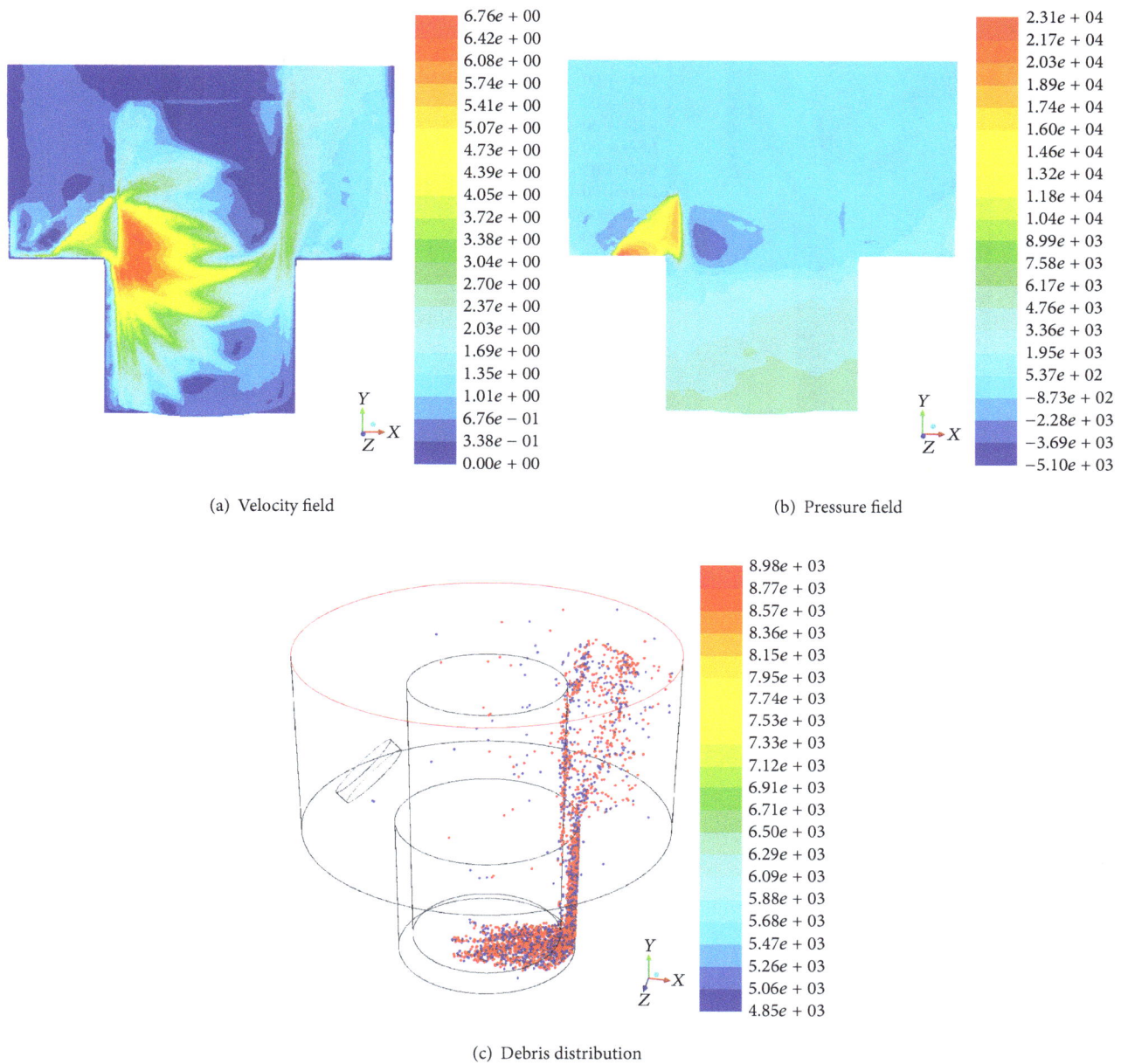

FIGURE 15: Schematics of gap flow field with flushing at 5 m/s and 2 mm deep hole.

impact on the debris distribution, and debris can hardly be removed.

5.3. Impact Analysis of Different Material Debris Distribution. Figure 23 illustrates the impact of flushing velocities and tool diameters on two kinds of debris distributions. As shown in Figure 23(a), from the view of debris distribution of every zone, the distribution rate of each zone for Ti to Cu is basically the same; Ti debris in each zone is approximately twice the quantity of Cu debris that is in accordance with settings. The quantity of removed Cu debris is a little less than that

of half of the Ti debris and the remaining Cu debris is a little larger than that of half of the Ti debris. For instance, in the case of 0.5 mm deep hole, 2 mm tool diameter, and 2 m/s flushing, the quantity percentage of Cu debris in S4 and S5 is 81.5% (S4 (2.5%); S5 (79.0%)), the quantity percentage of Ti debris in S4 and S5 is 84.0% (S4 (2.3%); S5 (81.7%)), and the probable reason is that the density of Cu is larger than Ti. In the case of 5 m/s, the removed percentage of Cu debris is 90.5% and Ti is 90.8%, and the impact of debris material is of limit with high velocity. The velocity impact to Cu and Ti alloy debris distribution is of similar effects. Thus

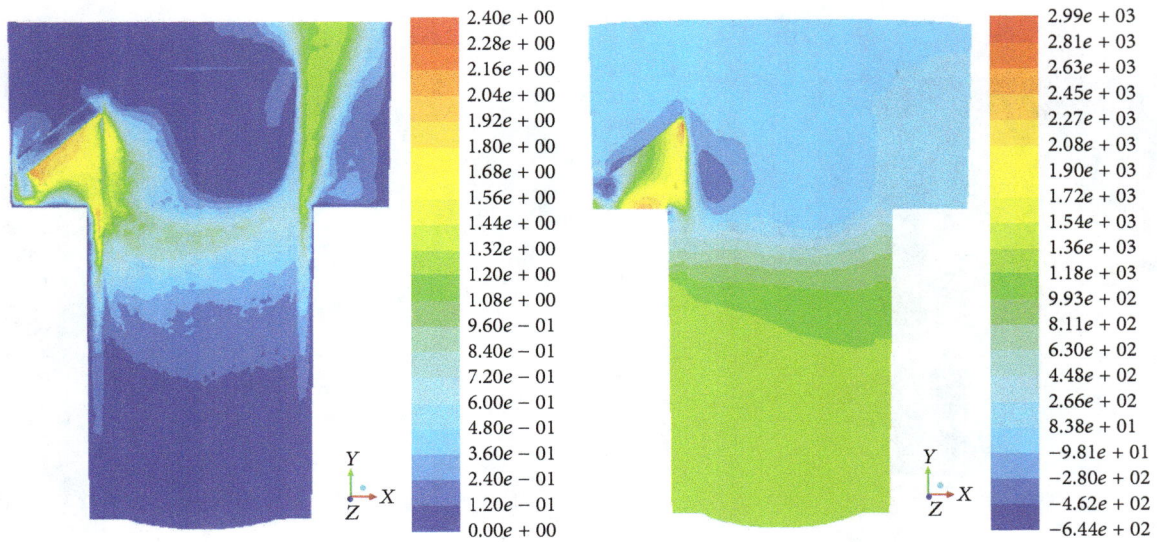

(a) Velocity field

(b) Pressure field

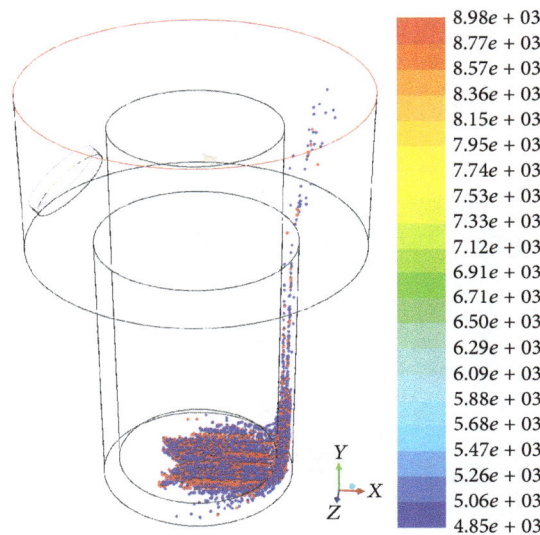

(c) Debris distribution

FIGURE 16: Schematics of gap flow field with flushing at 2 m/s and 3.5 mm deep hole.

the velocity has a bit influence on the debris distribution of different materials.

Figure 23(b) presents the copper and Ti alloy distribution with different tool diameters. As shown in Figure 23(b), different material debris distribution between two different tool diameters is not so obvious. For instance, in the case of 3.5 mm deep hole, 1 mm tool diameter, and 2 m/s flushing, the quantity percentage of Cu debris in S1 and S2 is 94.4% (S1 (44.6%); S2 (49.8%)) and the quantity percentage of Ti debris in S1 and S2 is 84.3% (S1 (44.2%); S2 (40.1%)). In the case of 3.5 mm deep hole, 2 mm tool diameter, and 2 m/s flushing,

the quantity percentage of Cu debris in S1 and S2 is 93.7% (S1 (18.7%); S2 (75.0%)) and the quantity percentage of Ti debris in S1 and S2 is 93.3% (S1 (18.2%); S2 (75.1%)). A conclusion can be made that, in the case of a large tool diameter, the quantity difference between Ti and Cu in S1 and S2 is around 10%, whereas, in the case of a small tool diameter, the quantity difference between Ti and Cu in S1 and S2 is less than 1%. The reason is that when the fluid flows through the gap with a large diameter, it suffers less resistance and has sufficient pressure to take away the debris from the gap. When the fluid flows through the gap with a small diameter, it suffers large

(a) Velocity field

(b) Pressure field

(c) Debris distribution

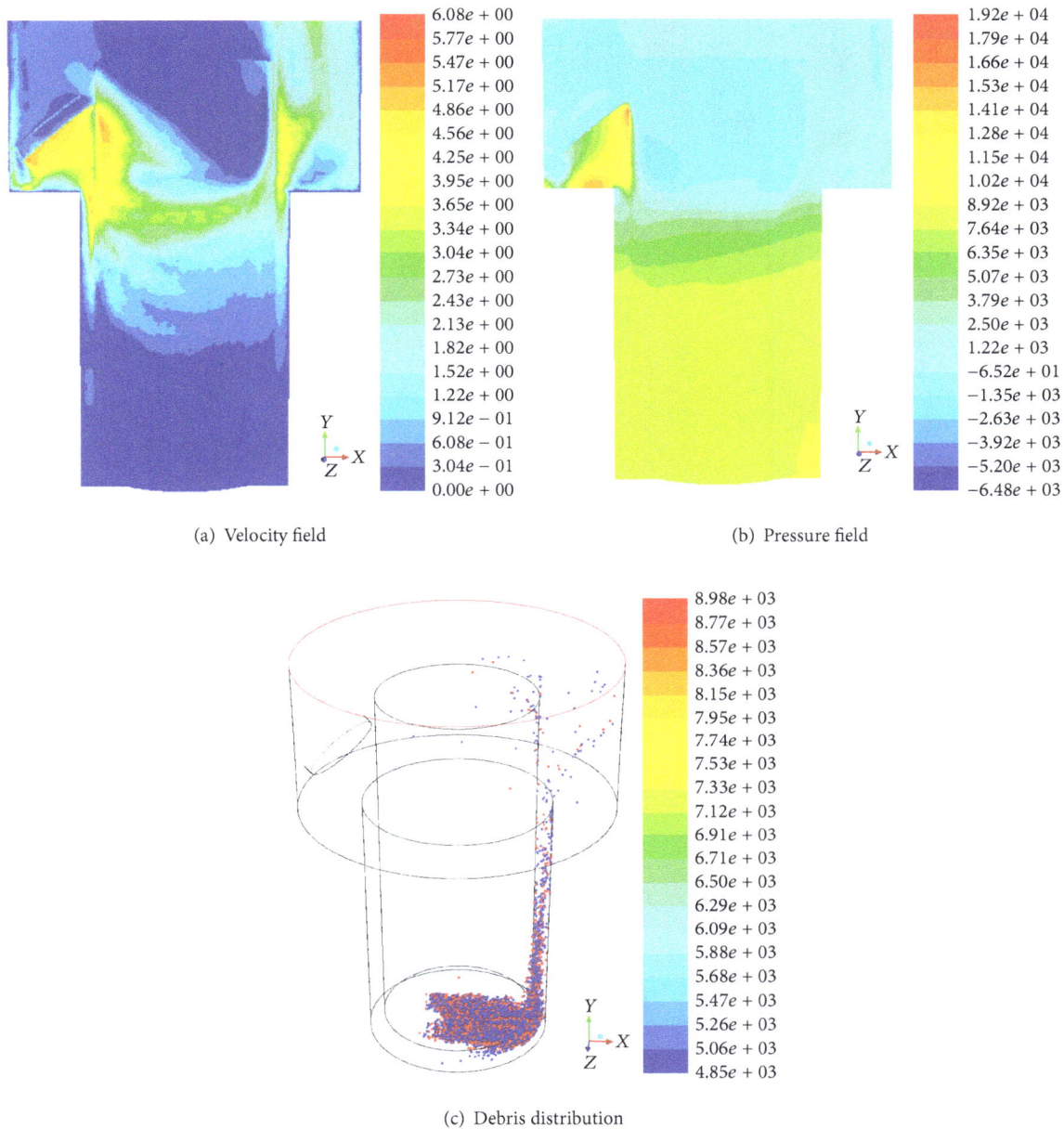

FIGURE 17: Schematics of gap flow field with flushing at 5 m/s and 3.5 mm deep hole.

resistance and has insufficient pressure to take away the debris from the gap.

6. Experiment Results and Discussion

The experiment is carried out on a self-made EDM machine. The detailed parameters applied are listed in Table 2. After the machining, the microscope is utilized to observe the cross sections of machined hole. Experimental results are shown in Figures 24 and 25.

Figures 24(a) and 24(b) present 0.5 mm deep hole and 2 mm deep hole without flushing. The cross section image

of hole is basically axial symmetric about the axis of the hole. The burn marks can be seen clearly, just as shown in the white cycle in Figure 24. The burn marks are generated by large quantity of debris remained at bottom gap. The existence of burn marks indicates that the fluid disturbance induced by tool movement can hardly remove the debris from the bottom gap, of which the concentrated discharge is created between the debris and the workpiece repeatedly. The concentrated discharge could cause the continuous melt and recast, which results in no material removed from the workpiece. The position of burn marks locates at the bottom of the hole, which corresponds well with the simulation results in

(a) Velocity field

(b) Pressure field

(c) Debris distribution

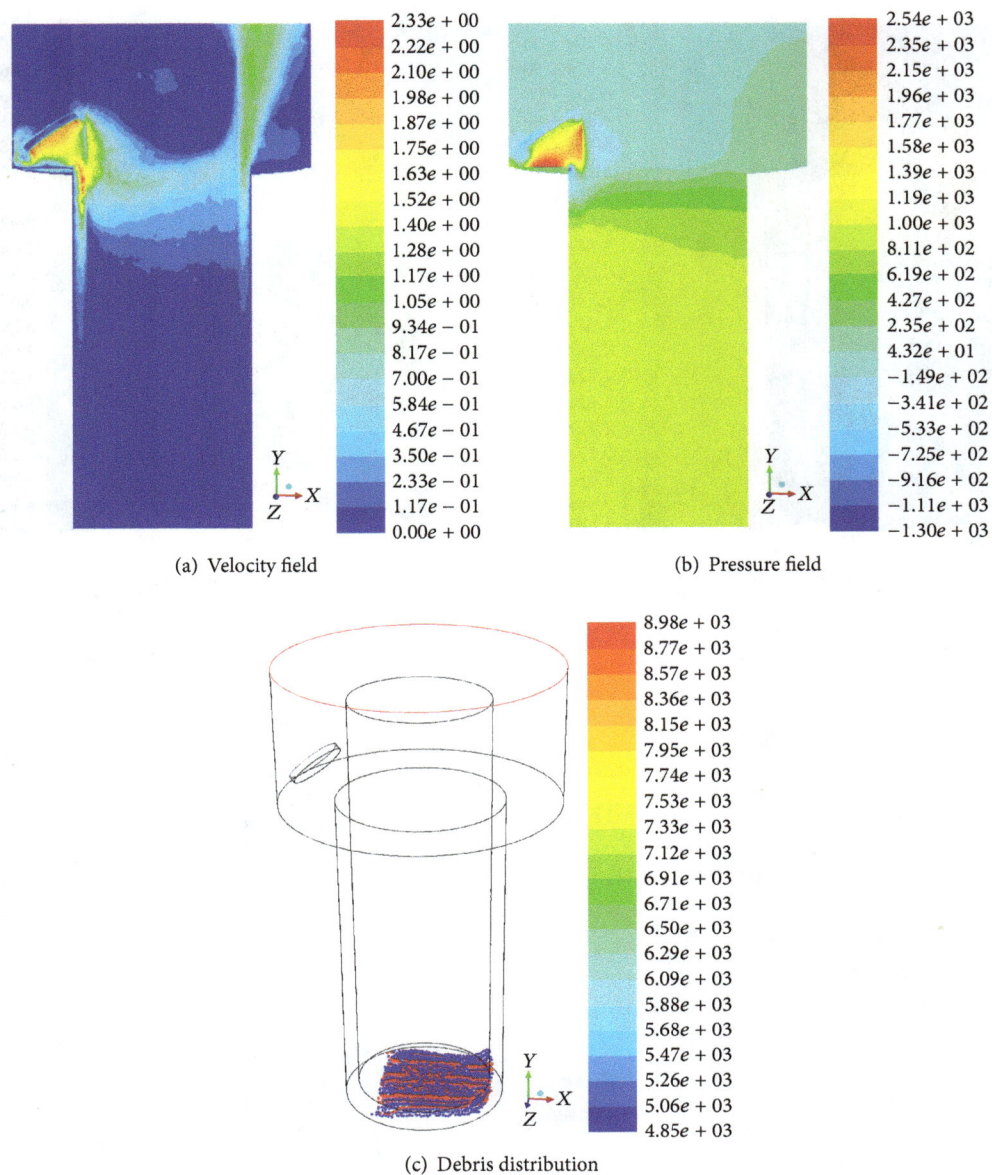

FIGURE 18: Schematics of gap flow field with flushing at 2 m/s and 5 mm deep hole.

TABLE 2: Continuous discharge simulating parameters.

Parameter	Description
Maintaining voltage (U)	15 V
Peak current (I_p)	0.8 A
Dielectric fluid	Deionized water
Tool material	Red copper
Workpiece material	TC4
Tool velocity	0.01 m/s
Tool diameter	2 mm
Depth (mm)	0.5, 2.0, 3.5, 5.0
Flushing velocity (m/s)	0, 2

Figures 4 and 5. With the increase of hole depth, the probability of debris removal decreases, which increases the burn mark area. Such results correspond well with the simulation results in Figures 4 and 5.

Figure 25(a) presents the result of 0.5 mm deep with side flushing. Obviously, Figure 25(a) has no burn mark on the machined surface, which means that the debris at bottom gap could be effectively flushed out. The area of burn mark in Figure 25(b) is smaller than that in Figure 24(b), which indicates that the debris is easily carried away by the fluid flushing. Besides, in the case of hole depth with 0.5 and 2.0 mm, with and without flushing, both cross sections of the walls are symmetric, which indicates that the debris distributed at the bottom is uniform.

(a) Velocity field

(b) Pressure field

(c) Debris distribution

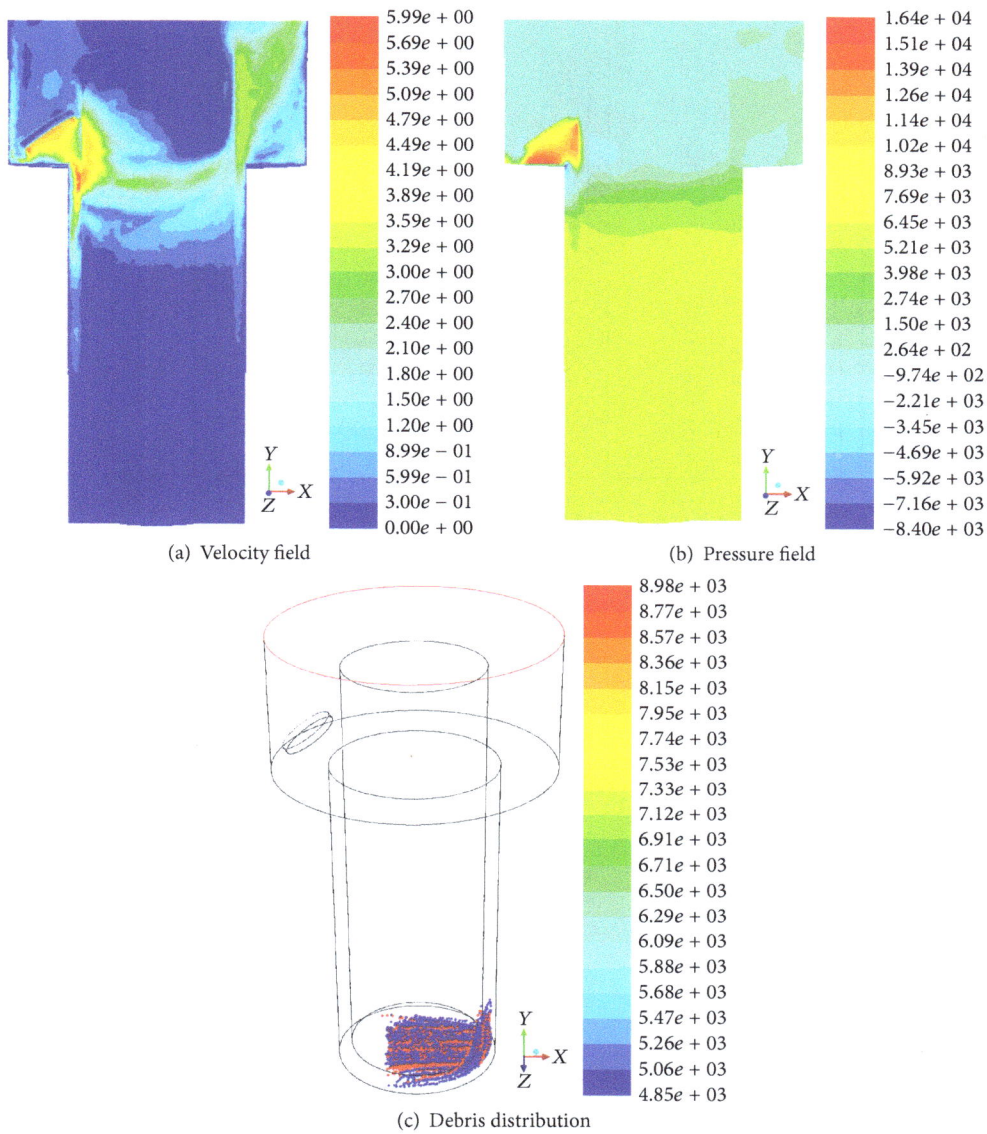

FIGURE 19: Schematics of gap flow field with flushing at 5 m/s and 5 mm deep hole.

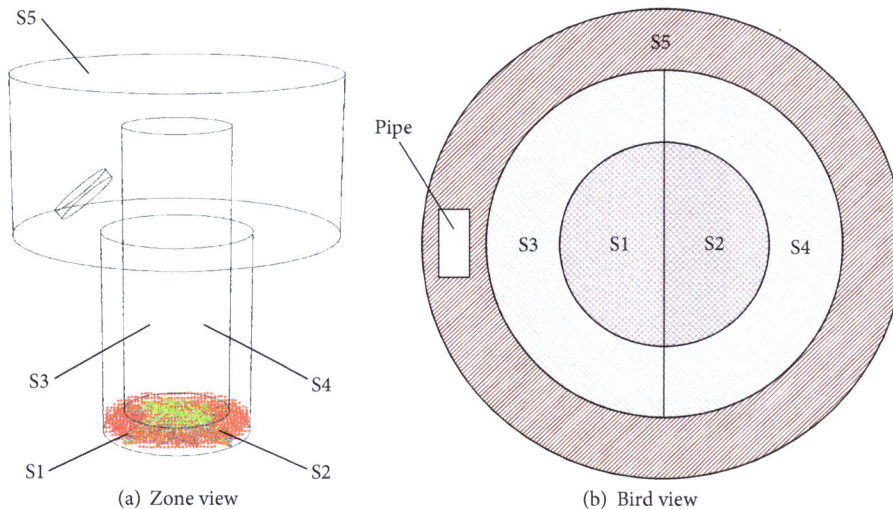

(a) Zone view

(b) Bird view

FIGURE 20: Schematics of debris domain (moves from the top to the bottom).

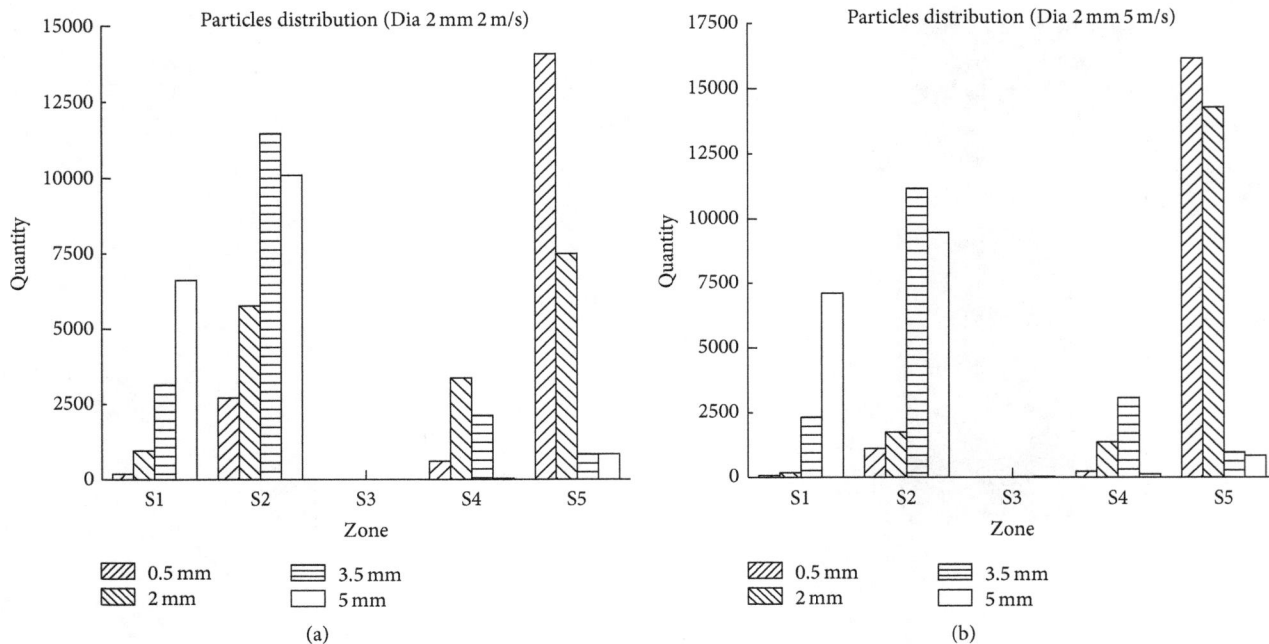

FIGURE 21: Impact of depths on the debris distribution.

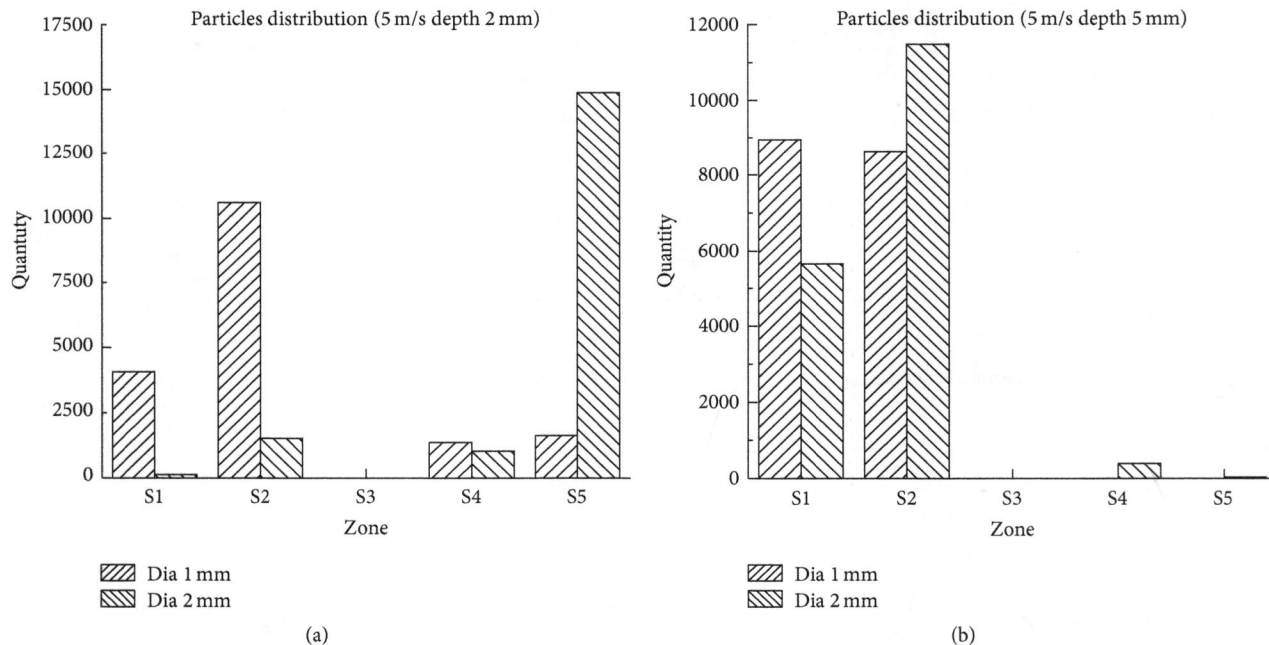

FIGURE 22: Impact of diameters on the debris distribution.

For 3.5 mm deep hole, the burn mark appears at the lower right corner of the hole, which illustrates that most debris accumulates at the lower right corner under the influence of flushing, as shown in Figure 25(c). The accumulated debris can increase the electrical conductivity of dielectric. So the

secondary discharges are easily created between the sides of tool and right wall before the energy is fully reloaded, which causes the burn marks at bottom of hole. Besides, the inclination of right wall of hole is larger than that of left one. During the process of debris flushed out from side gap, the

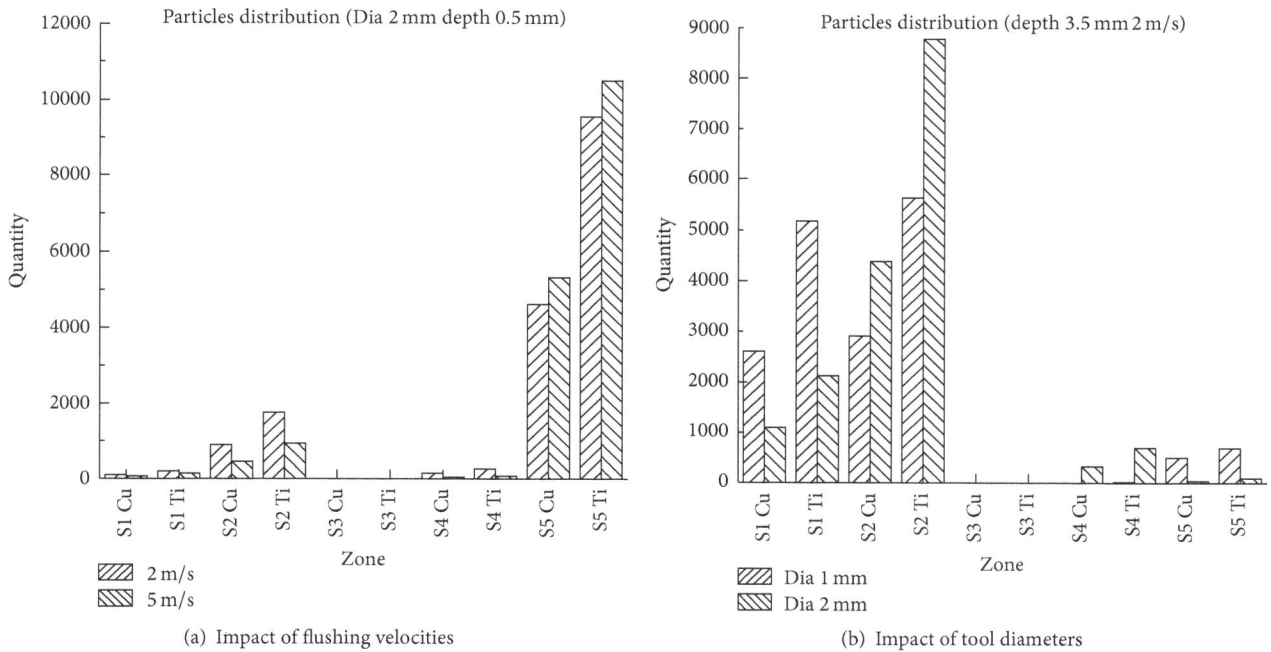

(a) Impact of flushing velocities

(b) Impact of tool diameters

FIGURE 23: Debris statistical distribution of Cu and Ti in different conditions.

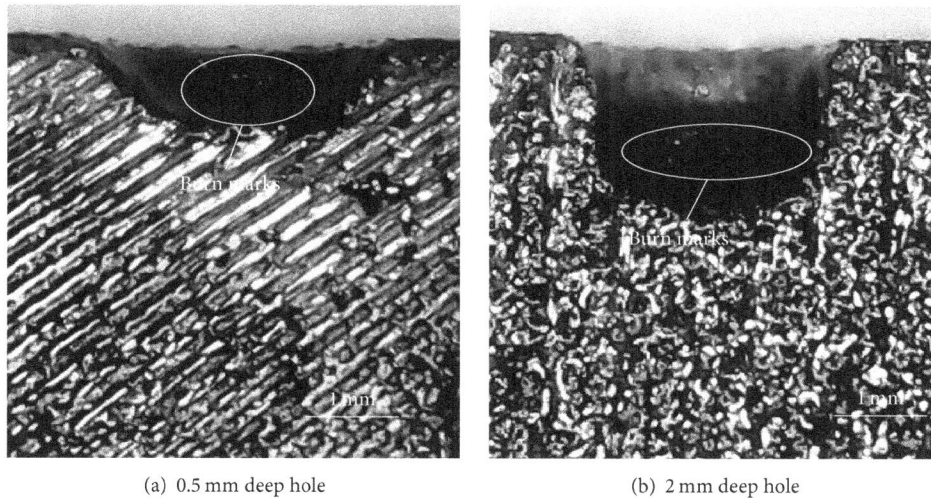

(a) 0.5 mm deep hole

(b) 2 mm deep hole

FIGURE 24: The cross section image of machined holes without flushing.

secondary discharge happens between the right side of tool and wall, which causes the unsymmetrical wall, just as shown in Figure 25(c). These experiment results correspond well with the mathematical model and distribution simulation results in Figures 16 and 17.

When processing 5 mm deep hole with 2 mm tool diameter, as shown in Figure 25(d), the upper wall of the hole has an obvious inclination on the right side, the reason is similar as above, but the lower part of two walls is nearly symmetrical. It is because the hole is too deep that nearly no fluid is injected into the deep gap and debris is distributed uniformly at the bottom. Therefore, the lower parts of walls are evenly worn out. This phenomenon is in accordance with the simulation results of Figures 18 and 19. In a word, if the ratio of depth to diameter exceeds 2, side flushing is not suitable for processing, as the debris removal rate is of limited effect, which can result in a very low machining efficiency; other flushing methods such as internal flushing or rotating tool machining for small hole EDM are recommended.

(a) 0.5 mm deep hole

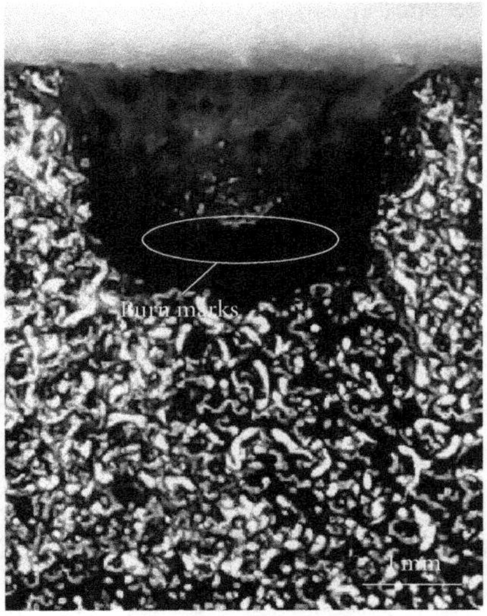

(b) 2 mm deep hole

(c) 3.5 mm deep hole

(d) 5 mm deep hole

FIGURE 25: The cross section image of machined holes with flushing.

7. Conclusions

This paper researches the effect law of machining parameters to the gap flow field and debris movement. Besides the disturbation induced by self-adaptive movement on the gap flow is considered. Based on the solid-liquid two-phase flow, a mathematical model of fluid in gap flow field and debris movement is derived. The methods of smoothing and remeshing are also considered. The real situation of continuous discharging and debris generating is taken into account, a simulation model of the small hole gap flow field in EDM and debris distribution was established, 3D model simulates the influence of tool movement, machining depth, flushing velocity, and tool diameter on interelectrode gap flow field, and debris distribution was analyzed; furthermore, the mathematical model was verified by the experimental results. The conclusions are listed as follows:

(1) During the machining, the self-adaptive movement of tool can generate disturbance to the machining region, which is positive for the debris removal. However, in the case of deep hole machining, the impact of flushing is of limited effect.

(2) The depth of the hole, flushing velocity and tool diameter influence the gap flow field and debris distribution. Normally, with the increasing depth of machining hole, the fluid velocity declines at the bottom gap, and the debris is not easily removed. With the increase of flushing velocity, the fluid at the bottom takes more debris away. A large diameter

is beneficial for reducing the resistance of fluid and debris removal. But when the depth of the machined hole increases to a certain degree, whose depth to diameter exceeds 3, the impact of flushing velocity and tool diameter on debris removal reduces obviously, which results in a lot of debris at the bottom. The side flushing is of limited effect.

(3) During the small deep hole machining process, the cross section appearance of side walls as well as positions of the burn mark is in accordance with the simulating results of debris distribution and movement, which proves the effectiveness of mathematical model and simulation model.

Conflicts of Interest

The authors declare that they have no conflicts of interest.

Acknowledgments

The financial support by National Natural Science Foundation of China under Grant no. 51405058, Program for Liaoning Excellent Talents in University under Grant no. LR2015012, and Talent Special Foundation of Dalian City under Grant no. 2016RQ054 is acknowledged.

References

[1] T. Muthuramalingam and B. Mohan, "A review on influence of electrical process parameters in EDM process," *Archives of Civil and Mechanical Engineering*, vol. 15, no. 1, pp. 87–94, 2015.

[2] J. DBhaghi, M. S. Bohat, V. P. Kalsi et al., "A review on effect of EDM parameters on die sinking EDM," *Trends in Mechanical Engineering & Technology*, vol. 5, no. 1, pp. 26–30, 2015.

[3] F. N. Leão and I. R. Pashby, "A review on the use of environmentally-friendly dielectric fluids in electrical discharge machining," *Journal of Materials Processing Technology*, vol. 149, no. 1-3, pp. 341–346, 2004.

[4] W. Koenig, R. Weill, R. Wertheim et al., "The flow fields in the working gap with electro-discharge-machining," *Annals of the CIRP*, vol. 25, no. 1, pp. 71–76, 1977.

[5] T. Masuzawa, X. Cui, and N. Taniguchi, "Improved jet flushing for EDM," *CIRP Annals—Manufacturing Technology*, vol. 41, no. 1, pp. 239–242, 1992.

[6] H. Takeuchi and M. Kunieda, "Relation between debris concentration and discharge machining," *Denki Kako Gakkaishi*, no. 12, pp. 17–22, 2007.

[7] P. Allen and X. Chen, "Process simulation of micro electro-discharge machining on molybdenum," *Journal of Materials Processing Technology*, vol. 186, no. 1-3, pp. 346–355, 2007.

[8] Y. Q. Wang, M. R. Cao, S. Q. Yang, and W. H. Li, "Numerical simulation of liquid-solid two-phase flow field in discharge gap of high-speed small hole EDM drilling," *Advanced Materials Research*, vol. 53-54, pp. 409–414, 2008.

[9] B.-C. Xie, Y.-K. Wang, Z.-L. Wang, and W.-S. Zhao, "Numerical simulation of titanium alloy machining in electric discharge machining process," *Transactions of Nonferrous Metals Society of China*, vol. 21, no. 2, pp. s434–s439, 2011.

[10] S. Cetin, A. Okada, and Y. Uno, "Electrode jump motion in linear motor equipped die-sinking EDM," *Journal of Manufacturing Science and Engineering, Transactions of the ASME*, vol. 125, no. 4, pp. 809–815, 2003.

[11] J. Wang, F. Z. Han, G. Cheng, and F. Zhao, "Debris and bubble movements during electrical discharge machining," *International Journal of Machine Tools and Manufacture*, vol. 58, no. 7, pp. 11–18, 2012.

[12] W. Kong, C. Guo, and X. Zhu, "Simulation analysis of bubble motion under ultrasonic assisted Electrical Discharge Machining," in *Proceedings of the 3rd International Conference on Machinery, Materials and Information Technology Applications (ICMMITA '15)*, vol. 1, pp. 1728–1731, Qingdao, China, November 2015.

[13] S. A. Mastud, N. S. Kothari, R. K. Singh, and S. S. Joshi, "Modeling debris motion in vibration assisted reverse micro electrical discharge machining process (R-MEDM)," *Journal of Microelectromechanical Systems*, vol. 24, no. 3, pp. 661–676, 2015.

[14] S. A. Mullya and G. Karthikeyan, "Dielectric flow observation at inter-electrode gap in micro-electro-discharge-milling process," *Proceedings of the Institution of Mechanical Engineers, Part B: Journal of Engineering Manufacture*, 2016.

[15] P. T. Eubank, M. R. Patel, M. A. Barrufet, and B. Bozkurt, "Theoretical models of the electrical discharge machining process. III. the variable mass, cylindrical plasma model," *Journal of Applied Physics*, vol. 73, no. 11, pp. 7900–7909, 1993.

[16] R. M. Bruce, F. Y. Donald, and H. O. Theodore, *Fundamentals of Fluid Mechanics*, John Wiley & Sons, 2009.

[17] S. Hayakawa, M. Yuzawa, M. Kunieda et al., "Time variation and mechanism of determining power distribution in electrodes during EDM process," *International Journal of Electrical Machining*, no. 6, pp. 19–26, 2001.

[18] Y. Liu, W. C. Zhang, S. F. Zhang, and Z. H. Sha, "The simulation research of tool wear in small hole EDM machining on titanium alloy," *Applied Mechanics and Materials*, no. 624, pp. 249–254, 2014.

13

Dynamical Analysis Applied to Passive Control of Vibrations in a Structural Model Incorporating SMA-SE Coil Springs

Yuri J. O. Moraes [ID],[1] Antonio A. Silva,[2] Marcelo C. Rodrigues,[1] Antonio G. B. de Lima [ID],[2] Rômulo P. B. dos Reis [ID],[3] and Paulo C. S. da Silva[2]

[1]Department of Mechanical Engineering, Federal University of Paraiba, SN/58.051-085 João Pessoa, PB, Brazil
[2]Department of Mechanical Engineering, Federal University of Campina Grande, Avenue Aprígio Veloso, 882/58.429-900 Campina Grande, PB, Brazil
[3]Department of Technology and Engineering, Federal University Rural of Semi-Árido, Avenue Francisco Mota, Costa e Silva, 572/59.625-900 Mossoró, RN, Brazil

Correspondence should be addressed to Yuri J. O. Moraes; yurijmoraes@gmail.com

Academic Editor: Wen Deng

Mechanical vibrations are severe phenomena of the physical world. These oscillations may become undesirable and may cause temporary and even irreversible damage to the system. There are several techniques to minimizing these vibration effects ranging from passive methods to the use of controllers with smart materials. In this sense, this study aims to analyze a passive vibration control system installed in a structure that simulates two-floor buildings. This system based on the incorporation of one SMA-SE (*Superelastic Shape Memory Alloys*) coil springs configuration for energy dissipation and the addition of damping. Modal analysis was performed using analytical, numerical, and experimental methods. In an experimental basis, response amplitudes were analyzed for free and forced vibrations in different configurations. As compared with the structure configuration with steel spring, the forced vibrations FRF (*Frequency Response Function*) analysis showed a reduction in displacement transmissibility of up to 51% for the first modal shape and 73% for the second mode in the SMA-SE coil spring configuration. As for damping, there was a considerable increase in the order of 59% in the first mode and 119% in the second, for the SMA-SE springs configuration.

1. Introduction

A periodic oscillation or mechanical vibration is a phenomenon defined as any movement that comes repeated after some time. Thus, the theory of vibration studies the oscillatory movements of the bodies and the forces associated with them. In general, a vibratory system alternates the transfer of its potential energy to kinetic energy. This system generally contains a means for storing potential energy, for example, a spring, another for storing kinetic energy, as a mass, and finally, one for gradually dissipating energy, called as damper [1, 2].

Under the viewpoints of mechanics, can be verified these effects easily in our everyday life, whether in the use of domestic appliances or the mining industry, among other

forms [3]. However, some physical phenomenon may become undesirable and may cause temporary or irreversible damage in a specific system, due to the malfunction, the progressive increase of noise, shortened life of its components, increased maintenance costs, and in the more severe cases with the own collapse or structural failure. Thus, in the case of an architectural project, taking into account not only the load exerted by the weight of the system is crucial. Also, the loads are derived from the conditions of the own use, such as the movement of people, automobiles, loads, among others. However, it is of equal importance to consider the effects provoked by natural phenomena, like winds, and waves of the sea [4–6].

Another type of excitation in structures is earthquakes, usually associated with a kind of action imposed by a base

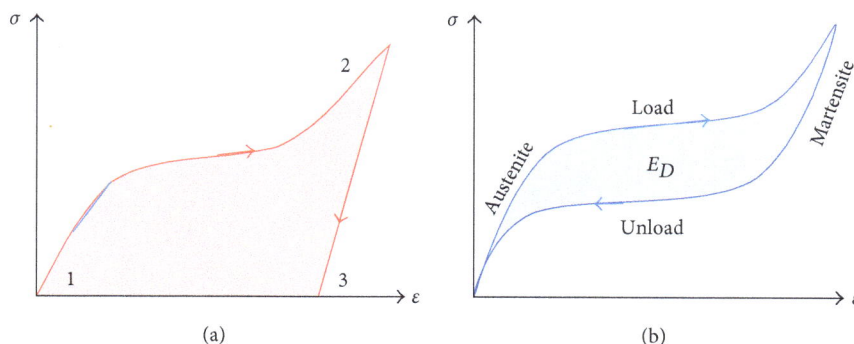

FIGURE 1: Representation of the two effects of an SMA. (a) SME and (b) SMA-SE.

excitation. Earthquake is related to the movements of the surface layers of the Earth, causing a deformation in the broad masses of rocks, which when broken, generate traveling waves in the Earth, provoking the earthquake [7, 8]. The seismic action depends on the mass of the structure, damping, and stiffness of its components. The behavior of a structure under earthquake conditions is a dynamic problem because the seismic movement causes forced vibration in it. Thus, several studies focus on the amplitudes and duration times of the vibration [9–11].

As a result, it is observed that many countries in Europe and Asia suffered the effects of natural disasters, with earthquakes contributing to more significant damage [12, 13]. In Brazil, the risks of a natural cause's shock are of remote chances, since there are no geological faults with sufficient dimensions to generate such an effect; however, specific human activities can also cause earthquakes, the so-called induced earthquakes. These vibrations are due to the construction of significant works, the actions of massive explosions, and the traffic of vehicles, which are associated with the human activity. The collapses and structural failures related to these earthquakes can arise because of the physical phenomenon well known as resonance, defined as the propensy of a system to oscillate at maximum amplitude under specific natural frequencies of the system. At these frequencies, even small forces can produce large-amplitude vibrations because the system stores sizeable vibration energy [14].

Techniques and models have been used to minimize the implications of these vibrations. The first studies with passive viscoelastic absorbers with a structural damping function date back to the 1950s [15–18]. In structural engineering, one of the first applications is related to the design of viscoelastic dampers used in the former WTC (*World Trade Center*) building, located in New York, USA [19], where 10,000 dissipators were used in each of the towers with the specific function of absorbing vibrations originated from dynamic actions of wind.

Currently, the materials most used for the needs imposed by the structural systems are the type with SME (*Shape Memory Effect*), where the main ones are the metallic alloys and polymers. In the metallic materials, the SME characteristic is the reaction to change in temperature, so-called SMA, and in some cases, to change under mechanical stress

known SMA-SE. These reactions are solid-state martensitic transformations [20–22]. Figure 1 shows a typical curve of these two-distinct phenomena. In Figure 1(a) is presented the SMA thermally active, in which the segments 1 and 2 establish that the material is submitted to an external load and modify its shape in the martensitic phase. After removing this external load, one small recovery segment occurred (2-3). To complete recovery, a change in temperature is necessary to transform the martensitic phase in the austenite phase. In the other hand, in SMA-SE the external load (Figure 1(b)) provokes a full-phase transformation from the austenite to the martensitic phase. Upon unloading, a reverse phase transformation is achieved. A complete load-unload cycle gives as result a hysteresis curve. The difference between the curves represents the dissipated mechanical energy.

In this way, the SMA used as an active actuator (requires an external source of energy) is inefficient because it is thermoactive: this means it has a low-frequency response up to 10 Hz. Further, the SMA-SE active per stress field can act as passive damping (without the need of external power source) capable of operating from low frequencies to frequencies above 10 Hz. In fact, the SMA-SE responds as fast as the change in strain submitted to stress field [23].

The use of these materials has greatly increased in the last decades, where many researchers have intensively carried out activities that aim to explore devices and applications that make use of them. In fact, the number of commercial applications is growing each year, with the largest market segment represented by actuators and motors. The global smart materials market in 2010 was approximately $19.6 billion, estimated at $22 billion in 2011, and more than $40 billion by 2016, with an annual growth rate of 12.8% between 2011 and 2016 [24–26].

A selection of SMA is available in the market, but only a few have developed on a commercial scale. From the discovery of Nitinol in 1963, many SMA's were investigated and adapted to specific requirements, such as modulus of elasticity and electrical resistivity, as for the use of sensors and actuators, for example. Nowadays, over 90% of all applications are based on NiTi, NiTiCu alloys, or NiTiNb alloys [27–30].

Then, based on the information related to SMA, this study aimed to perform a comparative analysis between the

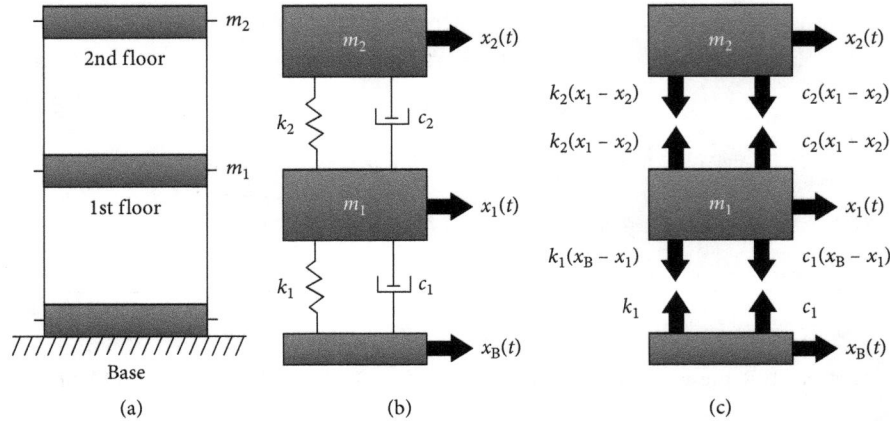

FIGURE 2: Representation of a 2DOF system with base excitation. (a) A simple model that simulates a two-floor building, (b) diagram of concentrated masses, and (c) free-body diagram of the structure.

use of elastic structural elements (steel springs) and elements of Nitinol alloy (SMA-SE springs) for the passive control of vibrations in structural systems. While the steel springs modify the structural parameters of stiffness and mass without incorporating any damping in the system and do not decrease the response amplitudes, the smart material increases the damping with only a small variation of these parameters and reduces the amplitudes in significant percentages, showing its superior efficiency and reliability.

This direct comparison is the main contribution of the research about previous works, showing the high possibility of the use of SMA-SE in dynamic systems with the aim of mitigates mechanical vibrations.

2. Methodology

This section describes a structural model that simulates two-floor buildings. The mathematical formulation is applied in an underdamped system with 2DOF (*Two Degrees of Freedom*) excited by the base. Thus, a numerical modal analysis of the system was performed without the SMA-SE absorbers, followed by the experimental stage, as a way of studying the structure with the elements of dissipation. Finally, a static characterization of the SMA-SE coil springs was performed.

2.1. Analytical Procedure. A free-body diagram illustrating the spring forces acting on each mass (Figure 2(a)) is illustrated in the Figures 2(b) and 2(c). The system will be analyzed when subjected to a forced excitation acting in the base. The constants c1 and c2 indicate the structural damping imposed on the system. More details can be found in the literature [31].

Summing the forces applied in each mass (in the horizontal direction) and applying Newton's law of equilibrium to the system yields (1).

$$m_1 \cdot \ddot{x}_1 - c_1(\dot{x}_B - \dot{x}_1) + c_2(\dot{x}_1 - \dot{x}_2)$$
$$- k_1(x_B - x_1) + k_2(x_1 - x_2) = 0, \tag{1}$$

$$m_2 \cdot \ddot{x}_2 - c_2(\dot{x}_1 - \dot{x}_2) - k_2(x_1 - x_2) = 0. \tag{2}$$

After a rearrangement of terms of (1) and (2), (3) is obtained. In this equation, "$f_1(t)$" is the input harmonic force imposed on basis of the structure, which is a function of the stiffness and damping of the first floor. The displacement, velocity, and "$f_2(t)$" are null.

$$[m_1 \cdot \ddot{x}_1 + (c_1 + c_2) \cdot \dot{x}_1 + (k_1 + k_2) \cdot x_1]$$
$$+ [-c_2 \cdot \dot{x}_2 - k_2 x_2] = [k_1 \cdot x_B + c_1 \cdot \dot{x}_B] = f_1(t),$$
$$[m_2 \cdot \ddot{x}_2 + c_2 \cdot \dot{x}_2 + k_2 \cdot x_2]$$
$$+ [-c_2 \cdot \dot{x}_1 - k_2 \cdot x_1] = 0 = f_2(t). \tag{3}$$

Modifying (3) into the domain of Laplace transforms, we assume that the initial conditions of the system are zero and that "$F_1(s) = \{k_1 \cdot [X_B(s)] + c_1 \cdot [s \cdot X_B(s) - x_B(0)]\}$." Now, taking the displacement values in evidence, that is, "$[A] \cdot [X(s)] = [F(s)]$," we can write

$$\begin{bmatrix} s^2 \cdot m_1 + s \cdot (c_1 + c_2) + (k_1 + k_2) & -s \cdot c_2 - k_2 \\ -s \cdot c_2 - k_2 & s^2 \cdot m_2 + s \cdot c_2 + k_2 \end{bmatrix}$$
$$\cdot \begin{bmatrix} X_1(s) \\ X_2(s) \end{bmatrix} = \begin{bmatrix} F_1(s) \\ 0 \end{bmatrix}. \tag{4}$$

For (4), four transfer functions were obtained in the Laplace domain "$X_1(s)/F_1(s)$, $X_1(s)/F_2(s)$, $X_2(s)/F_1(s)$ and $X_2(s)/F_2(s)$." For this case, $F_2(s)$ is always 0 since there is no other force acting in the system, other than the base excitation. Using Cramer's rule, one can solve the matrix equation for the displacements. The results are

$$X_1(s) = \left[\frac{s^2 \cdot m_2 + s \cdot c_2 + k_2}{\det(A)}\right] \cdot F_1(s) + \left[\frac{-s \cdot c_2 - k_2}{\det(A)}\right] \cdot F_2(s),$$

$$X_2(s) = \left[\frac{-s \cdot c_2 - k_2}{\det(A)}\right] \cdot F_1(s)$$

$$+ \left[\frac{s^2 \cdot m_1 + s \cdot (c_1 + c_2) + (k_1 + k_2)}{\det(A)}\right] \cdot F_2(s),$$

$$(5)$$

in which the "$\det(A)$" is given as follows:

$$\det(A) = \left[s^2 \cdot m_1 + s \cdot (c_1 + c_2) + (k_1 + k_2)\right]$$
$$\cdot \left[s^2 \cdot m_2 + s \cdot c_2 + k_2\right] - (-s \cdot c_2 - k_2)^2.$$

$$(6)$$

Since that, "$F_2(s) = 0$," we have only two transfer functions that govern our system "$[X_1(s)/F_1(s) \text{e} X_2(s)/F_1(s)]$." So, by multiplying these equations by the modal rigidities "k_1" and "k_2" respectively, we obtain the measured responses in the masses "m_1" and "m_2" simultaneously, as follows:

$$H_{11}(s) = \frac{X_1(s) \cdot k_1}{F_1(s)} = \left[\frac{s^2 \cdot m_2 \cdot k_1 + s \cdot c_2 \cdot k_1 + k_2 \cdot k_1}{\det(A)}\right],$$

$$(7)$$

$$H_{21}(s) = \frac{X_2(s) \cdot k_2}{F_1(s)} = \left[\frac{-s \cdot c_2 \cdot k_2 - k_2 \cdot k_2}{\det(A)}\right].$$

$$(8)$$

We admit that a system with two independent coordinates subjected to a primary harmonic excitation can be defined by two transfer functions by using the Laplace Transform method, as in (7) and (8).

To obtain the analytical response in the frequency domain, a simulation code in the *Matlab®* software was developed and can be seen in the literature [31].

2.2. Numerical Procedure.

For the numerical analysis of the system was developed a virtual structural model. With this model were determined the natural frequencies of the system with their respective modal shapes. In this analysis, the maximum and minimum displacements of amplitudes at each point of the structure can be visualized. For the development of the model, the design and modeling commercial software *AutodeskInventor®* version 2015 was used. For the modal analysis, was used the software *Ansys®* version 15.0 applying the FEM (*Finite Elements Method*).

The structural design in a computational environment was carried out with the objective of determining the geometric dimensions of the system. The idea is verifying the structural stability of the system via modal analysis.

The simulation is an essential stage of design since it has as objective to verify the dynamic behavior of the system, preventing failure in this process, which can compromise the physical integrity of the structure in operation.

In this process, the model was defined from the selection of the materials for each element that constitutes the structure, finishing with the choice of the control block for modal analysis. A mixed mesh with a predominance of hexahedral elements, containing 31,074 nodes and approximately 4,047 elements, was used, along with the selection of the "Base" floor as a fixed component of the system. In Figure 3 is illustrated the mesh generated before the simulation with the use of FEM.

2.3. Experimental Procedure

2.3.1. Structural Model and Absorber Elements. In this study, a structural model was designed and built in the LVI (*Vibration and Instrumentation Laboratory*) of the UFCG (*Federal University of Campina Grande, Brazil*). The prototype model consists of carbon steel beam for the composition of floors, rectangular stainless steel plates (AISI 304) for column composition, and as connecting elements Allen bolts of steel. Table 1 and Figure 4 summarize the number of elements, physical parameters, and dimensions of the prototype used in the fabrication of the structure.

Two structural configurations were used. The first one uses elastic elements of the type steel coil springs and the second with coil springs of superelastic effect (SMA-SE NiTi) at room temperature. In the second case, damping in the structure was due to the thermomechanical property of pseudoelasticity present in SMA-SE.

The SMA-SE coil springs were mechanically characterized by Instron 5582 at room temperature, using quasistatic load, with displacement control at 3 mm/min up to 445% strain. The thermal characterization was done by DSC (*Differential Scanning Calorimeter*) with 5°C/min from 100°C to −60°C, in order to verify the temperatures of the phase transformation.

The SMA-SE coil springs initially have orthodontic function and are commercialized by the Dental Morelli Company. Figure 5 shows the springs used in the study. In their original state, the SMA-SE coil springs are commercialized with different lengths between the eyelets (7.0, 9.0, 12.0, and 15.0 mm). However, the lengths "M7 = 7.0 mm" and "M12 = 12.0 mm" were selected due to better adaptation to the displacement amplitudes of the structure in the second and first floor, respectively. Thus, it is possible to obtain a higher efficiency of energy dissipation.

For other types and configurations of the elements incorporated, the useful length and the length between eyelets (L_u: useful length, L_o: length between eyelets, N_e: number of active turns, and V_m: volume of the useful material) were checked. These values are reported in Table 2.

For SMA-SE spring (M7), the number of active turns is seven, while in SMA-SE spring (M12), that number is twenty-six. Both have an initial turn angle equal to zero, initially closed. For the calculation of the linear deformation of the springs, the useful length of 2.5 mm was considered for M7 and 7.5 mm for M12 according to the point of crimping of the eyelets.

As for system tests incorporating SMA-SE elements, a spring type was attached for each floor, and calculated the total volume of useful material incorporated into the

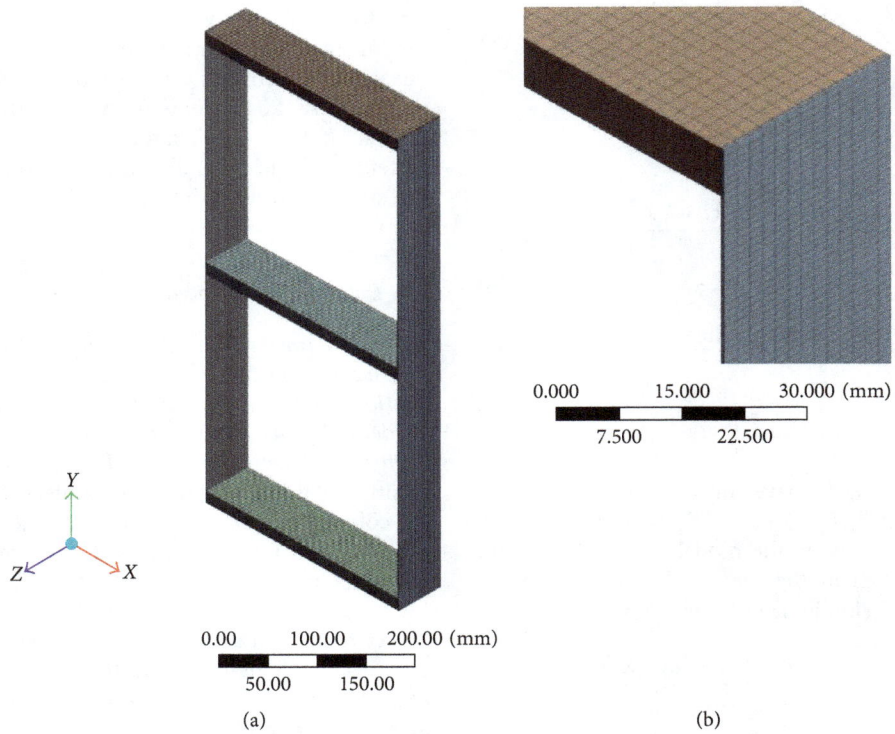

0.000   15.000   30.000 (mm)

7.500   22.500

0.00   100.00   200.00 (mm)

50.00   150.00

(a)   (b)

FIGURE 3: (a) The mesh of the physical model and (b) detail of the mesh in the columns.

TABLE 1: Description of the components used in the structural construction.

Description	Dimensions (mm)	E (GPa)	Mass (kg)	Amount (unit)
Steel beam 1020	$228.0 \times 50 \times 12.80$	200	1.110	3
Stainless plate 304	$501.6 \times 50 \times 0.97$	193	0.190	2
Allen bolt (3/4″)	9.5×15.50	200	0.004	24
Total	—	—	3.806	29

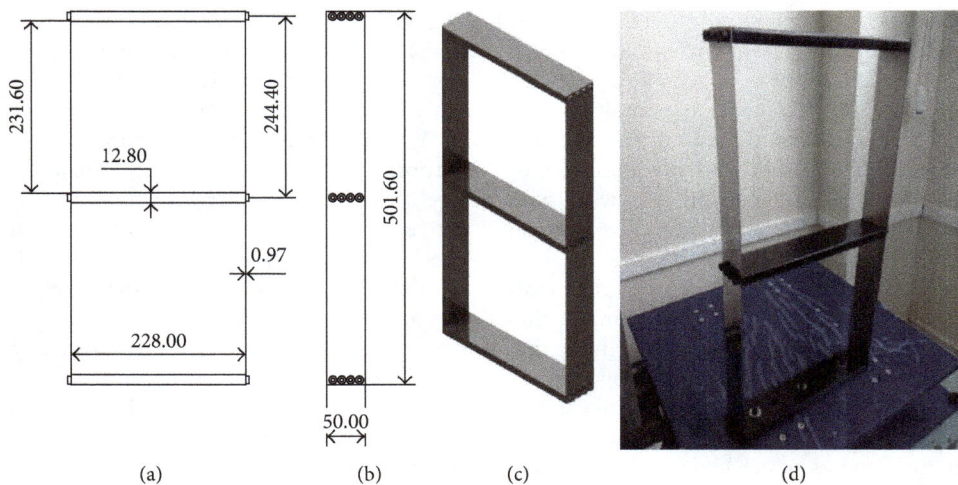

231.60

12.80

228.00

244.40

0.97

501.60

50.00

(a)   (b)   (c)   (d)

FIGURE 4: Technical drawing of the prototype. (a) Front view, (b) lateral view, (c) exploded view of the structure, and (d) real model on the shake table.

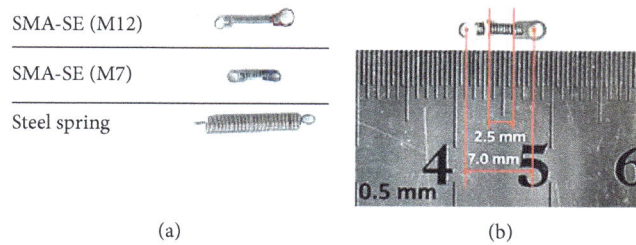

(a)   (b)

FIGURE 5: Absorber elements. (a) Springs used in the study and (b) length between eyelets and useful of the SMA-SE coil spring (M7).

TABLE 2: Dimensional parameters of the absorber elements.

Absorber elements	L_u (mm)	L_o (mm)	N_e (unit)	V_m (mm^3)
Steel spring	15.0	21.0	47	41.470
SMA-SE coil spring (M7)	2.5	7.0	7	2.494
SMA-SE coil spring (M12)	7.5	12.0	26	7.481

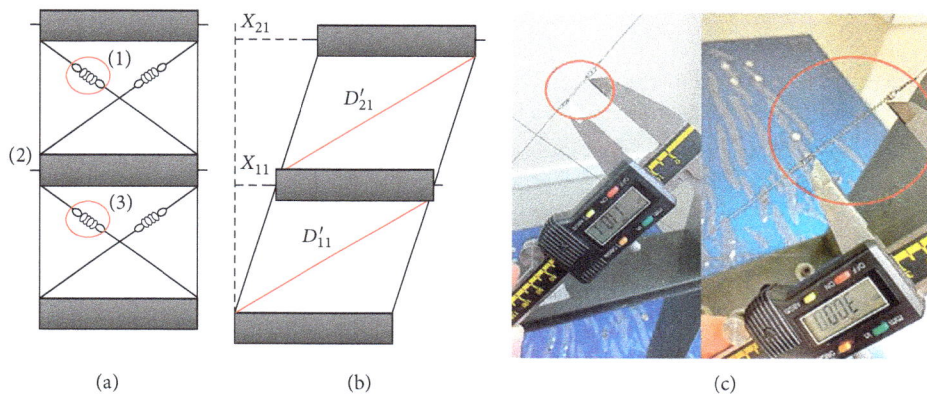

(a)   (b)   (c)

FIGURE 6: Coupling of the springs in the structure. (a) Illustrative representation of the arrangement of the elements, (b) procedure for measuring the deformation of the SMA-SE coil springs, and (c) calibration procedure for the springs.

structure. For the system configuration with SMA-SE springs, the entire material volume was about 88% less than in the case of steel springs, thus verifying that on a real scale, an optimization due to reduction in the material and space of the absorbers will be obtained.

The steel wires were received with an initial pre-tensioning to deform the SMA-SE springs at a certain level, allowing them to be in the superelastic region before performing dynamic tests. Following this recommendation was obtained this offset through measurements realized with an instrument used to measure displacements.

Assuming the springs have a much lower stiffness than steel wires, we can consider that the increase in damping imposed on the system must be associated with absorbers SMA-SE only. Figure 6(a) shows the connection of the steel wires in the system (1) and (3), which is fixed by anchor bolts (2). One can also observe the calibration procedure for the offset of the springs in Figure 6(c).

In this setup, when requesting the structure, it will move by expanding the vibration absorber element, occurring to the direct/martensitic phase transformation of the SMA-SE, and achieve the reverse/austenitic transformation when discharging the element, passing through the central point and forming the hysteretic energy dissipation loop.

To obtain the initial deformation, the forced vibration test of the model without the presence of absorbers was carried out, to determine the maximum deflection in each floor of the structure and the deformation in each diagonal.

Figure 6(b) shows the scheme for determining the parameters "D'_{11}" and "D'_{21}" of the first and second floor when it is subjected to the excitation relating to the first modal shape. After analysis of the FRF experimental curves, the displacement values "X_{1i}" and "X_{2i}" can be obtained for the maximum deflection at each floor, with "$i = 1, 2$" corresponding to the first and second modal shapes. These values can be obtained using (9) and (10):

$$X_{1i} = T_{d1i} \cdot A_{1i}, \qquad (9)$$

$$X_{2i} = T_{d2i} \cdot A_{2i}, \qquad (10)$$

where "T_{d1i}" and "T_{d2i}" represent the transmissibility of displacement with respect to the base and analogously "A_{1i}" and "A_{2i}" the input amplitude at the base for the excitation function. Assuming that "X_{1i}" has the same value for the

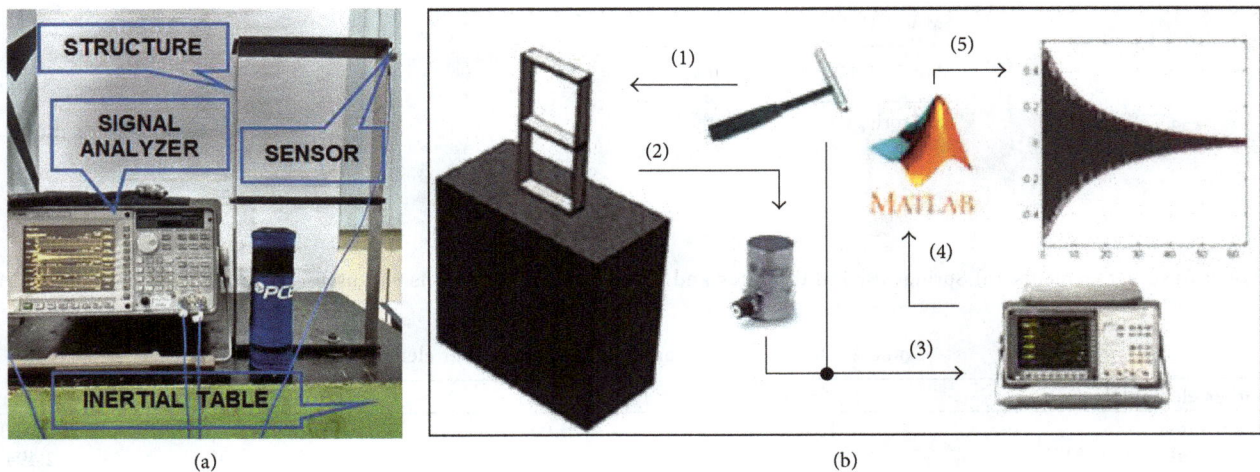

FIGURE 7: (a) Experimental setup of the structure tests in free vibration and (b) schematic representation of the experimental procedure in the free vibration tests.

entire thickness of the bar, the values of deformation in each diagonal are obtained in the condition of the first modal shape, using (11) and (12) as follows:

$$D'_{11} = \sqrt{(l_1)^2 + (X_{11} + P_1)^2}, \quad (11)$$

$$D'_{21} = \sqrt{(l_2)^2 + [(X_{21} - X_{11}) + P_2]^2}, \quad (12)$$

where, from the data of the technical drawing of the structure (Figure 4), $l_1 = l_2 = 231.60$ mm and $P_1 = P_2 = 228.00$ mm. Analogously, the values of these same parameters related to the second modal shape of the system can be found using (11) and (12), using $(i = 2)$. In this way, the strategy of using a minimum initial elongation must be adopted, ensuring that the maximum deformation does not exceed the elastic or final regime of the hysteresis loop.

2.3.2. Free Vibration Setup.

For the experimental tests in free vibration, was fixed the structure on the inertial table supported by vibra stops. The inertial table has a natural frequency lower than the primary frequencies of the system tested.

This experiment aims to determine the time response graph in which the amplitudes, regarding acceleration, and the period of vibration attenuation in the system are determined. The frequency spectrum was obtained in the cases where it is possible to visualize the natural frequency peaks and values of the structure to validate the analytical results. Figure 7(a) shows the actual assembly and the instrumentation used.

The setup of instrumentation is composed by an impact hammer (PCB® 086D05) which was used to generate a transient input signal and an accelerometer type sensor (PCB 353B01) to pick up the output signal. The data were collected by the signal analyzer, the (Agilent® Model 37670A). Following, they were manipulated in the Matlab software.

The tests were divided in three stages: no absorbers elements, with the steel springs, and with the SMA-SE coil

TABLE 3: Characteristics of the Shake Table II Quanser.

Dimensions	$61.0 \times 46.0 \times 13.0$ cm
Maximum displacement	± 7.6 cm
Maximum load	7.5 kg
Maximum mass	27.2 kg
Maximum acceleration	24.5 m/s^2
Maximum frequency	15.0 Hz

springs. In all situations was used the same instrumentation. The external force (impact hammer) focused on the second floor of the structure, along with the capture of the output signal (accelerometer) because they present less interference in the signals generated. Figure 7(b) shows the sequence of the experimental procedure.

2.3.3. Forced Vibration Setup.

For the experimental tests of forced vibration, an electromechanical machine Quanser® Shake Table II was used. Table 3 shows the main characteristics.

The shake table is managed by a computer and its QUARC® control software or the Matlab interface from Simulink®. The input and output signals of the structure were captured by the LVDT (Linear Variable Differential Transformer) type (WI/10 mm T) and (WA/20 mm L), both from the HBM® manufacturer and registered by the QuantumX® data acquisition system. The signal acquisition software is the CatmanEasy®, and data were processed on the Matlab software. Figure 8 shows the setup and instrumentation used in the experiments.

For the dynamic tests of the structure under forced vibration, behavior of the structure when subjected to harmonic excitation forces (sinusoidal function), using a peak input amplitude of 0.6 mm was studied.

Based on the choice of motion imposed on the system, it is possible to determine the FRF from the displacement measurements for each floor, extracting from these curves essential values in the analysis, such as the natural

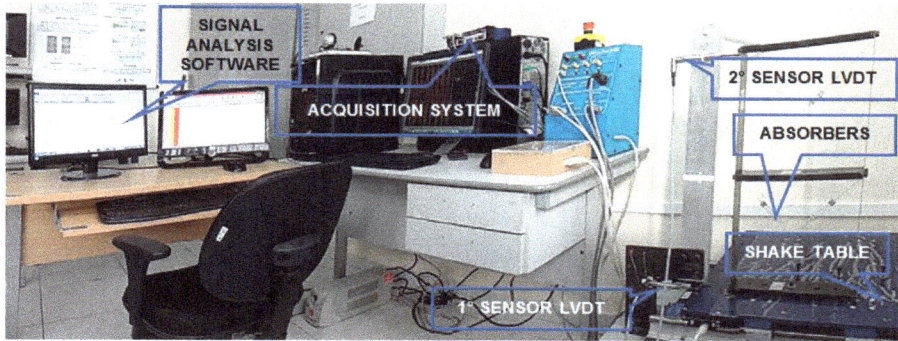

FIGURE 8: Experimental setup of the system under forced vibration.

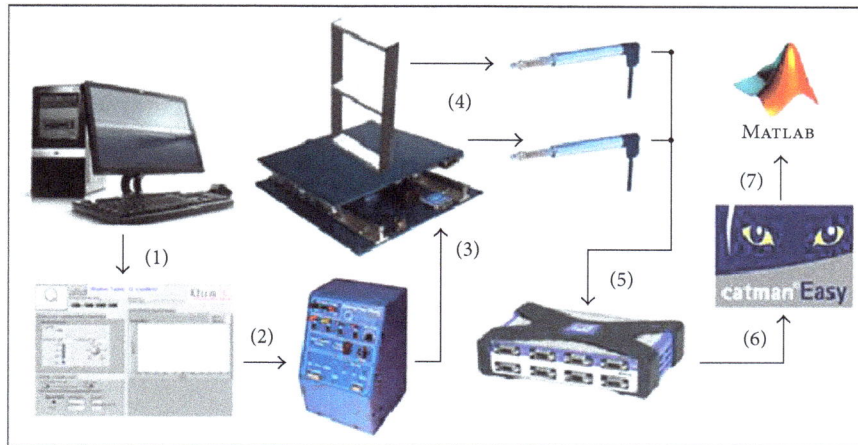

FIGURE 9: Representation of the experimental procedure under forced vibration.

frequencies, the transmissibility of the displacement, and the damping factors. Figure 9 shows the experimental sequence.

There are three stages for the dynamic test of the forced vibration structure: no absorber elements, with the steel springs, and with the SMA-SE coil springs, each subdivided in two measurements concerning the two floors of the structure. The measurements were carried out separately to avoid interference with the reading sensors, due to the small increase in the damping imposed by the friction between the moving parts of the sensor, perceived in previous tests.

All tests followed the same measurement setup and frequency range (1.0 to 11 Hz), by increments of the 0.05 Hz. For signal pickup, at each frequency was applied a hold time of 10 seconds for the system to stabilize. The signal pickup was performed at a rate of 50 Hz and a time period of 10 s, totaling 500 resolution points at each measurement. Therefore, each curve was generated with 201 pts, obtaining a good accuracy in the resonance peaks and increasing the precision of the damping values.

Initially, we cut off the offset value and calculated the FFT (*Fast Fourier Transformer*) for each of these signals and then obtained maximum peak-value. From these maximum peak data points, the FRF was constructed by dividing the output by the input values $[X(w)/Y(w)]$, with a resolution of 201 points, as previously described.

3. Results and Discussion

3.1. SMA-SE Coil Springs Characterization. Figure 10(a) shows the result from DSC of the M7 SMA-SE coil spring and the results from the Instron equipment for both M7 and M12 SMA-SE coil springs. The final temperature of austenite transformation (15.7°C) to the M7 coil spring is below room temperature, to confirm the superelastic behavior. This behavior was not verified to the M12 coil spring. Figure 10(b) illustrates the force versus displacement curves up to 445% of strain. For the M7 coil spring, the energy dissipated was 0.015 J and 0.026 J for M12 coil spring.

3.2. Predicted Results. For determination of the natural frequencies and modal shapes of the structure subjected to forced vibration, we used (7) and (8) and the experimental data of mass, damping, and stiffness. Figure 11 shows the analytical FRF curves of the system without the incorporation of absorbers, for the first and second floors, obtained from the transfer functions modeled by the Laplace transform theory described in Section 2.1. Table 4 displays the natural frequencies of the structure without absorbers.

Figure 12 shows the results of the numerical simulation (FEM) for the modal shapes of the structure without the incorporation of the absorber elements. It was verified that

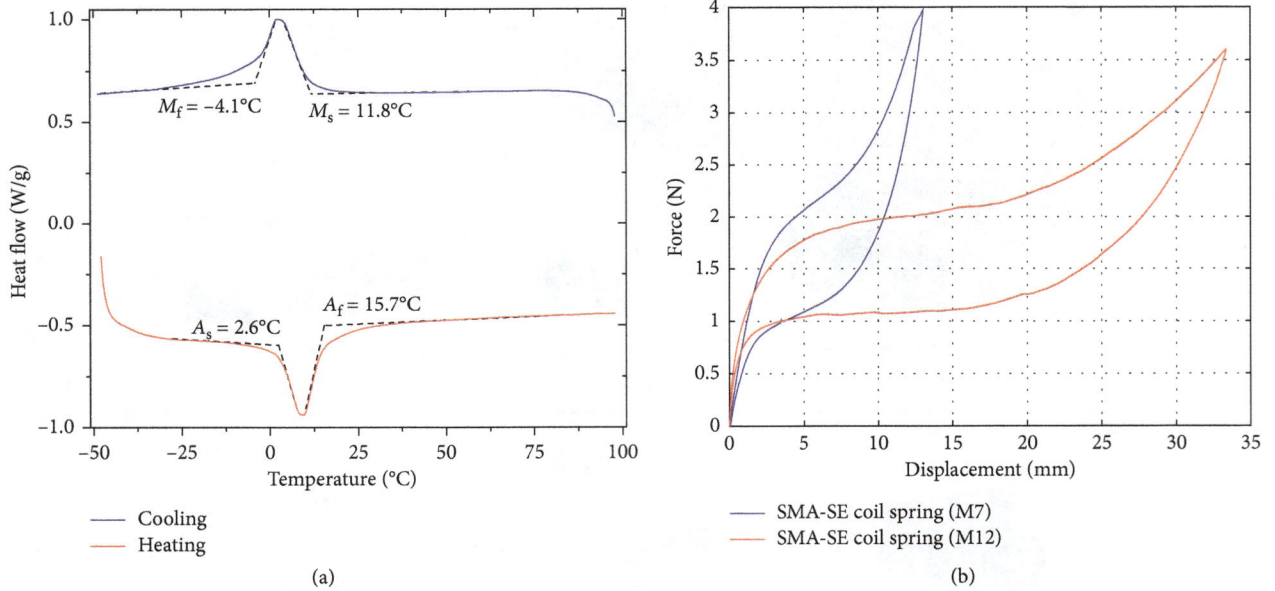

(a)   (b)

FIGURE 10: Characterization of SMA-SE absorbers elements. (a) DSC and (b) dynamical test at room temperature and quasistatic load.

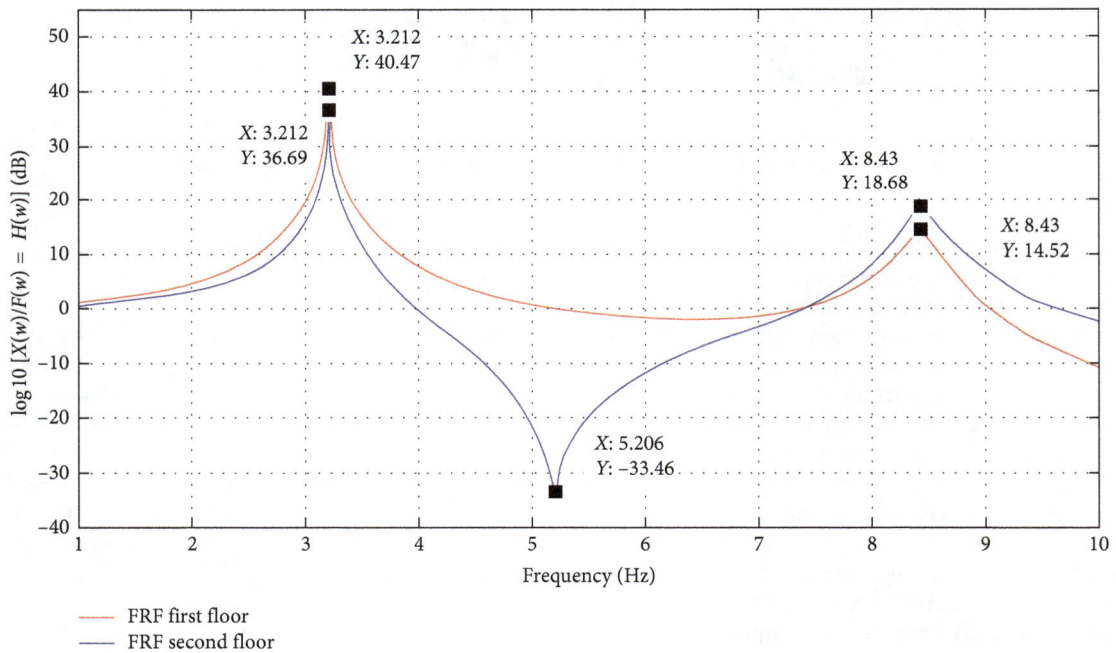

FIGURE 11: FRF analytical obtained by the system transfer functions.

TABLE 4: Natural frequencies of the structure without absorbers (free vibration).

Natural frequencies	A: analytical method (Hz)	B: numerical method (Hz)	C: experimental free vibration (Hz)	Relative error (%) $[(A - C)/A] \times 100$
1st	3.21	3.32	3.08	4.0
2nd	8.43	8.73	8.25	2.1

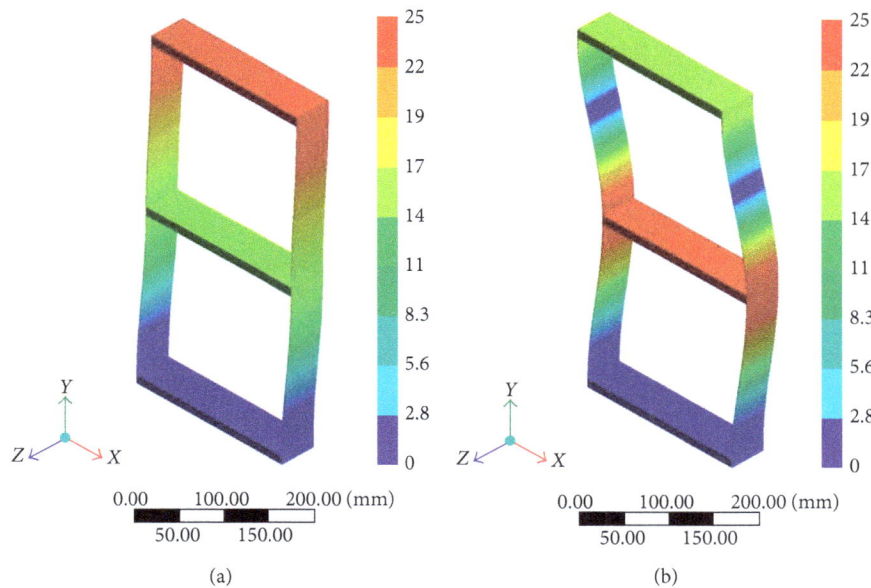

FIGURE 12: Modal shape obtained by numerical simulation. (a) First mode and (b) second mode.

the convergence analysis for the mesh size adopted validates the simulation. The modal shapes related to the two main frequencies in side view and their respective minimum and maximum displacement amplitudes are shown in the color bar. It is possible to notice the displacement imposed to the system, observing the position variations of the structure at rest when compared to the same one deformed, in its modal shapes of vibration.

According to the predicted results (Figure 11), we notice that they are in agreement with the numerical results (Figure 12) regarding natural frequencies, with errors from up the 3%. For the modal shapes, in the analytical response curve (Figure 11), it was observed that the highest transmissibility of displacements occurs on the second floor and the first mode. This fact was also observed in the numerical simulation (Figure 12(a)).

In the second modal shape, the analytical response curve (Figure 11) shows the greatest amplitude of displacement in the first floor of the structure; the same behavior was also obtained for the numerical simulation (Figure 12(b)), in which the largest displacements also occur on the first floor.

3.3. Experimental Results

3.3.1. Free Vibration. For validation of the predicted results, the transient responses of the structure in the free vibration tests (system acceleration responses) and their respective frequency spectrum FFT were obtained. Figure 13 presents the experimental results for the structure without absorbers and with the incorporation of two configurations, steel springs, and SMA-SE coil springs.

Note that a spring element made of steel stores elastic potential energy and returns a portion to the system in the form of kinetic energy. The difference between the stored energy and the energy recovered is due to the structural damping of the spring itself. However, the damping factor for the steel alloys is tiny and varying between 0.001 and 0.008 [32], so that the mechanical energy dissipation of these elements becomes insufficient to reduce the amplitudes of response. On the other hand, SMA-SE materials have a damping factor about 100 times higher than the steel, on the order of 0.1 to 0.2, and are known as Hidamets (*High Dampers Materials*). Therefore, when applying springs manufactured in SMA-SE, the system response suffers a reduction in amplitude [33].

From Figure 13, we can see that the mean peak amplitude achieved for all studied cases was approximately 0.45 g where "g" represents the gravity acceleration. The response in the time domain for the system without absorbers is very close to that obtained with the incorporation of steel springs (with stiffness similar to that for SMA-SE springs). This behavior confirms that elements of this type of material do not dissipate mechanical energy enough. In opposite, for the SMA-SE spring setup, greater energy dissipation is observed when compared to the previous cases.

Figure 14 illustrates the curves FRF of the system. This function was achieved by dividing the acceleration output signal and the force input signal from a predefined function in the dynamic signal analyzer. The response amplitude peaks are denoted by the behavior described in Figure 13. In this figure, we can see larger amplitudes for the system without the presence of actuators or with the incorporation of steel springs, and a considerable and significant reduction thereof in the SMA-SE coil springs configuration.

Table 4 shows the natural frequency values of the three methods of analysis adopted in this work. When the experimental results of the system in free vibration are compared with the analytical method, for the structure without absorbers, a variation of 4.0% is observed for the

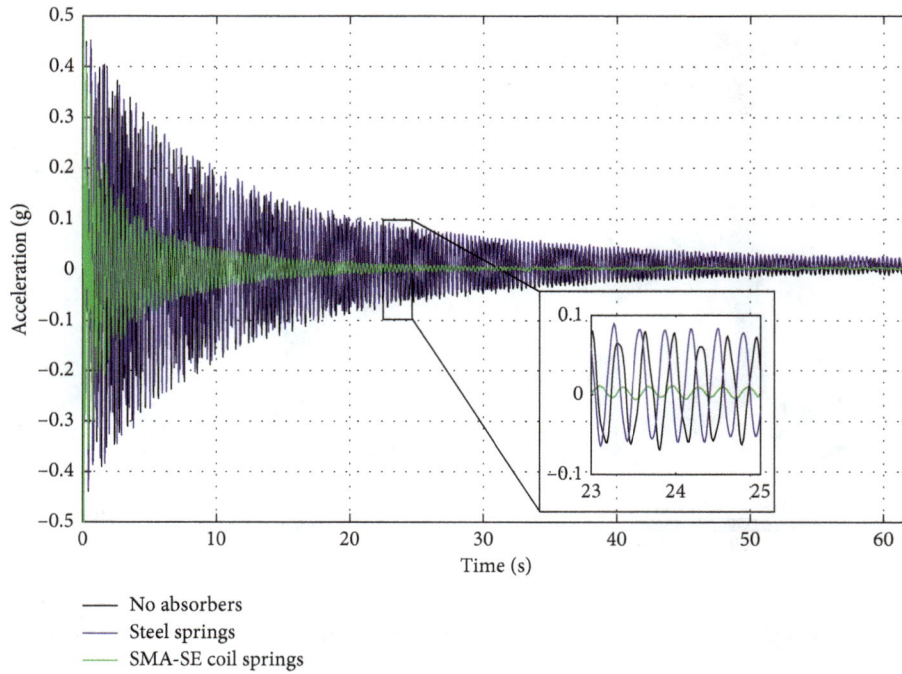

FIGURE 13: Time response for structure subjected to an impact in all configurations.

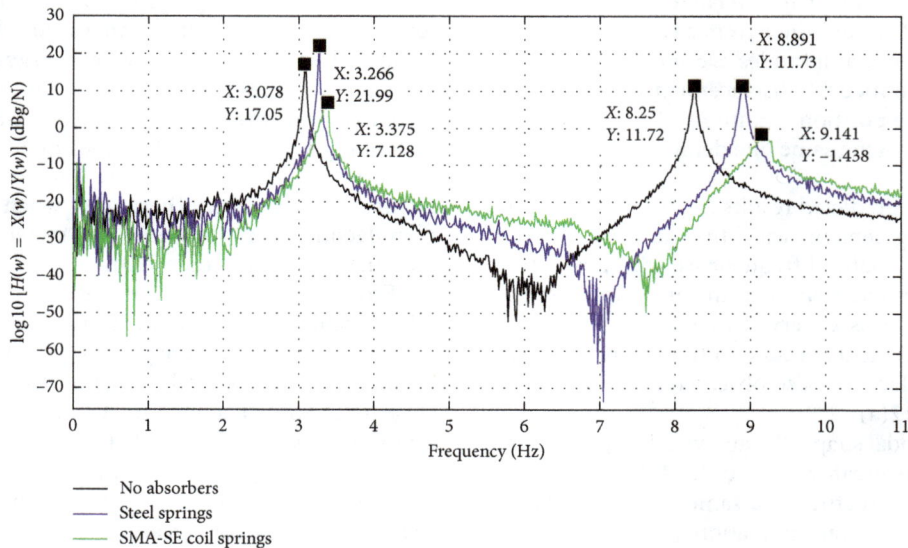

FIGURE 14: Experimental FRF of the system for the configurations adopted in this study.

TABLE 5: Natural frequencies and percentage reduction of amplitudes in free vibration tests.

Configurations	ω_{n1} (Hz)	$H(w)$ (g/N)	%R $[(B-C)/B] \times 100$	ω_{n2} (Hz)	$H(w)$ (g/N)	%R $[(B-C)/B] \times 100$
A: no absorbers	3.08	7.12	—	8.25	3.85	—
B: steel springs	3.26	12.58	—	8.89	3.86	—
C: SMA-SE coil springs	3.37	2.27	82	9.15	0.83	78

first natural frequency and of 2.1% for the second natural frequency.

The reason for this variation can be motivated by the consideration of perfect crimping adopted in the method, a fact that not is verified in the real experimental tests. The bolts make the connection between the floors and the columns, which decreases the stiffness of the model. This fact provokes a decrease in the natural frequencies of the system.

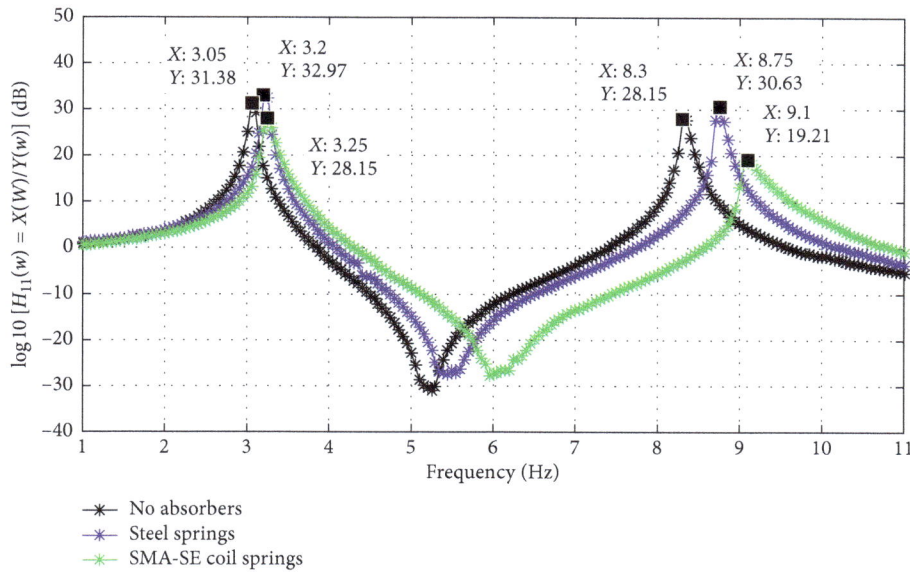

FIGURE 15: Experimental FRF of the first floor of the system.

Table 5 shows the values of the linear acceleration amplitudes. The small variation of natural frequencies of the system in comparison with the structure with steel springs and SMA-SE coil springs is observed. As already expected, an increase in global stiffness causes an increase in the natural frequency. The results of percentage reduction (%R) of acceleration amplitudes confirm the efficiency of the superelastic coil springs in the control of vibrations. In this configuration, reductions of 82% for the first natural frequency and 78% for the second when compared with the structure of steel springs were verified.

If we compare structural damping of the system without absorbers ($\xi = 0.0026$), with the system with steel springs ($\xi = 0.0022$), in both cases, damping factors of the same order of magnitude are observed, which are typical of conventional steel structures.

Concerning the values of attenuation of the of the acceleration signal in time, was verified that when the amplitudes reached a value of 0.03 g, a significant reduction of 72% for the case of the structure with SMA-SE springs compared to the system with steel springs.

The result of adding a steel spring in parallel is one increase in global stiffness. Just see (7) and (8); if "k" increases, the response amplitude increases too. So, add a steel spring without structural damping, and the response amplitude will increase. Also, the inclusion of a steel spring in a structure may characterize the addition of a secondary system with coupling by mass, and the possibility of a resonance phenomenon occurring and consequently the growth of the response amplitude.

3.3.2. *Forced Vibration.* For the experimental tests of the system subjected to forced vibration, the method established previously in the experimental methodology was followed, obtaining the curves of frequency response functions, in all configurations. Figure 15 illustrates the FRF obtained for the first floor.

Analyzing the curves in Figure 15, one can observe the same behavior already evidenced in the free vibration tests. For the case of the structure with steel springs, the displacement transmissibility peaks presented higher values for the two modal shapes. This fact indicates that elastic steel elements do not control or minimize vibrations in structures subjected to permanent excitations.

As expected, the FRF of the system with the SMA-SE coil springs presented a significant reduction in the transmissibility peaks, evidencing the high levels of mechanical energy dissipation. The behavior described here can also be visualized in an analogous way for the second floor of the structure, as evidenced in Figure 16.

Table 6 presents the values of natural frequencies and displacement transmissibility relation of the system, obtained from linear FRF. For the system with the SMA-SE coil springs, reductions in the transmissibility peaks "T_d" up to 51% in the first modal shape and up to 73% for the second shape were obtained.

Regarding the damping factor (ξ), the behavior of the system follows the same pattern perceived in previous analyzes. Table 7 highlights these rates for all modal shapes and their two floors. It seen that the system incorporated with steel springs does not add damping to the structure, presenting mostly values even lower than the structure without absorber. In the case of the system with the SMA-SE coil springs, there is a significant increase in the damping. There is an increase of up to 59% or approximately 1.6 times in the first modal shape and up to 119% or 2.2 times for the second mode.

4. Conclusions

This paper focuses on the passive control of vibrations using different configurations on the structural system: steel

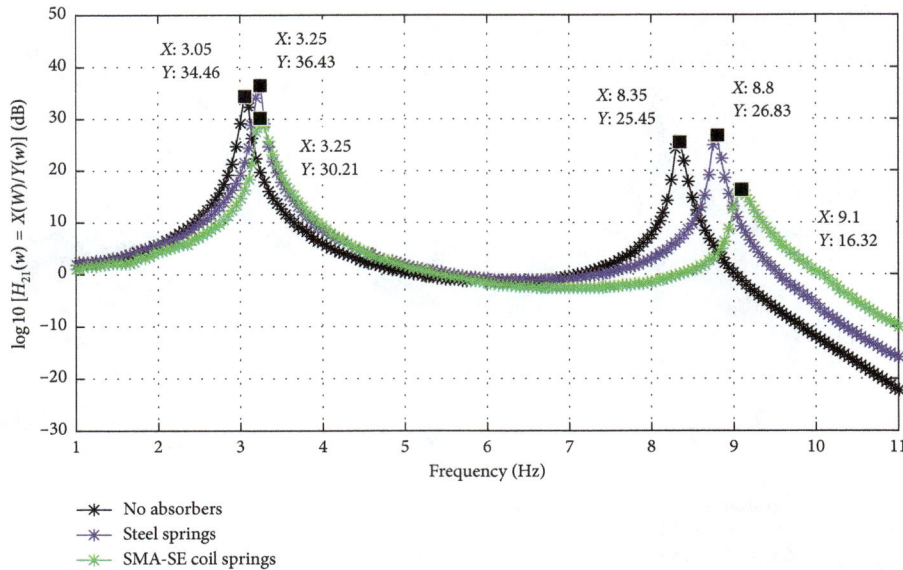

FIGURE 16: Experimental FRF of the second floor of the system.

TABLE 6: Natural frequencies and displacement transmissibility in forced vibration, linear scale.

Configurations	First floor						Second floor					
	ω_{n1} Hz	$H(w)_1$		ω_{n2} Hz	$H(w)_2$		ω_{n1} Hz	$H(w)_1$		ω_{n2} Hz	$H(w)_2$	
		T_d	%R^*		T_d	%R^*		T_d	%R^*		T_d	%R^*
A: no absorbers	3.0	37.1	—	8.3	26	—	3.0	52.8	—	8.3	19	—
B: steel springs	3.2	44.5	—	8.7	34	—	3.2	66.3	—	8.7	22	—
C: SMA-SE coil springs	3.2	25.5	43	9.1	9.1	73	3.2	32.4	51	9.1	6.5	70

%$R^* = [(B-C)/B] \times 100$.

TABLE 7: Structural damping of the system in forced vibration, linear scale.

Configurations	First floor				Second floor			
	ω_{n1} (Hz)		ω_{n2} (Hz)		ω_{n1} (Hz)		ω_{n2} (Hz)	
	ξ_1	%$\widehat{A}^*$	ξ_2	%$\widehat{A}^*$	ξ_1	%$\widehat{A}^*$	ξ_2	%$\widehat{A}^*$
A: no absorbers	0.0133	—	0.0067	—	0.0144	—	0.0065	—
B: steel springs	0.0142	—	0.0053	—	0.0123	—	0.0055	—
C: SMA-SE coil springs	0.0172	21	0.0116	119	0.0195	59	0.0115	109

%$\widehat{A}^* = [(C-B)/B] \times 100$.

springs and SMA-SE coil springs (NiTi Alloys). The results obtained by different methods (analytical, numerical, and experimental) were representative to compare the modal shapes. The proposed models were obtained in a simplified form and without absorber elements: however, the result confirms that the formulations represent the system as well. In the numerical analysis, good results were obtained using the Ansys in comparison with the analytical results, despite the limitation by the nonconsideration of the friction and the contact between the parts of the structure. A small relative variation up to 3.5% was obtained for the modal shapes, validating the modeling.

In an experimental modal analysis for the system in free vibration was verified the efficiency of the configurations that adopted the superelastic springs as a vibration absorber

element, obtaining significant reductions in the acceleration amplitudes, up to 82% for the SMA-SE coil springs configuration. Regarding the signal attenuation time, it observed that, for the structure subjected to a lateral impulse (impact hammer), the reductions were considerable. There was a reduction up to 72% in the SMA-SE coil spring configuration compared to the steel springs.

When analyzed, the system in forced vibration was verified that there was no significant variation in the natural frequencies for the configuration of steel springs when compared with the SMA-SE coil springs. It is perceived a variation of up to 4.6% for modal shapes. Thus, it is correct to say that amplitude reductions occur due to the increase in structural damping and energy dissipation. Regarding the displacement transmissibility, by analyzing the FRF, it can be seen that in comparison with the system with steel springs, the SMA-SE coil springs configuration showed a reduction up to 51% in the first modal shape and up to 73% for the second mode.

The damping factors ranging from (0.001 to 0.08) are within the range of expected values for structural damping. If we compare these factors for the SMA-SE spring configuration, there is an increase of up to 59% in the first modal shape and up to 119% in the second mode, in comparison with the worst case, the configuration of springs of steel. Thus, we stated that the objectives of the work were satisfactorily achieved, validating the use of SMA-SE in the passive control of vibrations in structures subjected to dynamic excitations.

Conflicts of Interest

The authors declare that there are no conflicts of interest regarding the publication of this paper.

Acknowledgments

The authors thank the Federal University of Campina Grande, LVI (*Laboratory of Vibrations and Instrumentation*) and the LaMMEA (*Multidisciplinary Laboratory of Materials and Structures Active*), of the Department of Mechanical Engineering/Federal University of Campina Grande, Brazil, for the field test. This work was supported by the financing granted in the form of scholarship and investment in scientific research fostered by CNPq (*National Council of Research and Development of Brazil*) through the projects 306732/2012-2 and 444039/2014-7.

References

[1] S. S. Rao, *Mechanical Vibrations*, vol. 4, Pearson Prentice Hall, São Paulo, SP, Brazil, 2008, ISBN 978-85-7605-200-5.

[2] A. K. Chopra, Dynamics of Structures, Vol. 2, Pearson Prentice Hall, Upper Saddle River, NJ, USA, 1995.

[3] L. N. F. France and J. Sotelo Jr., *Introduction to Mechanical Vibrations*, vol. 1, Edgard Blücher, São Paulo, SP, Brazil, 2006, ISBN 978-85-2120-338-4, in Portuguese.

[4] A. A. Correia, "System vibrations with 1DOF," in *Dynamics*, Higher Technical Institute, Nicosia, Cyprus, 2007.

[5] M. J. Brennan, "Some recent developments in adaptive tuned vibration absorbers/neutralizers," *Shock and Vibration*, vol. 13, no. 4-5, pp. 531–543.

[6] J. Liu and K. Liu, "A tunable electromagnetic vibration absorber: characterization and application," *Journal of Sound and Vibration*, vol. 295, no. 3–5, pp. 708–724, 2006.

[7] P. A. L. Yánez, "Seismic analysis of buildings by continuous environment technique," Ph.D. thesis, Structural Engineering, School of Engineering of São Carlos, Federal University of São Paulo (USP), São Carlos, SP, USA, 1992, in Portuguese.

[8] J. A. V. Veloso, "Man-induced earthquakes," We off the axis, Brasilia, Brazil, July 2016, in Portuguese, http://www.nosrevista.com.br/2012/09/18/terremotos-induzidos-pelo-homem/.

[9] F. L. B. Saavedra, "Comparative study in seismic analysis of building structures," M.S. dissertation, Civil Engineering Sciences, Federal University of Rio de Janeiro (UFRJ), Rio de Janeiro, RJ, Brazil, 1991, in Portuguese.

[10] M. Dolce and D. Cardone, "Mechanical behavior of shape memory alloys for seismic applications: austenite Ni-Ti wires subjected to tension," *International Journal of Mechanical Sciences*, vol. 43, no. 11, pp. 2657–2677, 2001.

[11] M. Dolce, D. Cardone, and R. Marnetto, "SMA re-centering devices for seismic isolation of civil structures," in *Proceedings of the SPIE*, vol. 4330, pp. 238–249, Chiba, Japan, July 2001.

[12] A. E. V. Lopes, "Seismic risk in Brazil and its impact on great works," *Journal of the Institute of Engineering*, no. 58, 2010, in Portuguese, https://www.institutodeengenharia.org.br/site/2010/05/02/edicao-58/.

[13] A. E. V. Lopes, *Seismic Intensities of Earthquakes: Formulation of Seismic Scenarios in Brazil*, vol. 91, pp. 90–102, USP Magazine, São Paulo, SP, Brazil, 2011, in Portuguese, 2011.

[14] D. Halliday and R. Resnick, *Fundamentals of Physics: Mechanics*, vol. 4, LTC, Rio de Janeiro, RJ, USA, 2008, ISBN 978-85-216-1605-4, in Portuguese.

[15] E. M. Kervin Jr., "Damping of flexural waves by a constrained viscoelastic layer," *Journal of Acoustical Society of America*, vol. 31, no. 7, pp. 952–962, 1959.

[16] D. Ross, E. E. Ungar, and E. M. Kervin Jr., "Damping of plate flexural vibrations by means of viscoelastic layer," in *Structural Damping*, ASME, New York, NY, USA, 1959.

[17] K. Williams, G. Chiu, and R. Bernhard, "Adaptive-passive absorbers using shape memory alloys," *Journal of Sound and Vibration*, vol. 249, no. 5, pp. 835–848, 2002.

[18] S. B. Choi and J. H. Hwang, "Structural vibrations control using shape memory actuators," *Journal of Sound and Vibration*, vol. 231, no. 4, pp. 1168–1174, 2000.

[19] P. Mahmoodi, "Structural dampers," *Journal of Structural Division*, vol. 95, no. 8, pp. 1661–1672, 1969.

[20] L. A. P. Semão, "Use of shape memory alloys in the control of vibrations in intelligent civil engineering structures," M.S. dissertation, Civil Engineering-Structures and Geotechnics, Faculty of Sciences and Technology, Nova de Lisboa University, Lisbon, Portugal, 2010, in Portuguese.

[21] M. Alam, M. Youssef, and M. Nehdi, "Utilizing shape memory alloys to enhance the performance and safety of civil infrastructure: a review," *Canadian Journal of Civil Engineering*, vol. 34, no. 9, pp. 1075–1086, 2007.

[22] T. W. Duerig, K. N. Melton, D. Stöckel, and C. M. Wayman, *Engineering Aspects of Shape Memory Alloys*, Butterworth-Heinemann Ltd, London, UK, 1990, ISBN 0-750-61009-3.

[23] D. C. Lagoudas, *Shape Memory Alloys: Modelling and Engineering Application*, D. C. Lagoudas, Ed., Springer, Houston, TX, USA, 2008.

[24] C. Menna, F. Auricchio, and D. Asprone, "Applications of SMA in structural engineering," in *Shape Memory Alloy Engineering*, Butterworth-einemann Limited, Oxford, UK, 2015, ISBN 978-0-08-099920-3.

[25] F. Auricchio, R. L. Taylor, and J. Lubliner, "Shape-memory alloy: macro modelling and numerical simulations of the superelastic behavior," *Computer Methods Applied Mechanics Engineering*, vol. 146, no. 3-4, pp. 281–312, 1997.

[26] J. Wilson and M. Wesolowsky, "Shape memory alloys for seismic response modification: a state-of-the-art review," *Earthquake Spectra*, vol. 21, no. 2, pp. 569–601, 2005.

[27] L. Lecce and A. Concilio, *Shape Memory Alloy Engineering: For Aerospace, Structural and Biomedical Applications*, Butterworth-Heinemann, Oxford, UK, 2015, ISBN: 978-0-08-099920-3.

[28] L. Delaey, R. V. Krishnan, and H. Tas, "Thermoelasticity, pseudoelasticity, and the memory effects associated with martensitie transforations: structural and microestructural changes associated with the transformations," *Journal of Materials Science*, vol. 9, no. 9, pp. 1521–1535, 1974.

[29] L. Delaey, R. V. Krishnan, and H. Tas, "Thermoelasticity, pseudoelasticity, and the memory effects associated with martensitie transforations: the macroscopic mechanical behavior," *Journal of Materials Science*, vol. 9, no. 9, pp. 1536–1544, 1974.

[30] K. Tanaka, S. Kobayashi, and Y. Sato, "Thermo-mechanics of transformation pseudo-elasticity and shape memory effect in alloys," *International Journal of Plasticity*, vol. 2, no. 1, pp. 59–72, 1986.

[31] Y. J. O. Moraes, "Dynamical analysis applied to control passive of vibrations in a structural device incorporating superelastic NiTi mini coils springs," M.S. thesis, Mechanical Engineering, Federal University of Campina Grande, Campina Grande, PB, Brazil, 2017, in Portuguese.

[32] C. E. Beards, *Structural Vibration: Analysis and Damping*, Halsted Press, New York, NY, USA, 1996.

[33] S.-H. Chang and S.-K. Wu, "Damping characteristics of SMA on their inherent and intrinsic internal friction," in *Handbook of Mechanics of Materials*, C.-H. Hsueh, S. Schmauder, C.-S. Chen et al., Eds., Springer Nature Singapore Pvt. Ltd., Singapore, 2018.

14

Microstructural Changes during High Temperature Service of a Cobalt-Based Superalloy First Stage Nozzle

A. Luna Ramírez,[1] J. Porcayo-Calderon,[2] Z. Mazur,[1] V. M. Salinas-Bravo,[1] and L. Martinez-Gomez[3,4]

[1]*Instituto de Investigaciones Eléctricas, Reforma 113, 62490 Cuernavaca, MOR, Mexico*
[2]*CIICAp, Universidad Autónoma del Estado de Morelos, Avenida Universidad 1001, 62209 Cuernavaca, MOR, Mexico*
[3]*Instituto de Ciencias Físicas, Universidad Nacional Autónoma de México, Avenida Universidad s/n, 62210 Cuernavaca, MOR, Mexico*
[4]*Corrosion y Protección (CyP), Buffon 46, 11590 Ciudad de México, DF, Mexico*

Correspondence should be addressed to J. Porcayo-Calderon; jporcayoc@gmail.com

Academic Editor: Amit Bandyopadhyay

Superalloys are a group of alloys based on nickel, iron, or cobalt, which are used to operate at high temperatures ($T > 540°C$) and in situations involving very high stresses like in gas turbines, particularly in the manufacture of blades, nozzles, combustors, and discs. Besides keeping its high resistance to temperatures which may approach 85% of their melting temperature, these materials have excellent corrosion resistance and oxidation. However, after long service, these components undergo mechanical and microstructural degradation; the latter is considered a major cause for replacement of the main components of gas turbines. After certain operating time, these components are very expensive to replace, so the microstructural analysis is an important tool to determine the mode of microstructure degradation, residual lifetime estimation, and operating temperature and most important to determine the method of rehabilitation for extending its life. Microstructural analysis can avoid catastrophic failures and optimize the operating mode of the turbine. A case study is presented in this paper.

1. Introduction

Gas turbines blades are manufactured mainly with nickel-based and cobalt-based superalloys. During the commercial operation of gas turbines, which are part of a power station, blades and other components of turbine are subject to natural wear and damage due to various causes which can interrupt continuous operation. The source of damage may be metallurgical or mechanical and is manifested in the equipment operation such as a decrease in the availability, reliability, and performance and an increase in the risk of failure. Also, after a prolonged service, moving blades and nozzles show a decrease in metallurgical characteristics, so the creep strength, fatigue, impact, and corrosion resistance decrease. There are different factors which influence lifetime of the main components of a gas turbine including design and operating conditions, but it is the latter that has an impact on the lifetime of these components. Generally, for most gas

turbines, operating conditions are very severe. The following factors have great effect: operation environment (high temperatures, fuel and air contamination, solid particles, etc.), high mechanical stresses (due to centrifugal forces, vibratory, and flexural stresses, etc.), and high thermal stresses (due to thermal gradients).

The phenomena described above do not operate in isolation; typically there are two or more factors being active simultaneously, causing reduction of blades or nozzle lifetime under the following damage mechanism [1, 2], that is, creep, thermal fatigue (low cycle fatigue), thermomechanical fatigue (high cycle fatigue), corrosion and oxidation, erosion, or foreign object damage (FOD).

The type of damage or degradation which occurs in gas turbine blades and nozzles after prolonged service mainly includes external surfaces damage (corrosion, oxidation, cracks, foreign object damage, erosion, and fretting) and internal damage of microstructure, such as γ' phase

154

Alloys: Metallurgy and Engineering

coarsening, grain growth, micro void growth in grain boundaries, carbide precipitation, and brittle phase formation.

Surface damage produces dimensional deterioration, generating loss of the blade/nozzle original dimension, resulting in increased stress and turbine efficiency reduction. During operation, the material microstructure is affected by high temperature combined with high stresses. However, the extent of deterioration differs due to the following factors: total service time and operation history (number of startups, shutdowns, and trips), gas turbine operation condition (temperature, rotational speed, mode of operation, i.e., base load or cyclic duty), and manufacturing alterations (grain size, porosity, alloy composition, heat treatment).

Then, a brief description of the Ni-base and Co-base base superalloys is given. These alloys are used in the manufacture of critical turbine components (moving blades, nozzles, combustors, and transition ducts) of stationary gas turbines. Fe based or Ni-Fe based superalloys are not mentioned because their use in gas turbine critical components is not common.

Ni-Based Alloys. The nickel-base alloys are the more complex and the most widely used for the hottest components of gas turbines (e.g., gas turbine first stage blades). In the heat treated condition superalloys represent a composite material consisting of several intermetallic phases linked by a metal matrix. The major phases present in these superalloys are [3] as follows: gamma matrix (γ), Ni-based austenitic phase (FCC), usually containing a high percentage of solid solution elements such as Co, Cr, Mo, and W; gamma prime (γ'), which is $Ni_3(Al, Ti)$ based intermetallic phase; Carbides, generally types M_6C and $M_{23}C_6$ which tend to precipitate into grain boundaries; topologically closed packed (TCP) type phases, such as σ, μ, and Laves, which precipitate after prolonged high temperature service.

These alloys can be classified into solid-solution hardened alloys and precipitation hardened alloys or gamma prime (γ') alloys. The former may be forged or cast, contain few elements forming γ' particles, but are hardened by refractory elements such as tungsten and molybdenum and carbide formation, and also contain Cr to impart corrosion resistance (oxidation) and Co to give microstructural stability. Precipitation hardened alloys can also be forged or cast. In addition to γ' particle formation as the main hardening mechanism also incorporates elements such as tungsten (W), molybdenum (Mo), tantalum (Ta), and niobium (Nb).

Co-Based Alloys. Cobalt-based superalloys (e.g., X 40, X 45, and FSX-414) are primarily used in the manufacture of all first stage nozzles and in some turbines are used in the last stage due to their good weldability and hot corrosion resistance. These alloys have higher strength at high temperatures than Ni-based alloys and also have excellent resistance to thermal fatigue, oxidation, and corrosion [4]. These alloys have cobalt as the principal alloying element, with significant amounts of nickel and chromium and smaller amounts of tungsten and molybdenum, niobium, tantalum, and sometimes iron. They are mainly hardened by carbide precipitation. Alloys hardening by carbide precipitation contain between 0.4 and 0.85% carbon. Such superalloys consist of an austenitic matrix (fcc)

and a variety of precipitated phases such as primary carbides (M_3C_2, M_7C_3, and MC) and coarse carbides ($M_{23}C_6$) and GCP types phases (geometrically compact phases) such as γ'' and η (Ni_3Al) and TCP (topologically close packed) type phases ($\sigma, \mu, R,$ or L) $(Cr, Mo)_x(Ni, Co)_y$ [5].

2. Superalloys Microstructural Degradation during High Temperature Service

There are several microstructural degradation mechanisms occurring in superalloys used in the manufacture of hot section components of gas turbines (nozzles, moving blades, and combustion chambers). The most common degradation mechanisms are aging and coarsening of the γ'-phase, transgranular precipitation growth of carbides in grain boundaries, brittle phases precipitation, and growth of cavities due to creep.

Coarsening and Aging of γ' Phase. The size and shape of the γ' phase in nickel-based superalloys are not stable after long periods of operation at high temperatures. However, after a heat treatment, this phase is very near to equilibrium with the γ matrix and therefore there is little additional precipitation or growth of this phase from the supersaturated matrix. Nevertheless some particles may grow by a diffusion mechanism [6]; that is, the average particle radius increases with aging time, t. This is represented by the following equation:

$$\bar{r}^3(t) - \bar{r}^3(0) = Kt, \qquad (1)$$

where $\bar{r}(0)$ is the average radius of the particle at $t = 0$, $\bar{r}(t)$ is the average radius of the particle in time t, and K is the kinetic constant which depends on temperature. Various studies [7, 8] on γ' growth phase in Ni-base superalloys and Fe-Ni-Al alloys have confirmed that growth obeys the law described in (1).

Changes in morphology of the γ' phase modify the mechanical properties of the material's component, since phase γ' is intended to act as a barrier to dislocation movement slowing creep; consequently resistance to this failure mechanism is greatly diminished [9]. In commercial superalloys, the γ' phase changes from spherical to cuboidal shape, although most of the particles have an intermediate form. Aging is revealed as an increase in average particle size. The γ' phase can be identified in the microstructure as particles whose shape is irregular and larger [10, 11]. The shape of this phase in a nondegraded and degraded condition can be observed in Figures 1(a) and 1(b).

Morphology and Degeneration of MC, $M_{23}C_6$, and M_6C Carbides. The role of carbides in superalloys is complex; carbides seem to prefer the grain boundaries as a site location in Ni-base superalloys, while in Co-base and Fe-base superalloys appear to precipitate intragranularly [3]. The most common carbides in all Ni, Fe-Ni, and Co-base superalloys are basic MC, $M_{23}C_6$, and M_6C and seldom M_7C_3 [12]. The most stable carbide found in Ni-base and Co-base superalloys is the MC type, where M represents the element Ti. A fraction

(a)   (b)

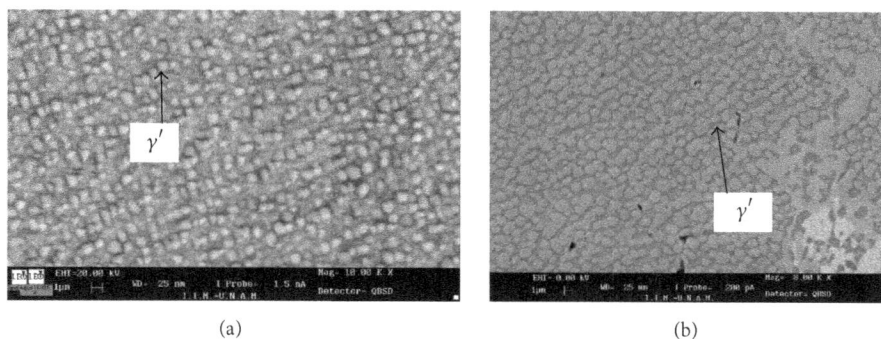

FIGURE 1: Blade root micrograph of (a) gamma prime (γ') without degradation and (b) gamma prime (γ') with moderate degradation after 24000 h in service, IN738LC superalloy.

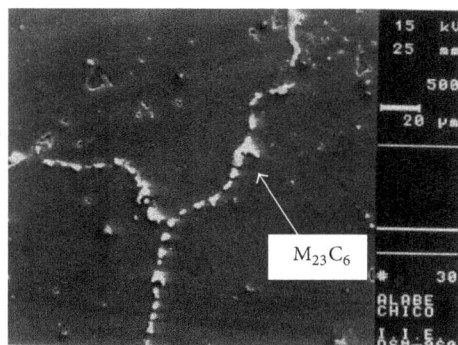

FIGURE 2: $M_{23}C_6$ carbides precipitated in grain boundaries, IN738LC superalloy, after 24000 h of service.

of Ti can be replaced by Nb, Ta, W, and Cr, depending on the alloy composition. In Co-based superalloys containing W, WC carbide is dominant [13]. This carbide generally has a pseudo cubic or script shaped figure; it precipitates as discrete particles distributed heterogeneously through the alloy in intragranular or transgranular locations. The source of carbon needed in the heat treatment of these alloys is taken from the WC. In the course of a prolonged service, MC primary carbides decompose into secondary carbides rich in chromium ($M_{23}C_6$). MC carbide decomposition occurs by diffusion of carbon into the γ matrix and γ' phase, resulting in the formation of $M_{23}C_6$ carbides near the matrix-interface [14], as shown in Figure 2. The MC decomposition can be stated by the reaction:

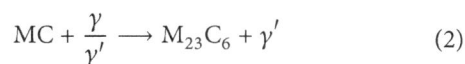

$$MC + \frac{\gamma}{\gamma'} \longrightarrow M_{23}C_6 + \gamma' \qquad (2)$$

or

$$(Ti, Mo)\, C + (Ni, Cr, Al, Ti)$$
$$\longrightarrow Cr_{21}Mo_2C_6 + Ni_3\,(Al, Ti) \qquad (3)$$

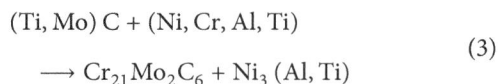

The above reaction occurs at a temperature of approximately 980°C (1800°F) but has also been observed at a temperature of about 760°C (1400°F) [15, 16]. The $M_{23}C_6$ carbide has a significant effect on the superalloys properties. Its critical

location (grain boundaries) increases the rupture strength inhibiting grain boundary sliding. However, failure by break may be originated by fracture of these particles.

Phase Topologically Close Packed (TCP). Superalloys have high levels of refractory elements such as Mo, W, Re, Ru, and Ta, in order to increase creep and crack resistance [17, 18]. These elements function as solid-solution enhancers of both the γ matrix and the γ' phase. Re is a strong hardener; it precipitates mainly in the γ matrix and apparently slows degeneration of the γ' phase. High amounts of refractory elements make the superalloy prone to form TCP phases, the σ phase being the most common in Ni-base superalloys [19]. It has been shown that the formation of these phases has a detrimental effect on the creep rupture life of superalloys. These phases increase the strain rate of both conventional and single crystal superalloys [20, 21]. Other detrimental effects on superalloys are a decrease in ductility, impact resistance, and thermal fatigue.

3. Case Study: Degradation in Service of a Gas Turbine First Stage Nozzle Segment

The nozzle segment (the complete wheel comprises 16 segments with two blades per segment) of the first stage of a gas turbine serves to rotate and direct the flow of hot gas to the rotating turbine with the most favorable incident angle. There is no centrifugal load on the nozzle segment. The combination of bending loads a thermal gradient caused by cooling of the nozzle results in high stationary operating stresses on the nozzle [22]. The first stage nozzle may experience damage mechanisms such as creep, fatigue-creep, oxidation, corrosion, and mechanical damage during its service life [23]. The microstructural evaluation is one of the most important tools in assessing the current condition of the nozzle segment for its correlation with the service conditions experienced by the component. The microstructural evaluation can point out strategies for repair and/or heat treatments for rejuvenation and recover the mechanical properties and extend the useful life of the alloy.

The evaluated component is a segment of the first stage nozzle of a 60 MW gas turbine; gas inlet temperature to

(a)

(b)

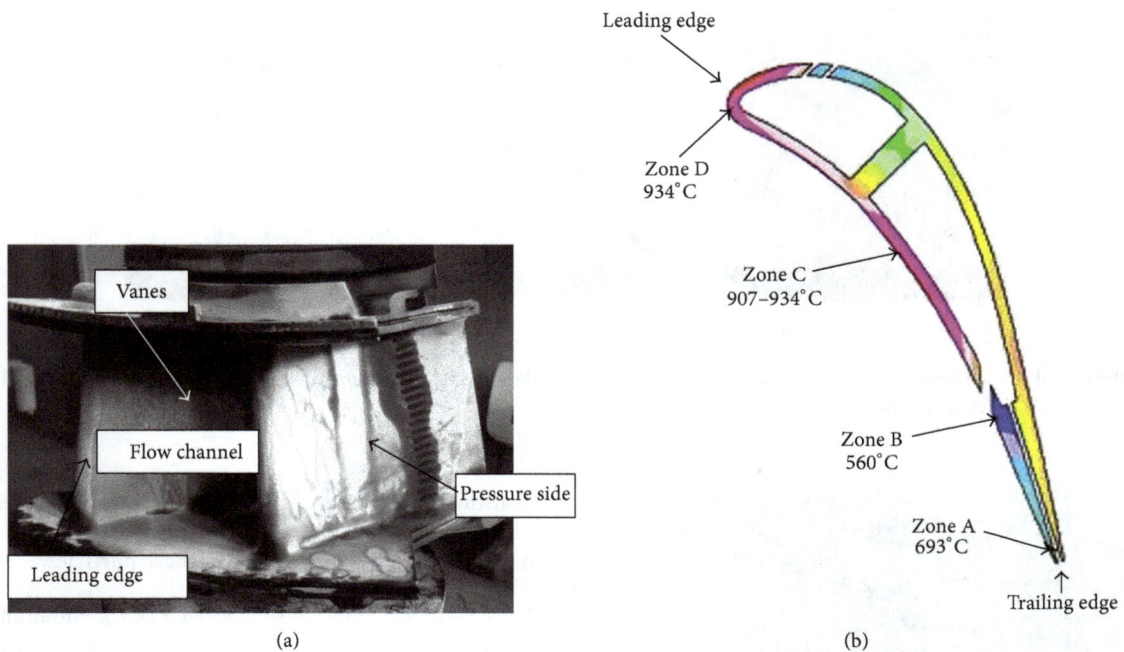

FIGURE 3: (a) General view of the nozzle vane. (b) Analysis regions and internal temperature distribution on the nozzle vane transversal section in the cutting plane at 50% height (section of maximum temperature).

TABLE 1: Chemical composition of FSX-414 superalloy (wt%).

Alloy	C	Cr	Ni	Co	W	Fe	B
FSX-414	0.25	29	10	52	7.5	1.0	0.01

the turbine is 1086°C. The full nozzle consists of 32 vanes and is cooled by air extracted from the compressor discharge. The microstructural evaluation was performed after 54,000 operating hours in mode of base load. The nozzle is made of a conventional cobalt-based FSX-414 superalloy by means of conventional investment casting (equiaxial grains) and without coating; its chemical composition is shown in Table 1. The gas turbine operates with natural gas. An overview of the nozzle segment (two vanes) is shown in Figure 3(a); the vanes have cooling passages on the pressure surface, and, in Figure 3(b), the different operating temperature zones are indicated on a section of the nozzle block. The maximum service temperature (Figure 3(b)) is recorded in the leading edge of the nozzle blade and the temperature distribution was obtained by numerical analysis using Computational Fluid Dynamics (CFD) with the code Star CD V 3150 [24]. Figure 4 shows some cracks detected near the cooling cavities of the nozzle, close to the trailing edge.

Microstructural Characterization of Nozzle Blade. The nozzle microstructure was evaluated at a zone corresponding to a height of 50% of the flow channel on the low and high temperature section. The characterization included grain size and carbide precipitation. In order to evaluate the extent of damage in the superalloy, the microstructure in the low temperature zone (zone B) was compared with the high

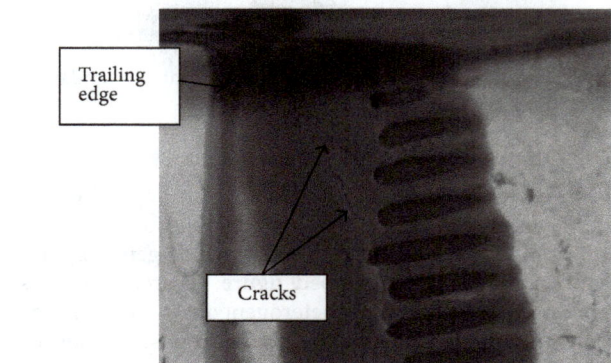

FIGURE 4: Cracks on the nozzle vane near the internal shroud close to the trailing edge.

temperature region (zone D). The microstructure of the low temperature zone can be taken as a reference or initial condition of the alloy, because at that temperature microstructural changes are insignificant. The microstructure of the low temperature zone is shown in Figure 5; this consists of equiaxed grains of the γ phase matrix (Figure 5(a)) and at higher magnification (Figure 5(b)) dispersed carbide particles in the grain boundaries and matrix can be observed. Figure 5(c) shows the unit area quantized to determine the percentage of precipitates. This microstructure is characteristic of cobalt-base superalloys [25–28].

Table 2 shows the grain size and the volume fraction at the different zones in the nozzle pressure side in cross section. Volumetric fraction of carbides in each area was determined

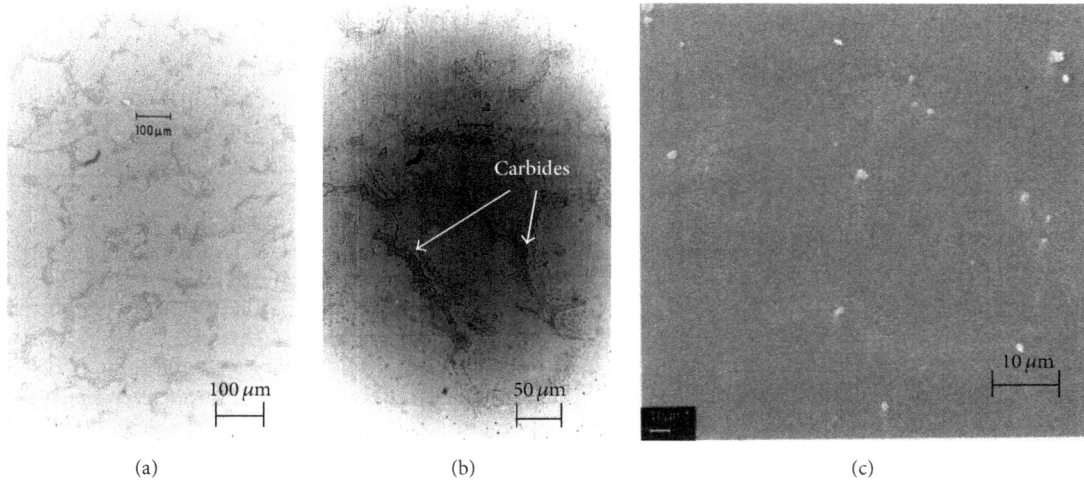

FIGURE 5: (a) General microstructure, (b) magnified microstructure of the low temperature zone (zone B), and (c) minor precipitation of carbide particles.

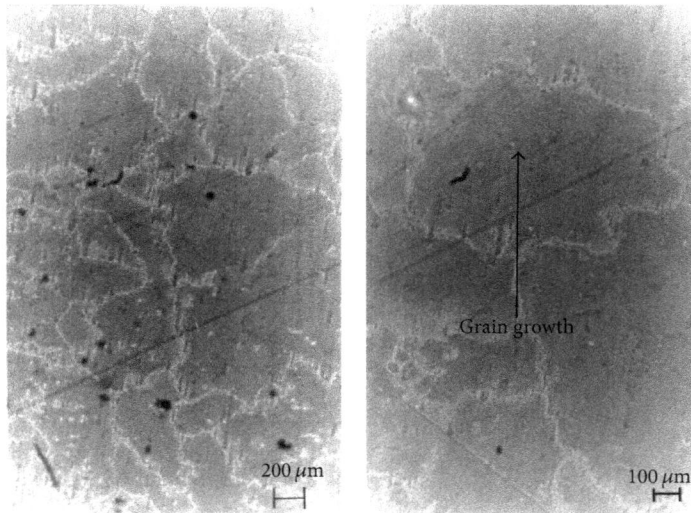

FIGURE 6: Grain growth in the high temperature (zone D).

TABLE 2: Quantitative microstructure in different zones of the nozzle vanes.

Microstructural parameter	Zone A	Zone B	Zone C	Zone D
	Temperature [°C]			
	693	560	907	934
Average grain size [μm]	313	54	401	531
Volume fraction of carbides [%]	7.36	0.72	10.22	12.96

taking into account the area ratio of carbides in μm^2/total measured area also in μm^2.

As shown in Table 2, the extent of deterioration of the superalloy (grain growth and higher amount of precipitates)

depends directly on the metal temperature. The micrographs in Figure 6 show a larger grain size for the area where the temperature is high (zone D), and the micrograph of Figure 7 shows a higher amount of precipitates (area D) compared to the "cold" or reference; see Figure 5. The grain size ratio between area A (693°C) and the high temperature zone D (934°C) is 0.6 and the grain size ratio between the reference area and the high temperature area is 0.1. The growth of grain size (coarse growth) is one of the main symptoms of microstructural worsening of nozzle's material. This is explained because the material is exposed to gas at high temperature and velocity.

The nozzle microstructural investigation revealed the presence of a continuous band of precipitated carbides in grain boundaries and a rise in the volume fraction of carbides up to 50%. This occurs because of the transformation of M_6C carbides to $M_{23}C_6$ carbides. Carbide transformation is

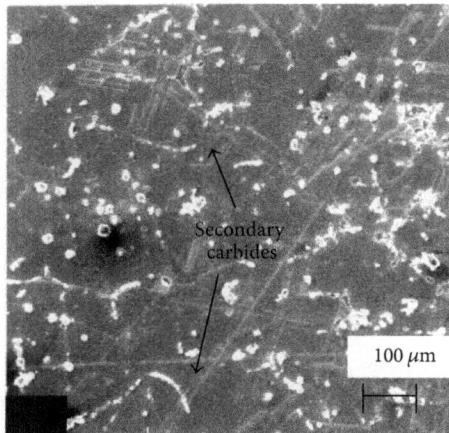

FIGURE 7: Precipitated carbides in grain boundaries of FSX-414 superalloy at intermediate temperature zone (area D).

encouraged by the high operating temperature of the nozzle, mainly at the blade leading edge; the latter type of secondary carbides takes place abundantly in the Co-base superalloys with more than 5% Cr [29]. Precipitation of these secondary phases in grain boundaries reduces material creep resistance.

This dense and continuous network of carbides observed mainly in the area D reduces the ductility and toughness of the alloy by up to 30% of its initial value and may facilitate initiation and propagation of cracks due to grain boundaries brittleness; all this leads to decreasing the useful life of the alloy. Additionally, such grain boundary precipitation reduces the material creep resistance [1, 27]. The microstructural characterization of FSX-414 superalloy revealed that the grain size increased considerably; see Figure 6. This may also reduce the material fatigue life [1, 30]. Also, the average grain size increment in the vane body reduces alloy fatigue life [1].

An analysis of thermal stress was performed which is not indicated in this work, but the results showed that the maximum tension stresses at steady-state were located near the cooling holes and blade profile on the pressure side of the nozzle.

In addition, because the gas turbine nozzle is a fixed component, its operational stresses are generated only by the gas flow pressure and by thermal loads due to temperature gradients through the nozzle elements. These stresses and temperature gradients cause fatigue damage during transient and steady-state operation; this thermal stress induces the initiation and propagation of cracks.

From the metallurgical evaluation carried out and cracks detected by nondestructive testing, the nozzle segment analyzed is a candidate for repair. It is noteworthy that there are virtually no limits for their rehabilitation by welding, because this component remains fixed during operation and is not exposed to concentrated mechanical stresses caused by the centrifugal force.

Repair may include use of conventional welding and/or brazing, subsequently applying a postweld heat treatment including a solution heat treatment at a temperature of 1150°C followed by rapid cooling and then an aging cycle at a

temperature of 980°C followed by cooling. In the event that the nozzle blocks have no coating and in order to decrease the effect of elevated temperature on the microstructure of the blades, the use of a thermal barrier coating (TBC) should be considered to improve corrosion and oxidation resistance.

4. Conclusion

A microstructural study to determine the extent of damage in terms of microstructural deterioration can be used to identify and therefore determine the type of repair (heat treatment, welding, conventional or brazing) to the nozzle block segment. Comparing deterioration parameters discussed above and the temperature distribution over the nozzle block, a direct relationship between the magnitude of damage of the superalloy and the metal temperature can be established. Therefore, metallurgical analysis of main components of a gas turbine is a very useful tool that provides information needed to make decisions about the possibility of repair, establish risk of fracturing and evaluate the operating conditions of the equipment. Consequently, metallurgical characterization should be incorporated into maintenance schedules. It is noteworthy to mention that any metallographic study should be complemented by a stress and temperature distribution analysis in order to corroborate or determine the nozzle failure mechanism.

Competing Interests

The authors declare that there is no conflict of interests regarding the publication of this paper.

Acknowledgments

Financial support from Consejo Nacional de Ciencia y Tecnología (CONACYT, Mexico) (Projects 196205, 159898, and 159913) is gratefully acknowledged.

References

[1] Z. Mazur, V. M. Cristalinas Navarro, A. Hernández Rossette et al., "Desarrollo de herramientas para predicción de vida útil residual de álabes de turbinas de gas," Internal Report II/43/11887/002/P/07, IIE, Cuernavaca, Mexico, 2003 (Spanish).

[2] X. Wu, W. Beres, and S. Yandt, "Challenges in life prediction of gas turbine critical components," Canadian Aeronautics and Space Journal, vol. 54, no. 2, pp. 31–39, 2008.

[3] C. T. Sims, N. Stoloff, and W. C. Hagel, Superalloys II: High-Temperature Materials for Aerospace and Industrial Power, Wiley-Interscience, New York, NY, USA, 1987.

[4] N. R. Muktinutalapati, "Materials for Gas Turbine," http://www.intechopen.com/download/pdf/22905.

[5] H. L. Bernstein, "Materials issues for users of gas turbine," in Proceedings of the 27th Texas AM Turbomachinery Symposium, College Station, Tex, USA, September 1998.

[6] A. Baldan, "Progress in Ostwald ripening theories and their applications to nickel-base superalloys. Part I: Ostwald ripening theories," Journal of Materials Science, vol. 37, no. 11, pp. 2171–2202, 2002.

[7] H. Li, L. Zuo, X. Song, Y. Wang, and G. Chen, "Coarsening behavior of γ particles in a nickel-base superalloy," *Rare Metals*, vol. 28, no. 2, pp. 197–201, 2009.

[8] H. J. Dorantes-Rosales, N. Cayetano-Castro, J. J. Cruz- Rivera, V. M. López-Hirata, JL. González-Velázquez, and J. Morena-Palmerín, "Cinética de engrosamiento de precipitados coherentes en aleaciones base Hierro," *Suplemento de la Revista Latinoamericana de Metalurgia y Materiales*, SI 2, pp. 637–645, 2009.

[9] J. Liburdi, P. Lowden, D. Nagy, T. R. De Priamus, and S. Shaw, "Practical experience with the development of superalloy rejuvenation," in *Proceeding of the ASME Turbo Expo, Power for Land, Sea and Air (GT '09)*, pp. 819–827, Orlando, Fla, USA, June 2009.

[10] P. S. Kotval, "The microstructure of superalloys," *Metallography*, vol. 1, no. 3-4, pp. 251–285, 1969.

[11] Z. Mazur, A. Luna-Ramírez, J. A. Juárez-Islas, and A. Campos-Amezcua, "Failure analysis of a gas turbine blade made of Inconel 738LC alloy," *Engineering Failure Analysis*, vol. 12, no. 3, pp. 474–486, 2005.

[12] Z. G. Yang, J. W. Stevenson, D. M. Paxton, P. Singh, and K. S. Weil, "Materials properties database for selection of high-temperature alloys and concepts of alloy design for sofc applications," Tech. Rep. PNNL-14116, U.S. Department of Energy under Contract DE-AC06-76RL01830, 2002.

[13] G. P. Sabol and R. Stickler, "Microstructure of nickel-based superalloys," *Phys Status Solidi*, vol. 35, no. 1, pp. 11–52, 1969.

[14] G. Lvov, V. I. Levit, and M. J. Kaufman, "Mechanism of primary MC carbide decomposition in Ni-base superalloys," *Metallurgical and Materials Transactions A: Physical Metallurgy and Materials Science*, vol. 35, no. 6, pp. 1669–1679, 2004.

[15] H. Kitaguchi, "Microstructure-property relationship in advanced Ni-based superalloys," in *Metallurgy—Advances in Materials and Processes*, Y. Pardhi, Ed., chapter 2, InTech, Rijeka, Croatia, 2012.

[16] V. P. Swaminathan and N. S. Cheruvu, *Condition and Remaining Life of Hot Section Turbine Components Is Essential to Insure Reliability*, Materials Engineering Department, Gas Turbine Materials, 1997.

[17] Z.-X. Shi, J.-R. Li, S.-Z. Liu, X.-G. Wang, and X.-D. Yue, "Effect of Ru on stress rupture properties of nickel-based single crystal superalloy at high temperature," *Transactions of Nonferrous Metals Society of China*, vol. 22, no. 9, pp. 2106–2111, 2012.

[18] A. C. Yeh and S. Tin, "Effects of Ru on the high-temperature phase stability of Ni-base single-crystal superalloys," *Metallurgical and Materials Transactions A: Physical Metallurgy and Materials Science*, vol. 37, no. 9, pp. 2621–2631, 2006.

[19] R. Darolia, D. F. Lahrman, and R. D. Field, "Formation of topologically closed packed phases in nickel-base single crystal superalloys," in *Superalloys 1988*, S. Reichman, D. N. Duhl, G. Maurer, S. Antolovich, and C. Lund, Eds., The Metallurgical Society, 1988.

[20] A. Volek, R. F. Singer, R. Buergel, J. Grossmann, and Y. Wang, "Influence of topologically closed packed phase formation on creep rupture life of directionally solidified nickel-base superalloys," *Metallurgical and Materials Transactions A: Physical Metallurgy and Materials Science*, vol. 37, no. 2, pp. 405–410, 2006.

[21] M. V. Acharya and G. E. Fuchs, "The effect of long-term thermal exposures on the microstructure and properties of CMSX-10 single crystal Ni-base superalloys," *Materials Science and Engineering A*, vol. 381, no. 1-2, pp. 143–153, 2004.

[22] R. Viswanathan, *Damage Mechanisms and Life Assessment of High Temperature Components*, ASM International, Materials Park, Ohio, USA, 1989.

[23] T. J. Carter, "Common failures in gas turbine blades," *Engineering Failure Analysis*, vol. 12, no. 2, pp. 237–247, 2005.

[24] *STAR-CD Version 3.15A, Methodology*, Computational Dynamics, 2002.

[25] H. Kazempour-Liacy, S. Abouali, and M. Akbari-Garakani, "Failure analysis of a repaired gas turbine nozzle," *Engineering Failure Analysis*, vol. 18, no. 1, pp. 510–516, 2011.

[26] R. Bakhtiari and A. Ekrami, "Microstructural changes of FSX-414 superalloy during TLP bonding," *Defect and Diffusion Forum*, vol. 312–315, pp. 399–404, 2011.

[27] H. L. Bernstein, R. C. McClung, T. R. Sharron, and J. M. Allen, "Analysis of general electric model 7001 first-stage nozzle cracking," *Journal of Engineering for Gas Turbines and Power*, vol. 116, no. 1, pp. 207–216, 1994.

[28] G. M. Shejale, "Metallurgical evaluation and condition assessment of FSX 414 nozzle segments in gas turbines by metallographic methods," *Journal of Engineering for Gas Turbines and Power*, vol. 133, no. 7, Article ID 072102, 2011.

[29] W. H. Jiang, X. D. Yao, H. R. Guan, Z. Q. Hu, and W. H. Jiang, "Secondary carbide precipitation in a directionally solified cobalt-base superalloy," *Metallurgical and Materials Transactions A*, vol. 30, no. 3, pp. 513–520, 1999.

[30] J. A. Daleo, K. A. Ellison, and D. H. Boone, "Metallurgical considerations for life assessment and the safe refurbishment and requalification of gas turbine blades," *Journal of Engineering for Gas Turbines and Power*, vol. 124, no. 3, pp. 571–579, 2002.

15

Numerical and Experimental Analyses of the Effect of Heat Treatments on the Phase Stability of Inconel 792

Maria M. Cueto-Rodriguez,[1] **Erika O. Avila-Davila** (ID),[1] **Victor M. Lopez-Hirata** (ID),[2] **Maribel L. Saucedo-Muñoz,**[2] **Luis M. Palacios-Pineda** (ID),[1] **Luis G. Trapaga-Martinez,**[3] **and Juan M. Alvarado-Orozco**[4]

[1]*Tecnológico Nacional de México/Instituto Tecnológico de Pachuca (DEPI), Pachuca de Soto, Hgo. 42080, Mexico*
[2]*Instituto Politécnico Nacional (ESIQIE), UPALM, Ciudad de México 07300, Mexico*
[3]*CIATEQ-Posgrado en Manufactura Avanzada, Querétaro, Qro. 76150, Mexico*
[4]*Centro de Ingeniería y Desarrollo Industrial (CIDESI), Querétaro, Qro. 76130, Mexico*

Correspondence should be addressed to Erika O. Avila-Davila; osirisavila77@yahoo.com.mx

Academic Editor: Amelia Almeida

A study about the precipitation and phase stability was carried out in an IN-792 superalloy used as a blade in a gas turbine. Microstructure analysis was conducted experimentally on three different cross sections of the blade designated as high temperature (HT), medium temperature (MT), and low temperature (LT). To identify the HT, MT, and LT sections, a numerical thermal analysis was performed using ANSYS software. To obtain the distribution gradient of temperature in the blade, the real conditions of operation in steady state of the gas turbine were considered. A numerical study about the occurrence of phases in the IN-792 superalloy was carried out with Thermo-Calc and TC-PRISMA software. The analysis of the as-cast IN-792 superalloy with Scheil-Gulliver equations permitted to explain the phase formation during the solidification process. The calculated time-temperature-precipitation (TTP) diagram explains consistently the precipitation process observed after two different heat treatment conditions applied experimentally and numerically to regenerate the original microstructure of the IN-792 superalloy. The experimental results were consistent with the calculated isoplethic and TTP diagrams. In terms of accuracy, the further development of the Thermo-Calc databases for thermodynamic calculations in superalloys is evident. It was possible to calculate precipitation temperatures and the local evolution of precipitated particles for two different heat treatment conditions.

1. Introduction

The Ni-based superalloys have excellent mechanical properties under high-load bearing at temperatures up to approximately 85% of their incipient melting point, as well as, good environmental resistance and metallurgical stability under service conditions from about 813 K (540°C) and, in some cases, up to 1473 K (1200°C) [1]. These superalloys have been used since the early 1940s to manufacture diverse structural components of aerospace and land-based gas turbine engines exposed to the highest temperatures during the operation stages (e.g., combustor, turbine arrangements, and in the final high pressure of the compressor) [2].

The current performance of gas turbine engines is the result of continuous improvements in different areas of engineering, including alloy design, casting technology, and coating methods [3–5]. Critical rotary and stationary components in current gas turbine engines are mainly manufactured by directional solidification via columnar-crystal-structure or single-crystal technology in order to reduce or remove the grain boundaries, introduce a preferred grain orientation, and hence, improve its mechanical properties [2]. The remarkable mechanical properties of Ni-based superalloys are attributable to their precise chemical balance that promotes a coherent precipitation of the intermetallic compounds, γ' ($L1_2 - Ni3Al$) and/or

γ'' (D0$_{22}$ – Ni$_3$Nb), in a matrix phase, γ(A1 – Ni). The precipitates of γ' are usually aligned along the elastically softest crystallographic <100> direction [2]. The mechanical strength of γ' reinforced Ni-based superalloys is maintained at high temperatures due to γ' coarsening resistance. This characteristic is favored by a low value of lattice misfit between the γ and γ' phases, which affects the interfacial energy between them [6].

The precipitation hardening of as-cast superalloys is generally carried out by a controlled heat treatment which may consist of a homogenization period followed by several aging steps. This treatment causes commonly a bimodal size distribution of γ' precipitates [2, 7, 8]. Additionally, it is known that superalloys are a good example of the more complex metallurgical systems due to the large number of alloying elements that compose them and promote the formation of other phases during its heat treatment and/or in-service operation. This fact makes difficult to determine what phases are stable after a heat treatment or during the component operation. For example, a study of the thermodynamic equilibrium for an IN-792 superalloy, using Thermo-Calc version S, showed the occurrence of γ, γ', M$_{23}$C$_6$ and σ phases in the calculated isothermal diagram at 973 K (700°C) [9]. It was concluded that even though the calculations predicted both σ-phase and M$_{23}$C$_6$ carbides to be thermodynamically stable, the experimental results showed occurrence of M$_{23}$C$_6$ carbides without σ-phase [9]. It is important to mention that an essential difference between this study and the analysis in [9] is the Thermo-Calc database used for the thermodynamic calculations. Another difference is that the blade in this study has an improved engineering design and cooling channels [2, 3]. This technological improvement increases the efficiency of the gas turbine and suggests that the blade operates under severe conditions of service, such as higher temperature [2]. Thus, the microstructural stability, which is related to the mechanical integrity of this component, acquires major technological and scientific interest. So, several experimental and theoretical efforts have been made to predict or estimate the phase evolution of superalloys under different aging conditions [7, 10–12]. Other phases like the topologically packed phases (TCP) can frequently appear into the microstructure of Ni-based superalloys. The TCP phases contain excessive amounts of Cr, W, and Mo that promote the precipitation of intermetallic phases enriched with these elements [2]. These phases have a high atomic density and some degree of nonmetallic behavior with directional bonding, as well as complex crystalline structures. The TCP phases have the general chemical formula A$_x$B$_y$, where A and B are transition metals. Some examples are the μ, σ, R, and P phases [2]. Thus, TCP phases are incoherent with the γ matrix phase and therefore do not contribute significantly to the mechanical strength. Additionally, different carbides and borides may be formed during processing or service, and their type and structure (i.e., MC, M$_6$C, M$_{23}$C$_6$, M$_7$C$_3$, and M$_3$B) depend on the alloy composition and thermal history to which Ni-based superalloy was subjected [13]. The continuous formation of these phases is recommended to be avoided on the grain boundaries, although there is little information about it. Their presence in these regions is due to an inadequate choice of processing temperatures or heat treatment of the superalloy to obtain the best mechanical properties [2, 11, 14, 15]. Then, the study of phases formation and their evolution is an important issue to be investigated for Ni-based superalloys. Recently, the computational thermodynamic, based on CALPHAD methodology, has emerged as an important alternative to analyze the phase stability and growth kinetics of precipitation in alloys. For instance, Thermo-Calc and TC-PRISMA have been employed to analyze the phase transformations of alloys [9,16–18]. Thus, the purpose of present work is to study experimentally and numerically the precipitation, coarsening process, and stability of the formed phases for an IN-792 superalloy. The analyzed blade was used in the first stage of a land-based gas turbine, and after, heat treatments were applied to regenerate the precipitation process in order to study their effect on the superalloy hardness.

2. Experimental Procedure

The IN-792 superalloy was obtained from a gas turbine first-stage blade, with a length of about 103 mm over the blade root, after 12,000 h of service. Chemical composition of the superalloy was determined by an atomic absorption spectrometer, Varian Spectr AA-220 FS. Microstructural characterization was performed at three transversal sections along the blade length, designated as high temperature HT~1211 K (938°C), medium temperature MT~1147 K (874°C), and low temperature LT~1091 K (818°C) located at 50, 90, and 10 mm from the blade root, respectively. Specimens were prepared metallographically and then etched with FeCl$_3$: 5 g, HCl: 2 mL, and ethanol: 100 mL [19] to be observed by optical microscopy (OM), with an Axio Observer D1m (Carl Zeiss) and by scanning electron microscopy (SEM), with a Jeol Thermo Scientific JSM-6300 at 20 kV, equipped with EDX analysis. Phase characterization was pursued by X-ray diffraction (XRD) with copper K$_\alpha$ radiation, with a Bruker D8 FOCUS X-ray diffractometer. Quantitative metallography of microstructure was conducted by image analysis using ImageJ software. Vickers microhardness was determined using a load of 50 g for 12 s.

Reheat treatments of IN-792 superalloy specimens were carried out only in the blade root considering, from the results of design and structural analysis of gas turbine blades, that this section is exposed to the lowest values of total deformation [20–22]. The conditions for heat treatments were as follows [23]: first, heating at 1393 K (1120°C) for 4h with an electric tubular furnace, then air cooling. Next heating at 1393 K (1120°C) for 4h followed by rapid air cooling until 1353 K (1080°C) and then aging at this temperature for 4h and final aging at 1118 K (845°C) for 24h followed by air cooling. Heat-treated specimens were previously encapsulated in quartz tubes, first evacuated to 10^{-2} Pa and then filled with pure Ar gas. These specimens were also characterized by OM, SEM, XRD, and Vickers microhardness test.

3. Numerical Procedure

3.1. Fluid Flow and Thermal Analysis. First, a computational fluid dynamics (CFD) calculation was carried out to obtain the temperature produced by the gas flow on the blade surface [24]. Then, these results were used as boundary conditions in a conduction heat transfer analysis to determine the temperature distribution inside of the blade. All computations were performed at steady state condition using the finite volume method and finite element method (FEM) implemented in ANSYS software. The CFD domain was considered between the end of the first stage nozzle and the beginning of the second stage nozzle, with the following boundary conditions: inlet and outlet static pressure of 1.43 MPa and 1.02 MPa, respectively; inlet and outlet flow temperature of 1316 K (1043°C) and 1135 K (862°C), respectively; and, a mass flow of 2.11 kg/s. It is important to mention that the cooling flow was also taken into account and the cooling flow inlet velocity was 256.4 m/s at 612 K (339°C). The high pressure turbine was rotating at an angular speed of 8405 rpm. The discretized CFD domain contained 287638 vertexes and 281100 cells with 7.82×10^{-4} m of minimum edge length. Additionally, the heat transfer domain by conduction was discretized with 32777 nonlinear elements [24, 25]. The condition of ceramic coating on the blade was not considered in this study. The attention was focused on the temperature distribution in three cross sections located at different blade heights (bh), measured from its platform. The first one was located near the platform (bh = 10 mm), the second one was situated at midspan (bh = 50 mm), and the third one was positioned near the blade tip (bh = 90 mm).

3.2. Thermo-Calc Analysis. Thermo-Calc, TC software 2016[a] version 8.0, [26] was used to determine the equilibrium and nonequilibrium phases for the IN-792 superalloy. The single point calculation, phase diagram, and property diagram modules were utilized to analyze the equilibrium phases based on the chemical composition of the alloy, temperature, and thermodynamic database TCNi8: Ni-Alloys v8.0 and MOBNi: Ni-Alloys v8.0. In contrast, the Scheil module was employed to determine the nonequilibrium phases expected for as-cast IN-792 superalloy. This module is based on the solution of Scheil-Gulliver equation [27]. The analysis of growth kinetic for precipitation was conducted using TC-PRISMA during aging of the IN-792 superalloy. The TC-PRISMA precipitation module [28] is based on Langer and Schwartz (LS) theory and uses the Kampmann and Wagner (KW) numerical method [29] to simulate the concomitant nucleation, growth, and coarsening of precipitates for multicomponent and multiphase alloy systems. The solution of equations for the LS theory enables to calculate the time evolution of the particle size distribution, mean radius, and the number density. The simplified growth model was used in precipitation simulations. The initial chemical composition for the precipitation analysis was assumed to be that corresponding to the maximum microsegregation of the γ phase dendritic structure calculated with TC using the Scheil-Gulliver equation. The concentrations of most of the alloying elements were slightly lower than their corresponding concentrations in the superalloy. The kinetic and thermodynamic data were acquired from the TC databases for Ni-based alloys [30]. The molar volumes of the matrix and precipitated phases for the superalloy were also calculated from the TC databases. The interface energy between the γ', $M_{23}C_6$, μ precipitates, and γ matrix phase were calculated by TC-PRISMA. Homogeneous nucleation was assumed for the precipitation simulation in the austenite matrix, which is referred as "bulk nucleation" in TC-PRISMA.

4. Results and Discussion

4.1. Thermal Analysis. Figures 1(a)–1(c) show the temperature distribution in the blade, obtained from FEM calculations. Frequently, these studies only show the temperature distribution on the blade surface [5, 22, 31, 32]. In this case, it was taken into account the cross section to analyze in order to obtain a good reference of the service temperature. The microstructural characterization was performed in specific locations, indicated with black square points, for high temperature (HT), medium temperature (MT), and low temperature (LT). The section with higher gradient of temperature corresponds to the section near the blade root where temperatures from 1168 K (895°C) to 994 K (721°C) were obtained (Figure 1(c)). However, it is clear that the highest temperature in this section is located in a small area near the trailing edge. Thus, the highest gradient of temperature in the blade can be considered that of the section at midspan (Figure 1(b)), where a substantial area of the section is exposed to a major gradient of temperature, from 1211 K (938°C) to 1053 K (780°C). It is important to mention that the operating parameters and material properties used in FEM simulations correspond to actual conditions of the gas turbine [25]. The substrate temperatures obtained by this computation are close to those obtained in similar works [5, 22, 32].

4.2. Thermodynamic Analysis of IN-792 Superalloy. The Thermo-Calc-calculated isoplethic phase diagram, plot of temperature (K) versus Al composition (wt.%), is shown in Figure 2 for the IN-792 superalloy. The real chemical composition is shown in Table 1.

The current Al content is about 3.2 wt.%, indicated in Figure 2 with a dotted line. If a very slow cooling process is analyzed for an IN-792 alloy with this aluminum content, from the liquid state, at 1643 K (1370°C) approximately, the first phases to be formed are carbides of the MC type and the γ matrix phase. It is well known that the Ni-based superalloys are characterized by having a hardening γ' phase [2], which is also observed in the diagram over 1373 K (1100°C). As the temperature decreases other phases, such as borides, carbides of $M_{23}C_6$ type, TCP phases, and Ni_3Ti phase, are formed. In addition, it is important to note that there is no a previous report about a pseudobinary phase diagram of the IN-792 superalloy determined by numerical simulation. All

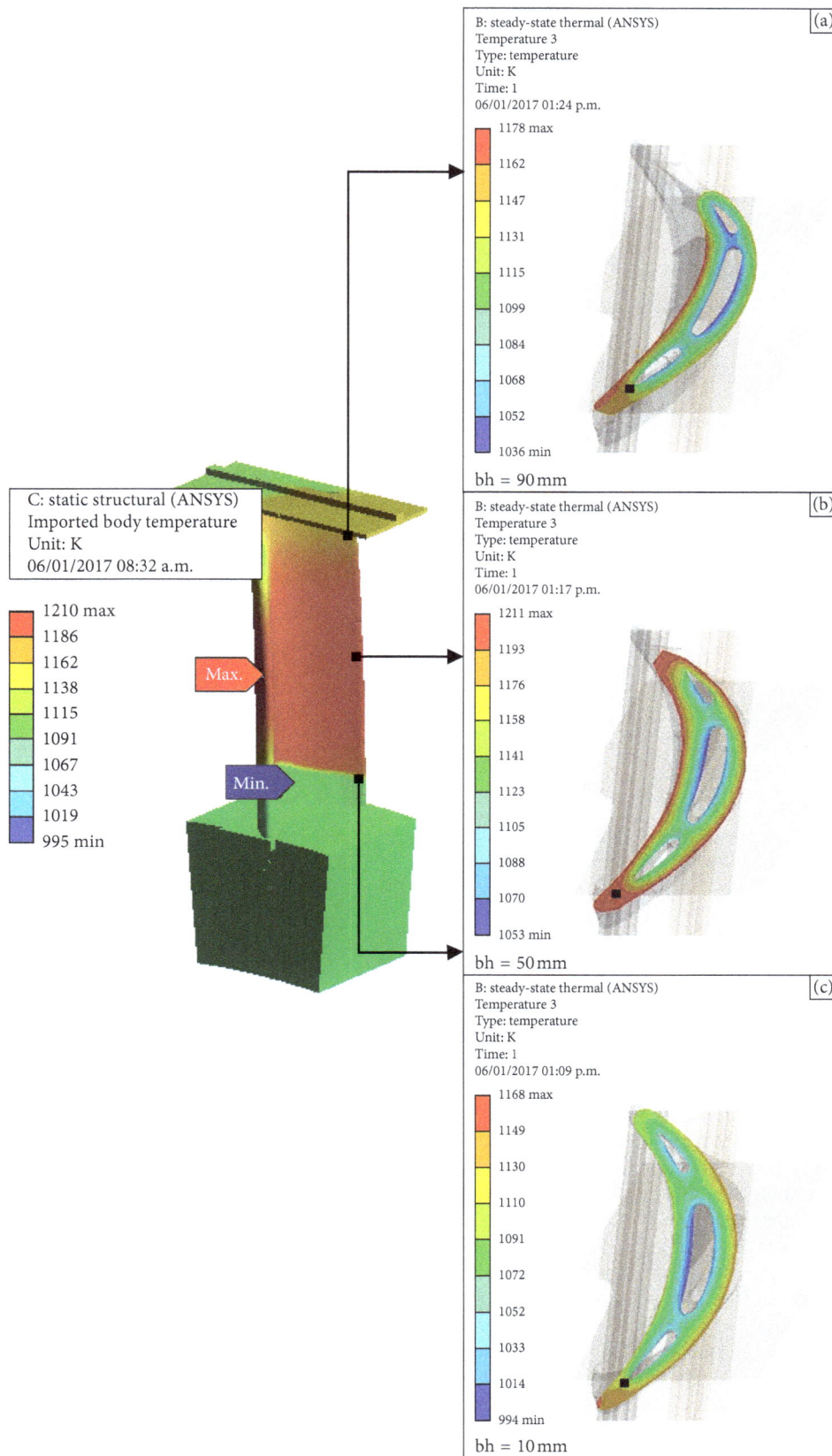

FIGURE 1: Calculated temperature distribution gradients of the blade for sections: (a) MT, (b) HT, and (c) LT.

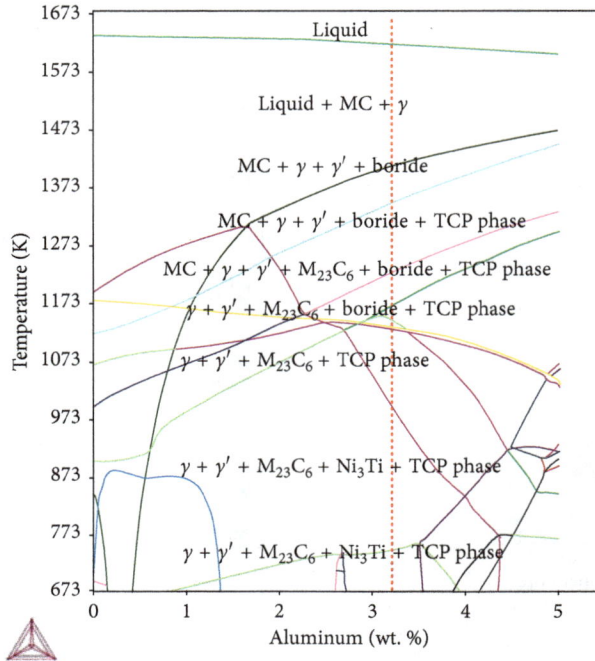

FIGURE 2: Isoplethic Ni-Al phase diagram of IN-792 superalloy.

TABLE 1: Real chemical composition (wt. %) of IN-792 superalloy.

Ni	Cr	Co	W	Al	Ti	Ta	Mo	Hf	Fe	C	Zr	V	Nb
Balance	11.7	9.0	3.5	3.2	3.8	3.14	2.0	0.49	0.26	0.15	0.05	0.05	0.01

phases shown in Figure 1, obtained by Thermo-Calc, are consistent with those reported in the literature for this type of Ni-based superalloys [2, 33]; however, their occurrence will depend on the real chemical composition of the superalloy.

4.3. Phase Analysis of As-Cast IN-792 Superalloy by Scheil-Gulliver Equation. Figure 3 shows the plot of temperature (K) versus mol fraction of solid determined using the Thermo-Calc Scheil module. This diagram represents an approximation of the solidification in nonequilibrium state based on the Scheil-Gulliver model. This model assumes a solidification process with no diffusion in the solid phase, a perfect mixing in the liquid phase (infinitely rapid diffusion in the liquid phase), and local equilibrium at the liquid/solid interface. It shows the phases in as-cast state for the IN-792 superalloy. Here, it is clear that the first solids to be formed are carbides of the MC type and γ matrix phase. According to the diagram, they start to be formed at 1623 K (1350°C), approximately. These results are similar to that obtained in [33] where temperatures of phase transformations were investigated in a cast polycrystal nickel alloy IN-792-5A experimentally, using the method of differential thermal analysis (DTA) under a cooling rate of 283 K (10°C)/min [33]. The hardening phase, $\gamma' - Ni_3Al$, and the η-Ni_3Ti phase appear near 1473 K (1200°C). It is important to say that η phase was suggested experimentally into the interval of temperatures of γ' precipitation [33]. Then, it was impossible

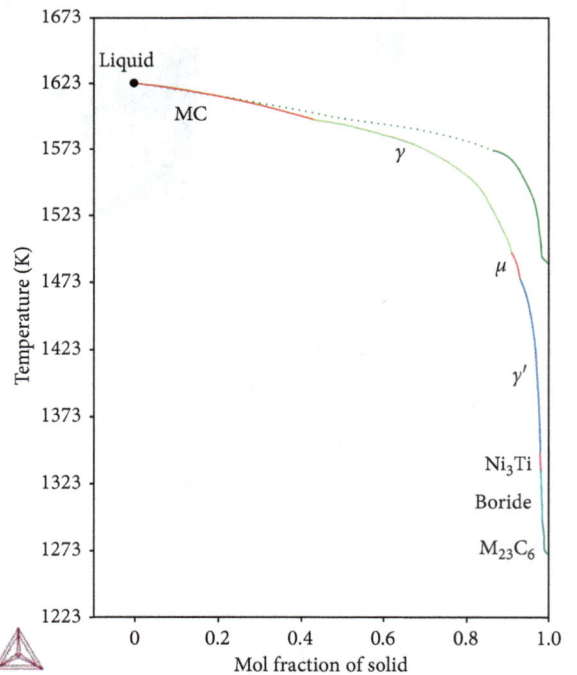

FIGURE 3: Temperature versus mol fraction of solid for the IN-792 superalloy.

to separate those thermal effects by DTA analysis and determine unequivocally temperatures of start and end of dissolution and precipitation of γ' and η phases. In this case,

Figure 3, it is possible to mention that γ' and η phases start their precipitation at 1477 K (1204°C) and at 1347 K (1074°C), respectively, and by comparison with [33], there is a difference of 279 K (6°C) and 278 K (5°C), respectively. The γ and γ' phases have been reported to be present as a result of a eutectic reaction, also. The γ' phase formed on this way is designated as primary γ' phase [2, 4]. The formation of μ phase in some Ni-based superalloys is unavoidable. It was shown that the μ phase can be formed in cast Ni-base superalloys when the amount of refractory elements, tungsten and molybdenum, is more than 3.5, and in these cases the μ phase will be the main TCP phase instead of the σ-phase [14]. So, the IN-792 superalloy of this study has a content of tungsten and molybdenum of about 5.5. The μ phase is based on the ideal stoichiometry A_6B_7. Examples of this phase are W_6Co_7 and Mo_6Co_7 [2]. Also, it has been determined that a primary μ phase may form during solidification, and secondary μ phase can precipitate during aged heat treatment, or during service temperature in a range since 1073 K (800°C) to 1413 K (1140°C) in the Ni-based superalloys [14]. At the end of the Scheil diagram, some borides and carbides, of the $M_{23}C_6$ type, are formed. These results are in agreement with those reported in the literature [2]. Besides, when the C content is in the range of 0.05 to 0.14 wt.% in a superalloy, fine dispersed MC and $M_{23}C_6$ carbides could be formed [13]. This particular type of carbides strongly influences the mechanical properties of Ni-based superalloys; for example, it is possible to improve the strength of this material by a controlled amount of particles of these carbides that promote an apparent inhibition of grain boundary slip. However, a failure in the superalloy eventually could occur due to a fracture in the particle or by decohesion of the carbide/matrix interface. The $M_{23}C_6$ carbides can also cause premature failure, but this behavior can be avoided by mean of a controlled heat treatment in the superalloy [2].

The Thermo-Calc-calculated chemical composition of the as-cast γ and γ' phases can be seen in Table 2. It is important to observe the proportion of the elements in the γ matrix phase. As expected, nickel is the element contained in major proportion, about 65.51 wt.%, and decreases to 60.02 wt.%. This behavior is similar to the aluminum content that decreases from 3.64 to 2.59 wt.%. Other elements are also present in γ phase, but they increase in content, chromium from 12.11 to 12.58 wt.% and cobalt from 7.88 to 10.14 wt.%. Besides, it is shown in Table 2, the nickel and aluminum content for γ' phase increase with respect to γ matrix phase from 62.19 to 65.83 and from 2.83 to 3.74 wt.%, respectively. The other elements of the γ' phase decrease in content for: chromium from 12.63 to 12.09 wt.% and cobalt from 9.98 to 8.5 wt.%. It is clear that the compositional ranges are similar between the phases γ and γ'. On the other hand, it is evident a good agreement of the chemical composition values of the elements shown in Table 2 with these values obtained experimentally for the real chemical composition shown in Table 1. Also, in Table 2, it can be observed that some elements promote a preferred substitution for nickel, like cobalt, or some elements promote the formation of γ' phase, like aluminum and nickel. In both cases, the values of

chemical composition for γ and γ' coincide with that reported in literature [2].

An EDX analysis was carried out by SEM to the sample exposed at MT, near of leading edge and of suction side of the blade (Figure 4). The spectrums obtained are shown only for the points 3-4 y 7-8, for simplicity. It is clear that the values of chemical composition are very closer between γ and γ' phases. However, it can be observed a larger content of nickel and aluminum for γ' phase than for γ matrix phase. This behavior is observed in an element-imaging SEM analysis of the IN-792 superalloy for the same section of the blade, MT (Figure 5). So, the experimental results are in agreement with the Thermo-Calc-calculated chemical composition of the as-cast γ and γ' phases.

4.4. Precipitation Analysis by TC-PRISMA. Figure 6 shows the time-temperature-precipitation (TTP) diagram for the IN-792 superalloy, determined by TC-PRISMA software. This diagram was calculated considering the chemical composition of as-cast γ' phase, shown in Table 2, and the following interfacial energy: 0.047, 0.17, 0.20, and 0.19 J/m^2 for the interfaces γ'/γ, μ/γ, $M_{23}C_6/\gamma$, and Ni_3Ti/γ, respectively. There is not a previous investigation about the TTP diagram for IN-792 superalloy, which is an important contribution because it is of great interest for the industry. The TTP diagram indicates that the isothermal aging at temperatures between approximately 1198 K (925°C) and 1323 K (1050°C) causes the following precipitation reaction:

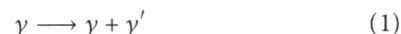

$$\gamma \longrightarrow \gamma + \gamma' \qquad (1)$$

It is important to notice that the precipitation of γ' phase occurs very fast. This precipitate is known as secondary γ' precipitates.

In contrast, the precipitation sequence for aging at temperatures of about 873 K (600°C) and 1173 K (900°C) is as follows:

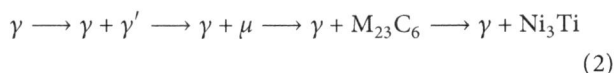

$$\gamma \longrightarrow \gamma + \gamma' \longrightarrow \gamma + \mu \longrightarrow \gamma + M_{23}C_6 \longrightarrow \gamma + Ni_3Ti \qquad (2)$$

That is, γ' phase is the first precipitate to be formed and then μ, $M_{23}C_6$, and Ni_3Ti or η phases appear successively as the aging time progresses.

It is important to mention that it was determined that η phase is found at the grain boundaries in Ni-based alloys when the Ti/Al ratios are higher than about 2.64 and after prolonged exposure at temperatures higher than 1073 K (800°C) [34]. It is considered as deleterious phase because it usually poses negative impact on the mechanical properties of superalloys [34]. The Ti/Al ratio in this superalloy is of about 1.19. Thus, it would be not a surprise in this study not to observe this structure in the samples reheat treated to regenerate the precipitation process. But, the formation of η phase in the blade used, after 12,000 h of service, could be possible.

The occurrence of $M_{23}C_6$ carbides were not experimentally detected in samples of IN-792-5A superalloy [33]. So, it would be not a surprise if this structure is not present in the samples re-heat-treated of this study due to the

TABLE 2: Thermo-Calc-calculated chemical composition (wt. %) of γ and γ' phases for as-cast Ni-based superalloy.

Phase	Ni	Cr	Co	W	Al	Ti	Ta	Mo	C	Nb
γ	65.51–60.02	12.11–12.56	7.88–10.14	2.89–4.70	3.64–2.59	2.06–4.72	2.08–4.69	2.05–2.88	0.01–0.042	0.005–0.0007
γ'	62.19–65.83	12.63–12.09	9.96–8.50		2.83–3.74					

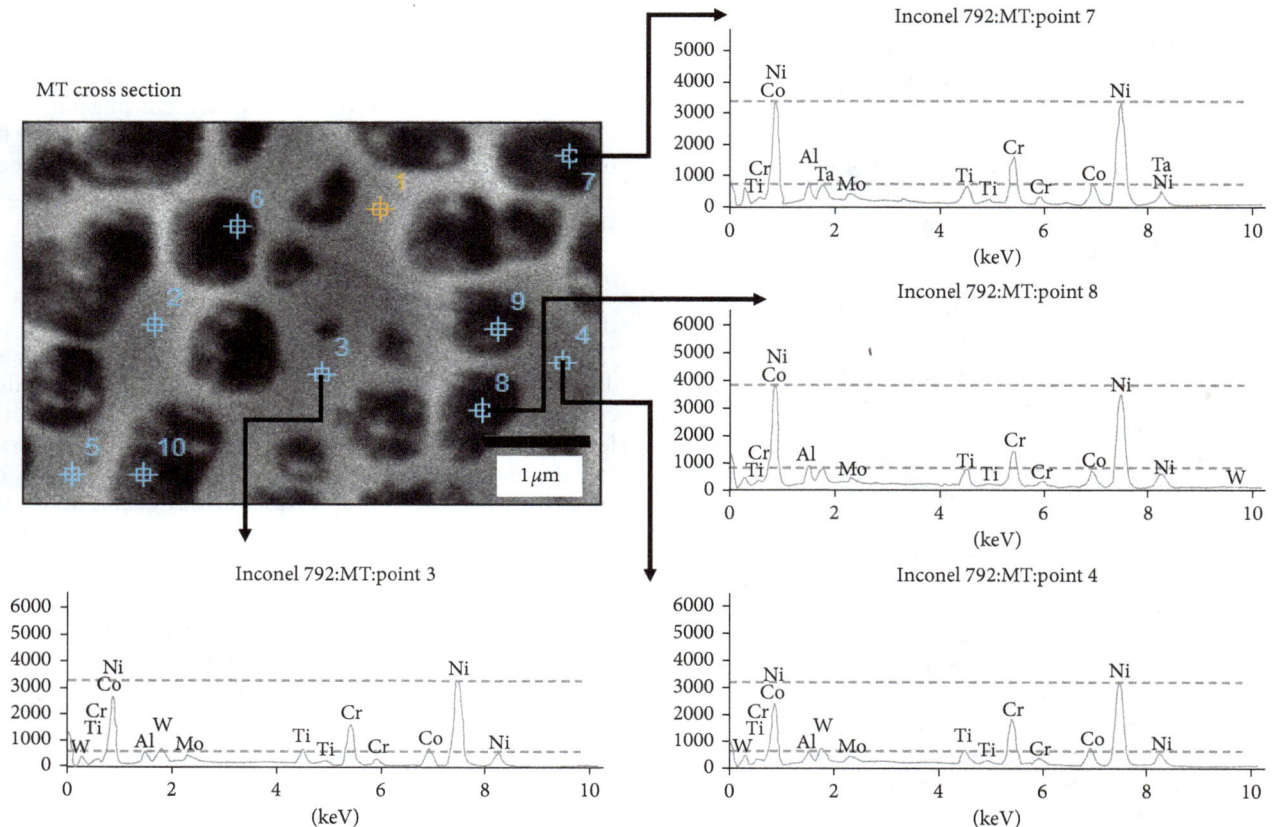

FIGURE 4: Analysis by EDX of the sample exposed at MT, near leading edge and suction side of the blade.

conditions of the heat treatments carried out. However, the formation of $M_{23}C_6$ carbides in the blade used, after 12,000 h of service, could be possible due to the decomposition of MC carbides that usually take place in Ni-based superalloys [2].

The TC-PRISMA simulation of heat treatment for the IN-792 superalloy was conducted considering the typical conditions: solution treatment at 1393 K (1120°C) for 4 h followed by rapid air cooling, then isothermal aging at 1353 K (1080°C) for 4 h followed by air cooling, and finally isothermal aging at 1118 K (845°C) for 24 h followed by air cooling. Thus, two different simulations were conducted to analyze the precipitation process during the heat treatment for IN-792 superalloy. The first treatment consisted of solution treating at 1393 K (1120°C) and subsequently isothermal aging at 1353 K (1080°C) for 4 h. The second treatment involved the solution treating at 1393 K (1120°C) and subsequently a double isothermal aging until 1118 K (845°C) for 4 h. Figures 7(a) and 7(b) show the plot of mean radius of precipitates (nm) versus time (s) determined by TC-PRISMA software for the former and latter treatments, respectively. Figure 7(a) shows that the precipitated corresponds to γ' phase. It is noted the presence of three stages:

the first of them shows a continuous increase in radius with time up to about 165 nm after 100 s of aging. This corresponds to the nucleation and growth stage. Then, there is clearly a plateau, which indicates that the precipitate growth stops and thus, the coarsening stage is reached. This behavior has been observed in a previous study where it was identified that precipitate coarsening commences at the beginning of the second aging step [18]. In contrast, Figure 7(b) indicates that the precipitation of several phases, $M_{23}C_6$, γ', μ, and Ni_3Ti, take place during aging at this condition. Furthermore, the radius of $M_{23}C_6$ precipitates is larger than that of the γ' phase. As expected, the radius of γ' precipitates is larger for aging at 1353 K (1080°C) than that at 1118 K (845°C). This fact suggests that the size distribution of precipitates is a bimodal type for the complete treatment.

In addition, Figures 8(a) and 8(b) illustrate the plot of precipitate volume fraction as a function of time for the same conditions of Figures 7(a) and 7(b). The volume fraction of the aging at 1353 K (1080°C) is much lower, about 18%, than that of aging at 1118 K (845°C), approximately 42%. These figures also show that the γ' phase is the first to be formed at both heat treatment conditions. These

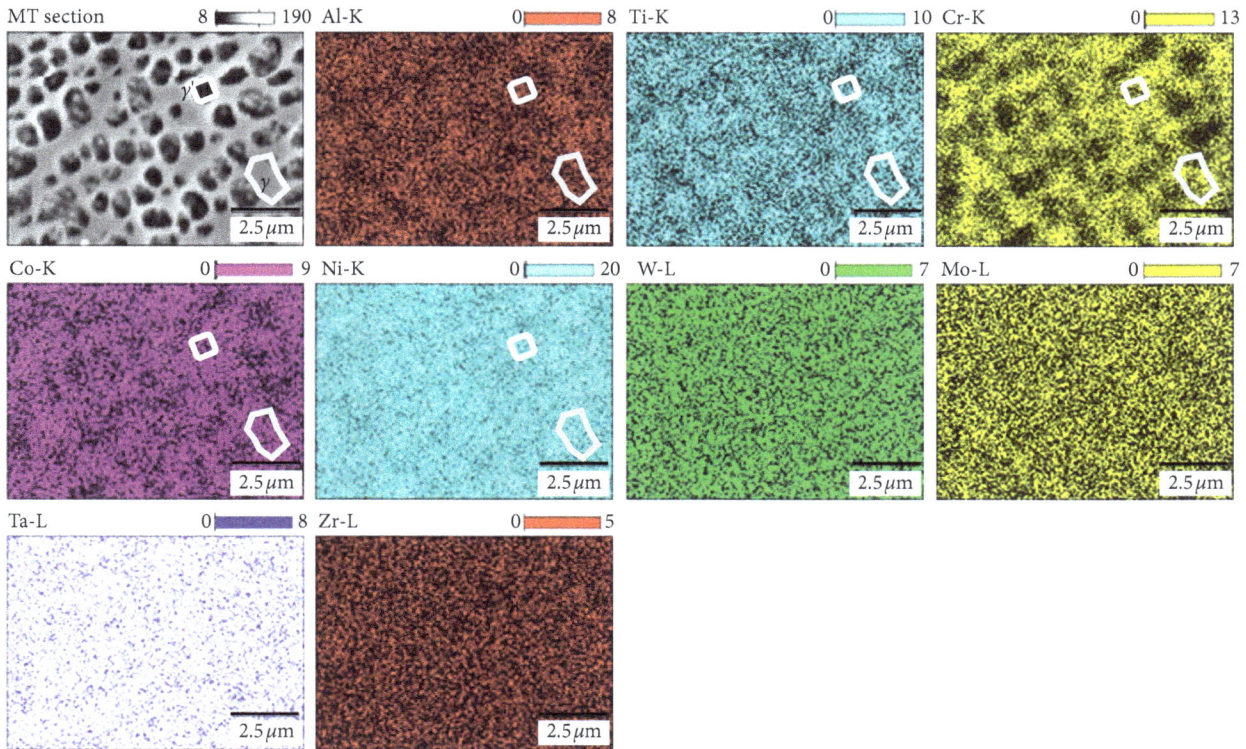

FIGURE 5: Element-imaging SEM analysis of IN-792 superalloy for MT section.

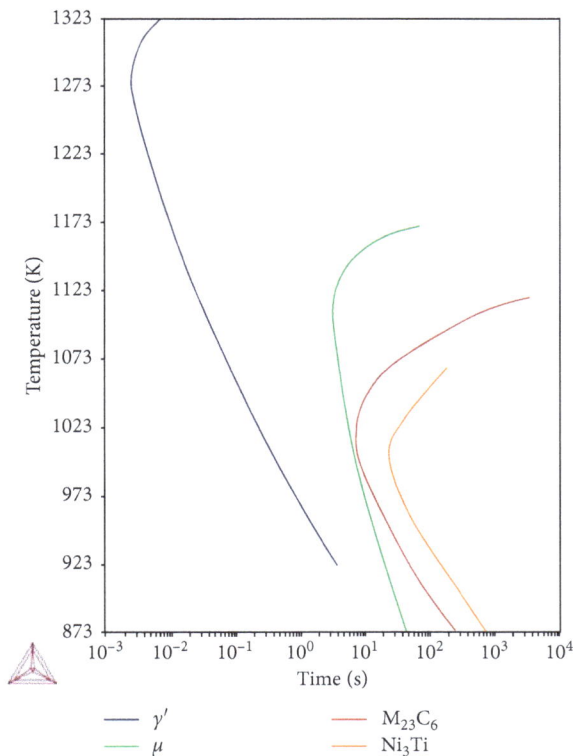

FIGURE 6: Time-temperature-precipitation diagram for IN-792 superalloy.

results, Figures 7(a), 7(b), 8(a), and 8(b), are similar to those obtained in [35]. In that experimental study, it was determined the effect of the solution temperature on the microstructure of a Ni-based superalloy and they observed a primary γ' fraction of about 7.4% at solution temperature for subsolvus condition. No primary γ' fraction was identified for supersolvus condition. Also, it was determined the precipitation of secondary γ' particles from the matrix during quenching for different heat treatment conditions of aged [35]. The size of γ' precipitates increased as the cooling rate decreased, and different conditions of cooling in their aging treatments led to a bimodal precipitation of secondary γ' particles [18, 35]. So, it is suggested that the results obtained in Figures 7(a) and 8(a) could be referred as a unimodal precipitation of γ' particles. Same behavior is suggested for the results obtained in Figures 7(b) and 8(b). It is important to mention that there is a previous study where the simulation of γ' precipitation kinetics was carried out in a commercial Ni-based superalloy with two simulation tools, TC-PRISMA and Pan-Precipitation [18]. In that analysis, it was predicted that monomodal γ' size distributions were consistent with their experimental observations where the equilibrium volume fraction γ' obtained would virtually be established at the beginning of the second aging step [18]. Thus, this simulation results allow to know the local evolution of precipitated particles in components under different conditions of heat treatments.

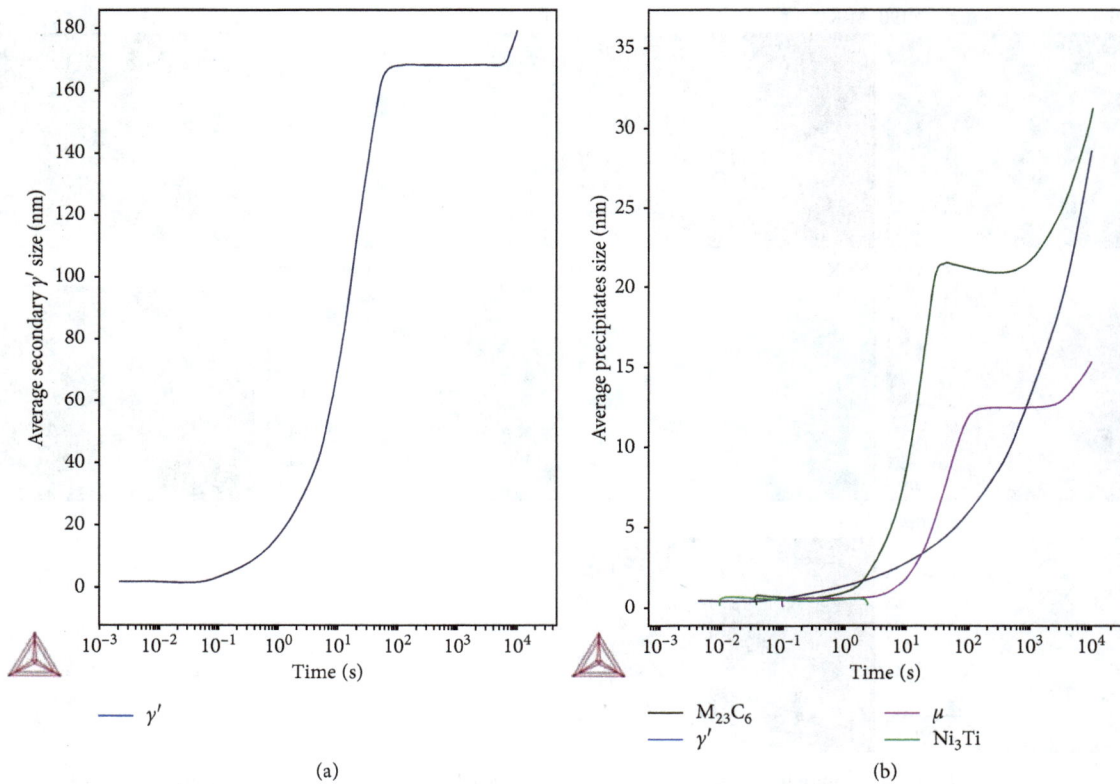

(a)

(b)

FIGURE 7: Mean radius of precipitates versus time for the conditions of heat treatment of (a) isothermal aging until 1353 K (1080°C) and (b) double isothermal aging until 1118 K (845°C).

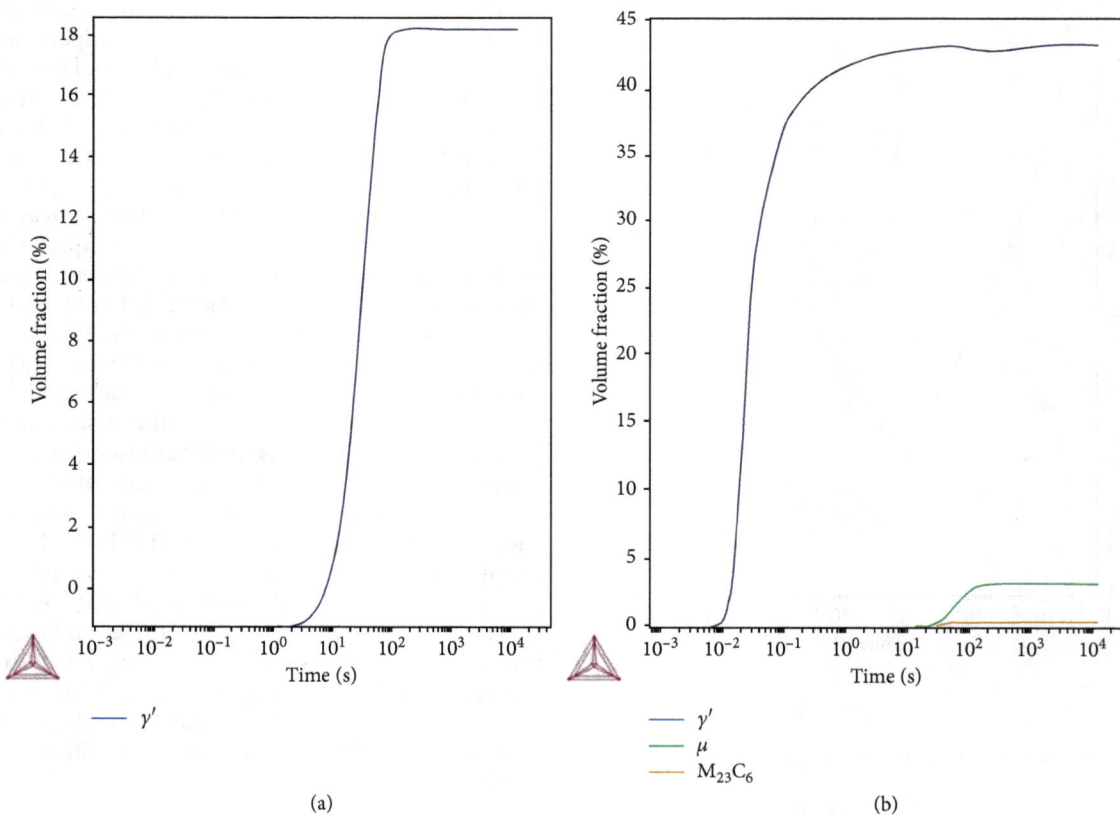

(a)

(b)

FIGURE 8: Volume fraction of precipitates versus time for the conditions of heat treatment of (a) isothermal aging until 1353 K (1080°C) and (b) double isothermal aging until 1118 K (845°C).

4.5. Microstructural and Mechanical Characterization of In-Service Operation Blade for Power Plant Turbine. In this study, the macroscopic analysis of IN-792 showed that this superalloy was obtained by a directional solidification process (SD). The results of the microstructure revealed by SEM are shown in Figures 9(a)–9(c) for the three sections: MT: 1147 K (874°C); HT: 1211 K (938°C); and LT: 1091 K (818°C), respectively. It can be observed the different formed phases: γ matrix (light gray), γ' precipitate (dark gray), and carbides of the MC type (white). The MC carbides, with a size of approximately $10\,\mu m$, are clearly observed in Figures 9(a)–9(c). These are the first phase to be formed during the solidification, as shown in Figure 3. They are composed of Ti and Ta, as indicated in Figure 10, which corresponds to the element-imaging SEM analysis of the IN-792 superalloy, in addition to $M_{23}C_6$ carbides (black) near the MC carbides, due to the decomposition of MC carbides that usually take place in Ni-based superalloys, as previously mentioned. This may cause the release of carbon that reacts in several ways and leads to the formation of $M_{23}C_6$ according to the following reactions [2]:

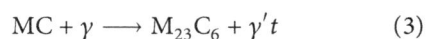

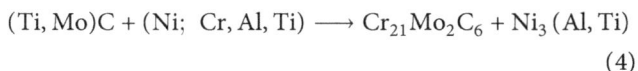

$$MC + \gamma \longrightarrow M_{23}C_6 + \gamma' t \qquad (3)$$

$$(Ti, Mo)C + (Ni;\ Cr, Al, Ti) \longrightarrow Cr_{21}Mo_2C_6 + Ni_3 (Al, Ti) \qquad (4)$$

These reactions promote the precipitation of carbides mainly on grain boundaries and, in many cases, the γ' phase generated by reaction (3) covers the surface of MC carbides [2]. This is in agreement with the MC carbides observed in Figures 9(a)–9(c) which are covered by γ' phase.

In addition to, a different coarse microconstituent with an irregular plate shape was observed in Figure 9(c), indicated with black arrows. This constituent may correspond to a topologically packed phase (TCP). It is known that the TCP phases are formed in Ni-based superalloys that have high content of elements such as Cr, Mo, W, or Re which promote precipitation of rich intermetallic phases in these elements [2]. The real chemical composition of IN-792 superalloy indicates a high Cr content, 11.7 wt.%, Mo and W, with 2.0 and 3.5 wt.%, respectively. This may cause the formation of the μ phase according to Thermo-Calc and TC-PRISMA results, previously presented. Also, it was shown in Figure 8(b) a low value of volume fraction for this precipitate. Additionally, particles of the γ' phase can be accumulated and coarsened around the μ phase. Thus, the identification of μ phase is difficult by SEM [14].

The eutectic γ and γ' lamellar constituent is noted in Figures 9(a)–9(c). This eutectic reaction was identified at 1504 K (1231°C) [33]. The aging process at temperatures between 1118 K (845°C) and 1353 K (1080°C) is expected to cause the precipitation of the γ' phase from the γ phase of as-cast superalloy. These precipitates have a cuboid morphology as can be observed in Figures 9(a)–9(c). They are also aligned on the softest crystallographic direction of the fcc γ phase, <100>. The average radius of γ' particles, determined on different SEM micrographs, was HT: 338.5 nm; MT: 320.4 nm; and, LT: 328.5 nm. It is evident that the highest service temperature promotes the largest size of γ' particles. However, the smallest radius of precipitates was not observed for LT section. This seems to be attributable to the occurrence of rafting of γ' precipitates in MT section due to the presence of highest value of total deformation in this section and to high mechanical stress during in-service operation of the blade.

The XRD pattern for superalloy, after in-service operation, is shown in Figure 11(a). Most of the diffraction peaks for γ' and γ phases are overlapped because of the similar lattice parameter; however, the superlattice diffraction peaks of γ' phase are clearly detected at low 2θ angle. The element-imaging SEM analysis of Figure 10 revealed the presence of $M_{23}C_6$ carbides, mainly composed of Cr and Mo, and it is confirmed in Figures 11(a) and 11(b). This fact is consistent with literature [2] where it is reported that M_6C carbide is favored to be formed instead of $M_{23}C_6$ carbide for nickel-based superalloys where Mo and W content is in the range of 6–8 at. %, which is different from 1.21 at. % Mo and 1.11 at. % W for the current alloy composition.

Figure 11(a) confirmed the presence of Ni_3Ti or η phase, predicted by Thermo-Calc and TC-PRISMA software. It was expected to take place due to the in-service operation conditions of the blade. The presence of η phase identified in this investigation is similar to that of the results obtained in [33, 34]. However, η phase was not detected by SEM.

It is important to mention that μ particles were not identified in previous studies about precipitation kinetic of IN-792 superalloys [9, 33]. This can be attributed to little differences of chemical composition and to the different bases of thermodynamics data used in this study.

The Vickers microhardness of the blade for each section, HT, MT, and LT, was determined to be about 478, 434 and 453 VHN, respectively. This difference in hardness can be attributed to the γ' phase average size. Besides, the highest value hardness was observed to occur when a high volume fraction of MC and $M_{23}C_6$ carbides were present, distributed intra- and intergranularly in the γ matrix phase.

4.6. Microstructural and Mechanical Characterization of the Reheat-Treated Blade. Figures 12(a)–12(f) show the SEM micrographs for the base of the blade with the following heat treatment conditions: heating at 1393 K (1120°C) for 4 h and subsequently air cooling (Figures 12(a)–12(c)), heating at 1393 K (1120°C) for 4h followed by rapid air cooling until 1353 K (1080°C), and then aging at this temperature for 4 h and final aging at 1118 K (845°C) for 24 h followed by air cooling (Figures 12(d)–12(f)). The first heat treatment caused the partial dissolution of the γ' phase because of the decrease in its volume fraction. That is, the γ' particles were not observed well distributed on the γ matrix, and its average radius decreased, approximately 312 nm. Furthermore, the crystallographic alignment between the γ and γ' phases is not clearly observed in Figures 12(a)–12(c). Additionally, the morphology of γ' precipitates is rounded instead of cuboidal. This result suggests that the γ' phase was not completely dissolved which is in agreement with the pseudobinary phase diagram of Figure 2 where no γ single

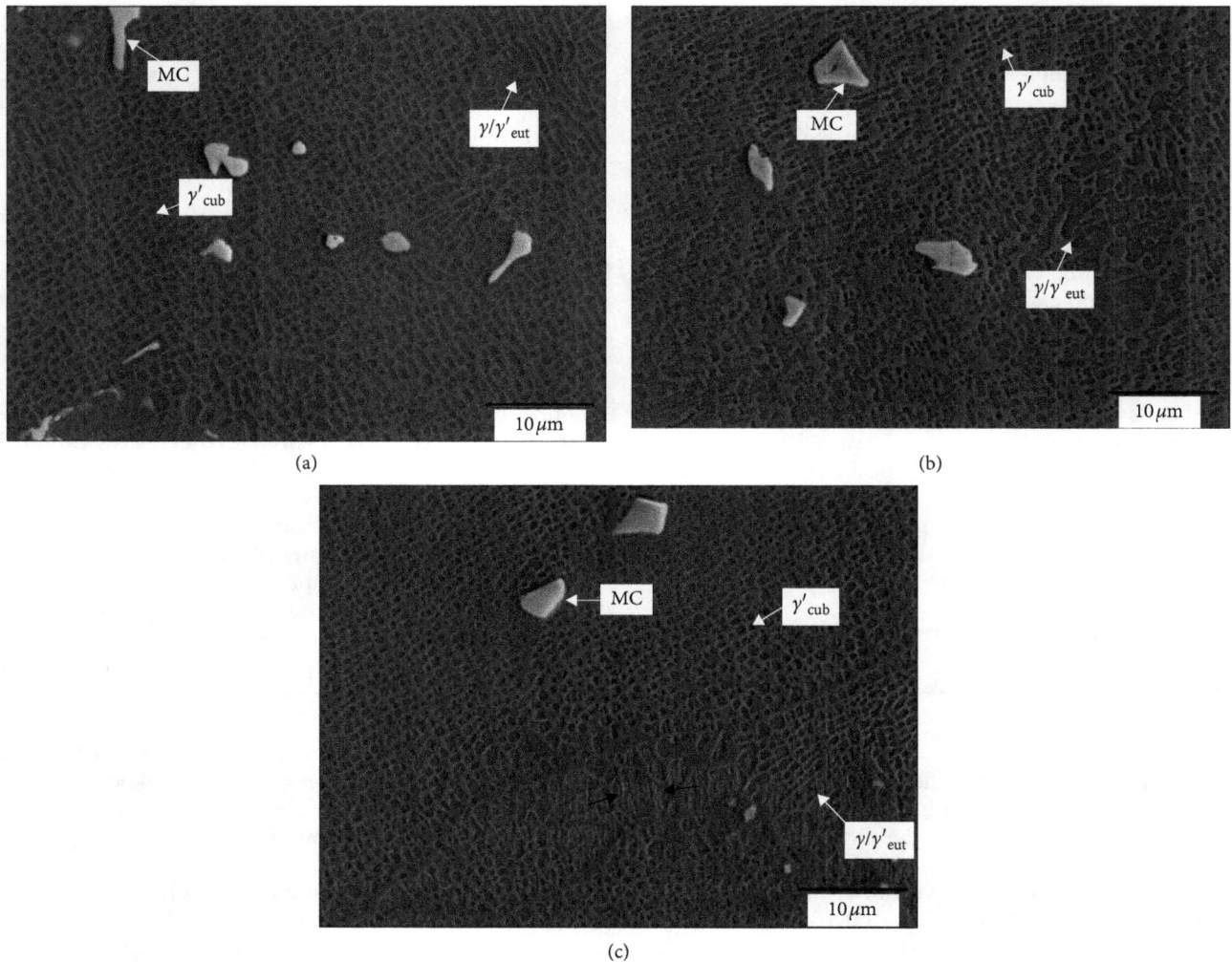

FIGURE 9: SEM micrographs of IN-792 superalloy for (a) MT: 1147 K (874°C); (b) HT: 1211 K (938°C); (c) LT: 1091 K (818°C).

phase region is observed for 3.2 wt.% of aluminum. The eutectic constituent of the γ and γ' phases is still present, and it seems that this treatment caused no change in them. This is in good agreement with the formation temperature, 1504 K (1231°C), of γ/γ' eutectic reported in literature [33]. The following aging at 1353 K (1080°C) for 4 h and aging at 1118 K (845°C) for 24 h have no effect on γ/γ' eutectic (Figures 12(d)–12(f)). MC carbides seem to be unaffected for both treatments (Figures 12(a)–12(f)). However, $M_{23}C_6$ carbides were dissolved apparently, which is consistent with the phase diagram obtained by Thermo-Calc that shows no $M_{23}C_6$ carbide presence at 1393 K (1120°C), solution treating temperature.

Furthermore, the XRD pattern, as shown in Figure 13(a), for the specimen with the first heat treatment shows only the presence of γ and γ' phases. The hardness for this section of the in-service operation specimen was 453 VHN in comparison to 452 VHN for the specimen with first treatment. That is, there is almost no change in hardness. In contrast, the specimen with the second treatment, complete heat treatment, shows the presence of very small spherical γ' precipitates, < 100 nm, located

among the original γ' precipitates as a result of aging at 1118 K (845°C). In addition, the XRD for this specimen, as shown in Figure 13(b), again indicates only the presence of the γ and γ' phases.

In a previous experimental study [36], it was analyzed the as-cast microstructure of an IN-792 superalloy, obtained by conventional cast process. They identified γ matrix phase, γ' phase, γ/γ' eutectic, and carbides. Then, after a heat treatment of solution treated at 1393 K (1120°C), they concluded that the MC carbides and rose-shaped γ/γ' eutectic partly dissolves into γ matrix phase and $M_{23}C_6$ carbide forms. In this study, $M_{23}C_6$ carbides were not observed after a heat treatment of solution treated at 1393 K (1120°C). It can be a consequence of the differences in chemical composition between the alloys studied. Then, after two-stage aging process with 4 h at 1353 K (1080°C) and 24 h at 1118 K (845°C) in [36], they identified γ' phase arranged regularly and the reprecipitation of profuse fine γ' throughout γ matrix channels. These results are in good agreement with the experimental results of this study.

The hardness for this complete heat treatment was 410 VHN, lower than the first treatment. This can be attributed

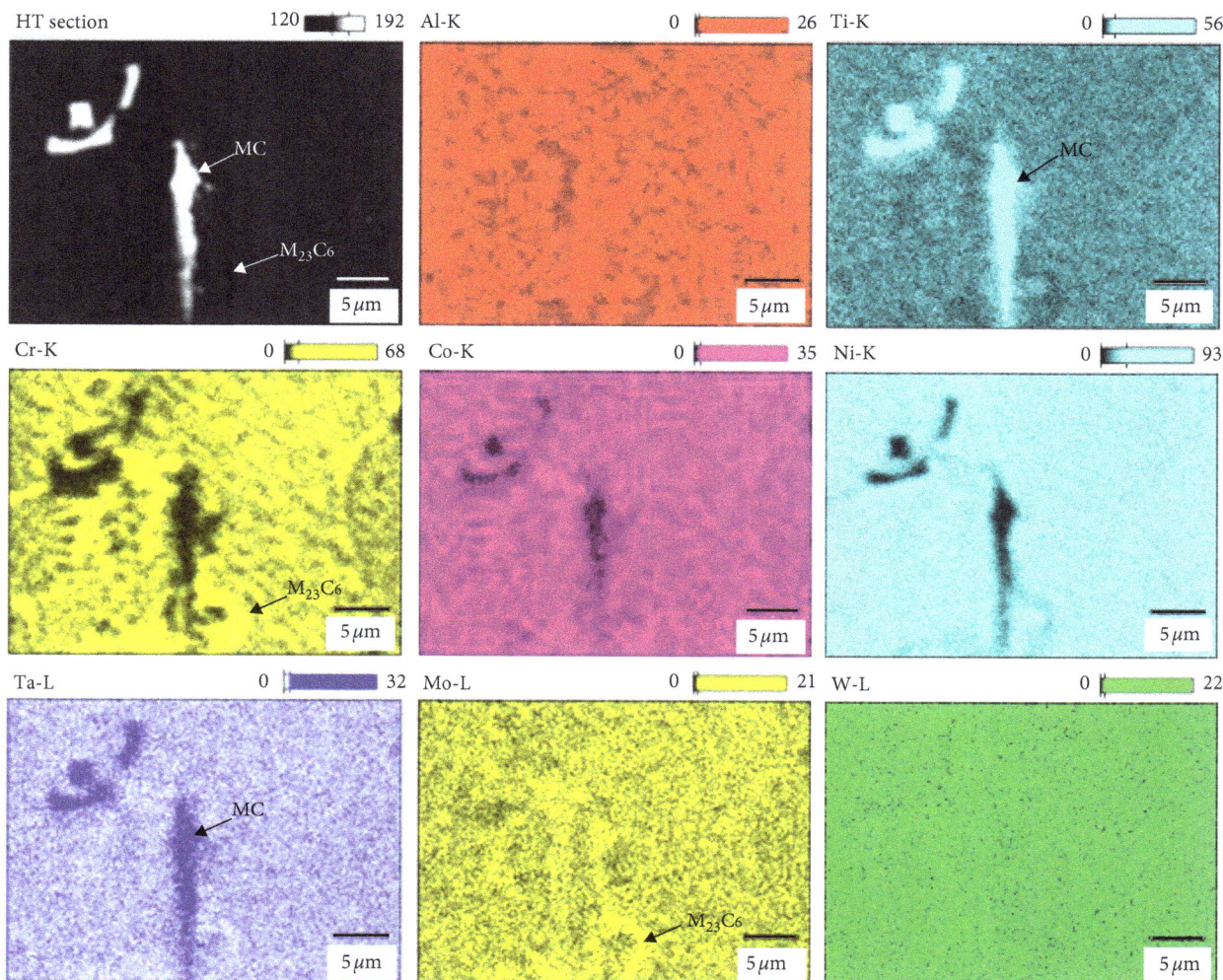

FIGURE 10: Element-imaging SEM analysis of IN-792 superalloy for HT section.

to the loss of solute in the γ matrix phase because of the precipitation of small γ' precipitates and to the dissolution of $M_{23}C_6$ carbides in the IN-792 superalloy, after applying the reheat treatments.

The precipitation of small γ' particles are in good agreement with the TTP diagram of Figure 5 determined by TC-PRISMA. Like in [36], in this study was not observed cold crack formation at grain boundaries for a solution temperature of 1393 K (1120°C). Finally, for both heat treatments, the heating at 1393 K (1120°C) is not enoughly high for dissolving the γ/γ' eutectics and the carbides of the MC type in the microstructure of the IN-792 superalloy. On the other hand, the complete heat treatment promotes a distribution more regular and homogeneous of γ' precipitates, and their size and morphology become finer and cuboidal, respectively, because of the repreciptation γ' particles throughout γ matrix phase.

5. Conclusions

Experimental and numerical analysis of the microstructure evolution for the IN-792 superalloy, extracted from a gas turbine blade, was conducted, and the conclusions are as follows:

(a) The Thermo-Calc analysis based on the equilibrium and nonequilibrium diagrams permitted to explain the phase formation during the solidification of the superalloy. The MC carbides were first formed during the solidification, and then the eutectic constituent, composed of the γ and γ' phases, was formed. These calculated results are in good agreement with the phases observed microscopically in the superalloy.

(b) The calculated time-temperature-precipitation diagram shows that if an isothermal aging is carried out in the IN-792 superalloy at temperatures between approximately 1198 K (925°C) and 1323 K (1050°C), the following precipitation reaction occurs: $\gamma \longrightarrow \gamma + \gamma'$, and the growth kinetic of precipitation of the γ' phase is very fast. The precipitation sequence for aging at temperatures of about 873 K (600°C) and 1173 K (900°C) is as follows: $\gamma \longrightarrow \gamma + \gamma' \longrightarrow \gamma + \mu \longrightarrow \gamma + M_{23}C_6 \longrightarrow \gamma + Ni_3Ti$. All these precipitated phases are in agreement with those phases observed experimentally from the

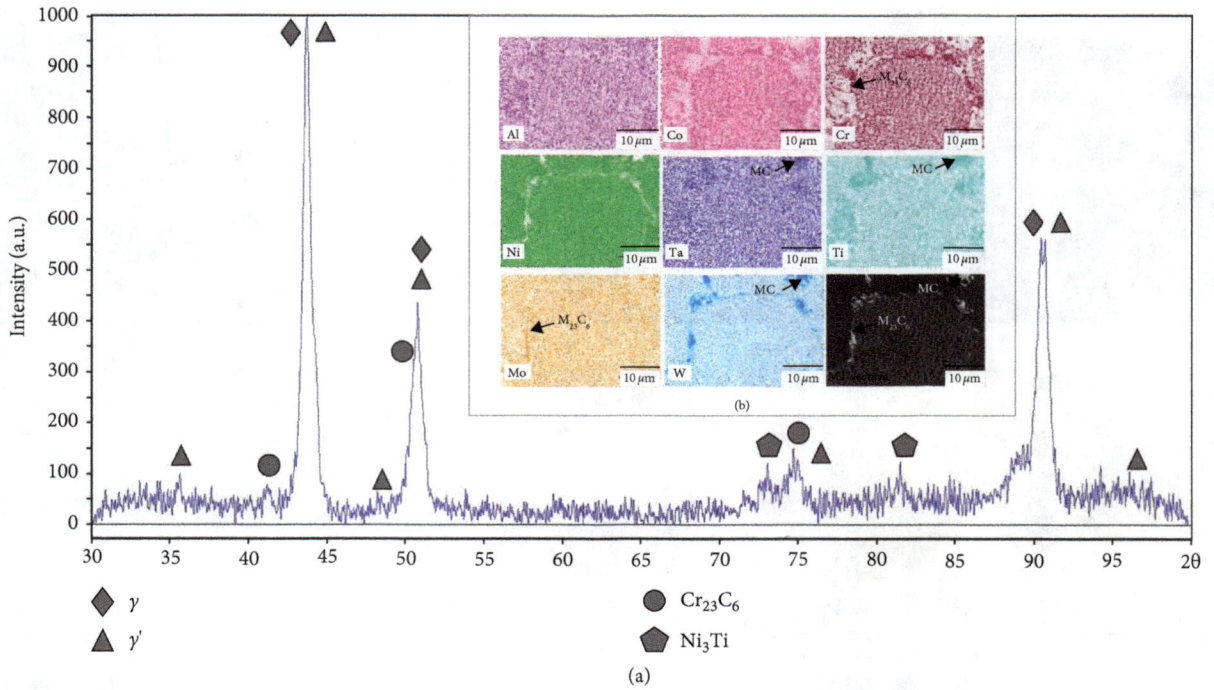

FIGURE 11: Characterization of the IN-792 superalloy. MT section by (a) XRD pattern and (b) element-imaging SEM analysis.

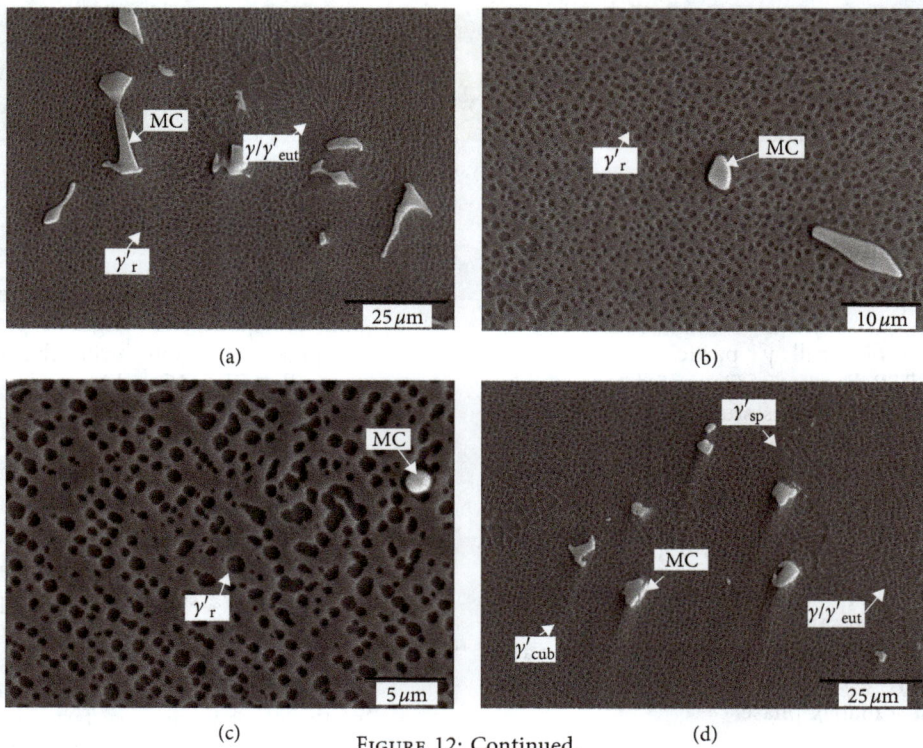

FIGURE 12: Continued.

(e)   (f)

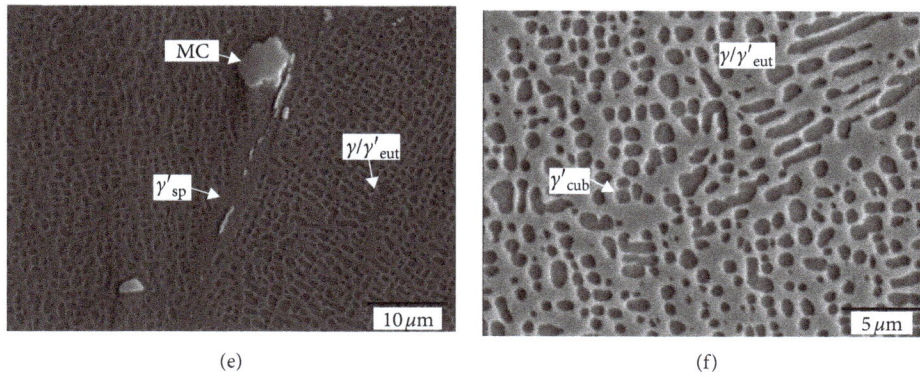

FIGURE 12: SEM micrographs of the reheat treatment blade for the conditions: (a–c) solution treating and (d–f) complete heat treatment.

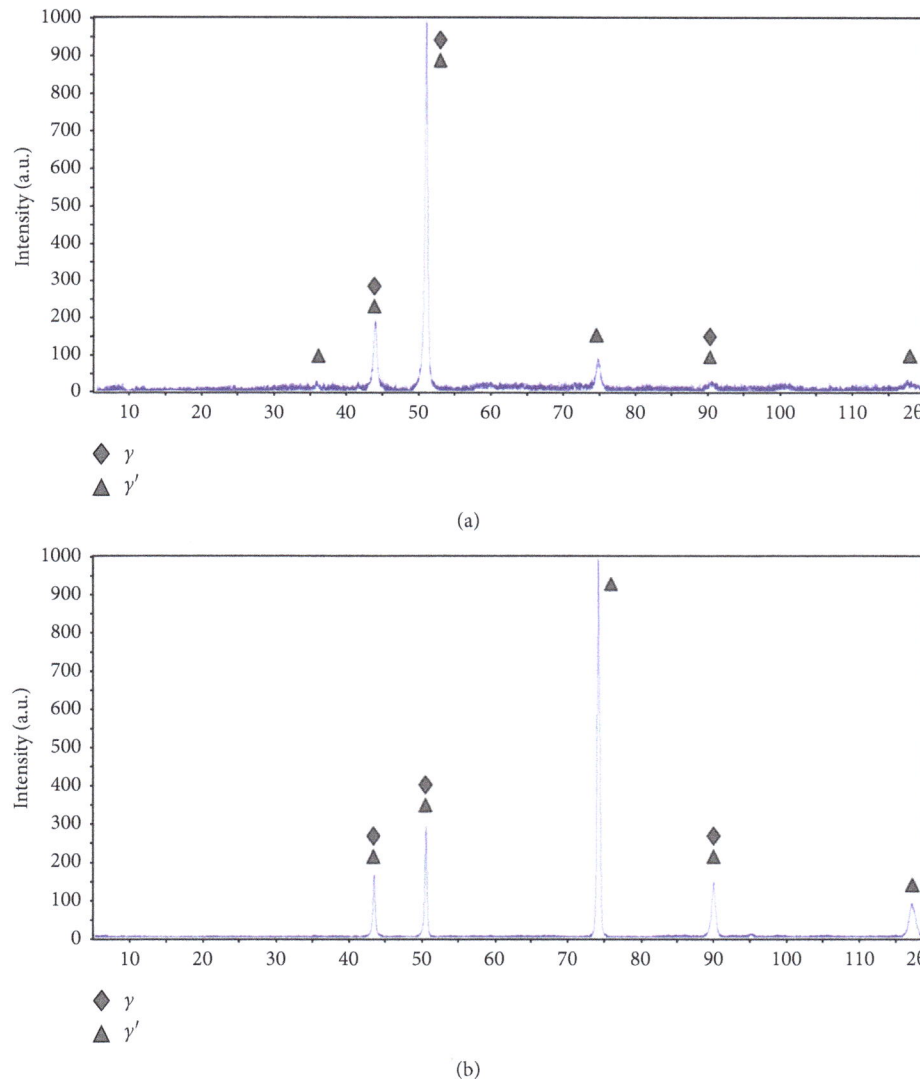

(a)

(b)

FIGURE 13: XRD patterns of the IN-792 superalloy with the following heat treatment conditions: (a) solution treating and (b) complete heat treatment.

microstructural characterization of in-service operation blade.

(c) In the case of both conducted reheat treatments, no effect of dissolution was observed on the γ/γ' eutectic and MC carbides. The $M_{23}C_6$ carbides were dissolved, apparently. This is consistent with the phase diagram obtained by Thermo-Calc that shows no $M_{23}C_6$ carbide presence at the solution treating temperature, 1393 K (1120°C). For the condition of heating at 1393 K (1120°C) for 4 h and subsequently air cooling, it was observed partial dissolution of the γ' phase and the precipitation of the γ' phase with a morphology round instead of cuboidal. In the case of the complete heat treatment, the fine secondary γ' precipitates showed a more regular distribution and bimodal type. These results are consistent with the calculated TTP diagram.

(d) The difference in the hardness values for the sections' in-service operation can be attributed as a consequence of the γ' average size and due to the high volume fraction of MC and $M_{23}C_6$ carbides, considering their intra- and intergranular distribution in the γ matrix phase. Nevertheless, it was not obtained an evident change in hardness for the specimen with the first treatment. In contrast, the specimen with a complete heat treatment showed hardness lower than the first treatment. This can be a consequence of the loss of solute in the γ matrix phase due to the reprecipitation of small γ' particles and of the dissolution of $M_{23}C_6$ carbides in the IN-792 superalloy, after applying the reheat treatments.

(e) The η phase was detected in the samples for the sections in-service operation, experimentally. In contrast, it was not identified in the samples reheat-treated. Then, for this IN-792 superalloy, its presence is due to the prolonged exposure at high temperatures of the blades. Similarly, the μ phase was predicted to appear in the IN-792 superalloy by Thermo-Calc and TC-PRISMA software. Thus, it is expected that the coarse microconstituent with an irregular plate shape observed in the LT section of the blade, after in-service operation, corresponds to the μ phase. However, there is good agreement between the numerical and experimental results obtained.

Data Availability

The Thermo-Calc and TC-PRISMA data used to support the findings of this study were supplied by Thermo-Calc software under perpetual license and so cannot be made freely available. Also, the Thermo-Calc and TC-PRISMA data used are mentioned in references [26, 28, 30] within the article. Previously reported computational fluid dynamics (CFD) data were used to support this study and are available in reference [24], within the article. These prior studies were used to calculate the service temperatures inside the blade.

Conflicts of Interest

The authors declare that there are no conflicts of interest regarding the publication of this paper.

Acknowledgments

The authors wish to acknowledge the financial support from CONACYT and from TecNM (6554.18-P project).

References

[1] J. R. Davis, *ASM Specialty Handbook: Heat-Resistant Materials*, ASM International, Materials Park, OH, USA, 1st edition, 1997.

[2] R. C. Reed, *The Superalloys Fundamentals and Applications*, Cambridge University Press, New York, NY, USA, 1st edition, 2006.

[3] T. M. Pollock and S. Tin, "Nickel-based superalloys for advanced turbine engines: chemistry, microstructure, and properties," *Journal of Propulsion and Power*, vol. 22, pp. 361–374, 2006.

[4] B. Du, J. Yang, C. Cui, and X. Sun, "Effects of grain size on the high-cycle fatigue behavior of IN792 superalloy," *Materials and Design*, vol. 65, pp. 57–64, 2015.

[5] L. Yang, Q. X. Liu, Y. C. Zhou, W. G. Mao, and C. Lu, "Finite element simulation on thermal fatigue of a turbine blade with thermal barrier coatings," *Journal of Materials Science & Technology*, vol. 30, no. 4, pp. 371–380, 2014.

[6] R. C. Ecob, R. A. Ricks, and A. J. Porter, "The measurement of precipitate/matrix lattice mismatch in nickel-base superalloys," *Scripta Metallurgica*, vol. 16, no. 9, pp. 1085–1090, 1982.

[7] A. Sato, J. J. Moverare, M. Hasselqvist, and R. C. Reed, "On the mechanical behavior of a new single-crystal superalloy for industrial gas turbine applications," *Metallurgical and Materials Transactions A*, vol. 43, no. 7, pp. 2302–2315, 2012.

[8] R. W. Kozar, A. Suzuki, W. W. Milligan, J. J. Schirra, M. F. Savage, and T. M. Pollock, "Strengthening mechanisms in polycrystalline multimodal nickel-base superalloys," *Metallurgical and Materials Transactions A*, vol. 40, no. 7, pp. 1588–1603, 2009.

[9] K. V. Dahl and J. Hald, "Identification of precipitates in an IN792 gas turbine blade after service exposure," *Practical Metallography*, vol. 50, no. 6, pp. 432–450, 2013.

[10] S. Gorgannejad, E. A. Estrada Rodas, and R. W. Neu, "Ageing kinetics of Ni-base superalloys," *Materials at High Temperatures*, vol. 33, no. 4-5, pp. 291–300, 2016.

[11] I. Lopez-Galilea, J. Koßmann, A. Kostka1 et al., "The thermal stability of topologically close-packed phases in the single-crystal Ni-base superalloy ERBO/1," *Journal of Materials Science*, vol. 51, no. 5, pp. 2653–2664, 2016.

[12] P. Wangyao, N. Chuankrerkkul, S. Polsilapa, P. Sopon, and W. Homkrajai, "Gamma prime phase stability after long-term thermal exposure in cast nickel based superalloy, IN-738," *Chiang Mai Journal of Science*, vol. 36, pp. 312–319, 2009.

[13] J. Yang, Z. Qi, M. Ji, X. Sun, and Z. Hu, "Effects of different C contents on the microstructure, tensile properties and stress-rupture properties of IN792 alloy," *Materials Science and Engineering: A*, vol. 528, no. 3, pp. 1534–1539, 2011.

[14] K. Zhao, Y. H. Ma, L. H. Lou, and Z. Q. Hu, "μ phase in a nickel base directionally solidified alloy," *Materials Transactions*, vol. 46, no. 1, pp. 54–58, 2005.

[15] A. S. Wilson, "Formation and effect of topologically close-packed phases in nickel-base superalloys," *Materials Science and Technology*, vol. 33, no. 9, pp. 1108–1118, 2017.

[16] Q. Chen, K. Wu, G. Sterner, and P. Mason, "Modeling precipitation kinetics during heat treatment with calphad-based tools," *Journal of Materials Engineering and Performance*, vol. 23, no. 12, pp. 4193–4196, 2014.

[17] O. Prat, J. García, D. Rojas, J. P. Sanhueza, and C. Camurri, "Study of nucleation, growth and coarsening of precipitates in a novel 9%Cr heat resistant steel: experimental and modeling," *Materials Chemistry and Physics*, vol. 143, no. 12, pp. 754–764, 2014.

[18] M. G. Fahrmann and D. A. Metzler, "Simulation of γ' precipitation kinetics in a commercial Ni-base superalloy," *JOM*, vol. 68, no. 11, pp. 2786–2792, 2016.

[19] G. F. Vander Voort, G. M. Lucas, and E. P. Manilova, "Metallography and microstructures of heat-resistant alloys," in *ASM Handbook: Metallography and Microstructures*, vol. 9, pp. 820–859, ASM International, Materials Park, OH, USA, 2004.

[20] S. Somashekar, R. Prem Chand, K. M. Chandrashekar, and D. Sachin, "Design and structural analysis of turbine blades," *International Journal for Research in Applied Science & Engineering Technology (IJRASET)*, vol. 5, pp. 1572–1578, 2017.

[21] N. L. Sindhu and N. Chikkanna, "Design and analysis of gas turbine blade," *International Journal for Research in Applied Science & Engineering Technology (IJRASET)*, vol. 5, pp. 1097–1104, 2017.

[22] M. Tofighi Naeem, N. Rezamahdi, and S. A. Jazayeri, "Failure analysis of gas turbine blades," *International Journal of Engineering Research & Innovation*, vol. 1, pp. 29–36, 2009.

[23] J. R. Davis, *ASM Specialty Handbook: Nickel, Cobalt, and their Alloys*, ASM International, Materials Park, OH, USA, 2000.

[24] Online World Academy of Science, "Engineering and technology (WASET)," September 2016, https://waset.org/Publication/fatigue-life-consumption-for-turbine-blades-vanes-accelerated-by-erosion-contour-modification/10962.

[25] Maintenance Manual MFT4-CID/LF, vol. 1-2.

[26] Online Thermo-Calc, "Thermodynamic equilibrium calculations," November 2016, http://www.Thermo-Calc.com/index.html.

[27] D. A. Porter and K. E. Easterling, *Phase Transformations in Metals and Alloys*, Chapman and Hall, London, UK, 2nd edition, 1992.

[28] Online TC-PRISMA, "Thermodynamic and diffusion calculations," December 2016, http://www.Thermo-Calc.com/index.html.

[29] G. Kostorz, *Phase Transformations in Materials*, Wiley-VCH, Weinheim, Germany, 2001.

[30] Online database Thermo-Calc, "Thermodynamic equilibrium calculations," November 2016, http://www.thermocalc.com/media/23648/tcni8.pdf and http://www.thermocalc.com/media/23658/mobni8.pdf.

[31] E. Poursaeidi, A. Kavandi, K. Vaezi, M. R. Kalbasi, and M. R. Mohammadi Arhani, "Fatigue crack growth prediction in a gas turbine casing," *Engineering Failure Analysis*, vol. 44, pp. 371–381, 2014.

[32] B. Deepanraj, P. Lawrence, and G. Sankaranarayanan, "Theoretical analysis of gas turbine blade by finite element method," *Scientific World*, vol. 9, no. 9, pp. 29–33, 2011.

[33] S. Zlá, J. Dobrovská, B. Smetana, M. Žaludová, V. Vodárek, and K. Konečná, "Thermophysical and structural study of IN 792-5a nickel based speralloy," *Metalurgija*, vol. 51, pp. 83–66, 2012.

[34] Y. Xu, L. Zhang, J. Li et al., "Relationship between Ti/Al ratio and stress-rupture properties in nickel-based superalloy," *Materials Science and Engineering: A*, vol. 544, pp. 48–53, 2012.

[35] D. Locq, L. Nazé, J.-M. Franchet et al., "Metallurgical optimisation of PM superalloy N19," in *Proceedings of Eurosuperalloys 2014*, p. 6, Giens, France, May 2014.

[36] J. Yang, Q. Zheng, H. Zhang, X. Sun, H. Guan, and Z. Hu, "Effects of heat treatments on the microstructure of IN792 alloy," *Materials Science and Engineering: A*, vol. 527, no. 4-5, pp. 1016–1021, 2010.

16

Prediction of Cutting Conditions in Turning AZ61 and Parameters Optimization using Regression Analysis and Artificial Neural Network

Nabeel H. Alharthi,[1] Sedat Bingol,[2] Adel T. Abbas ⬥,[1] Adham E. Ragab ⬥,[3] Mohamed F. Aly ⬥,[4] and Hamad F. Alharbi ⬥[1]

[1]Department of Mechanical Engineering, College of Engineering, King Saud University, P.O. Box 800, Riyadh 11421, Saudi Arabia
[2]Department of Mechanical Engineering, Dicle University, Diyarbakir 21280, Turkey
[3]Department of Industrial Engineering, College of Engineering, King Saud University, P.O. Box 800, Riyadh 11421, Saudi Arabia
[4]Department of Mechanical Engineering, School of Sciences and Engineering, American University in Cairo, AUC Avenue, P.O. Box 11835, New Cairo, Egypt

Correspondence should be addressed to Adel T. Abbas; atabbas1954@yahoo.com

Academic Editor: Fernando Lusquiños

All manufacturing engineers are faced with a lot of difficulties and high expenses associated with grinding processes of AZ61. For that reason, manufacturing engineers waste a lot of time and effort trying to reach the required surface roughness values according to the design drawing during the turning process. In this paper, an artificial neural network (ANN) modeling is used to estimate and optimize the surface roughness (R_a) value in cutting conditions of AZ61 magnesium alloy. A number of ANN models were developed and evaluated to obtain the most successful one. In addition to ANN models, traditional regression analysis was also used to build a mathematical model representing the equation required to obtain the surface roughness. Predictions from the model were examined against experimental data and then compared to the ANN model predictions using different performance criteria such as the mean absolute error, mean square error, and coefficient of determination.

1. Introduction

Magnesium alloys are often used in many industrial applications such as the manufacturing of several components used in the aerospace and modern automobiles industry. Also, magnesium block engines have been widely used in some high-performance vehicles. In those applications, the final surface roughness of machined components is playing a major factor in the acceptance of those parts.

Many researchers have investigated the optimization of cutting parameters for the prediction of surface roughness as a key performance measure. Asiltürk used ANN (artificial neural network) and MRM (multiregression models) to predict the surface roughness of steel AISI 1040. They developed their own models and used ANN to optimize the cutting parameters formulating the surface roughness as objective function. They used cutting speed, feed rate, depth of cut, and nose radius as optimized parameters. Surface roughness is characterized by the mean (R_a) and total (R_t) values of the recorded roughness at different locations on the produced surface. They conducted many experiments, each with a different set of the cutting parameters, and the corresponding R_a and R_t values were reported. Obtained results were then used to train an ANN model. Mean squared error of approximately 0.003% was achieved which outperforms error rates reported in the early literature and are claimed to be suitable for robust prediction of the surface roughness in industrial settings [1].

Another approach can be found in the work of Mokhtariet Homami et al. [2], and they employed a design of experiment

(DOE) technique based on a full factorial design to determine the number of experiment and the corresponding parameters. They represented their results in a statistical analysis, and they used ANN to model the system. Optimization was done using genetic algorithm (GA). The conclusion of their work was that the main factors affecting the flank wear and the produced surface roughness are the feed rate, nose radius, and approach angle, while the cutting speed had the major effect on flank wear. Optimized values of the cutting conditions were attained and showed a significant reduction in the surface roughness values.

Jafarian et al. [3] used GA and particle swarm optimization (PSO) techniques to determine optimal cutting parameters in turning operations with a multiobjective optimization aiming to minimize surface roughness and cutting forces and maximize tool life. They discussed their results claiming that training ANN using GA gave superior results than those reported in the literature with high accuracy and gives the flexibility of analyzing the effect of each parameter separately on the output.

The PSO technique was also utilized in the work of Karpat and Özel [4], and they used the Pareto optimal frontier to select optimized parameters to maximize material removal rate (MRR) without affecting the induced stresses or the final surface finish of the produced components. They obtained good results making use of dynamic-neighborhood PSO approach in solving complex turning optimization problems.

A different approach was utilized in the work of Natarajan et al. [5] to test the reliability of ANN in the prediction of surface roughness values when machining Brass C26000 material in dry cutting condition on a CNC turning machine. Surface roughness has been measured and compared to the experimental data and concluded that ANN can be implemented reliably and accurately to predict surface roughness in turning operations of Brass C26000 material.

The applicability of radial base function (RBF) neural networks was investigated in the work of Pontes et al. [6] to predict surface roughness in turning processes of SAE 52100 hardened steel. Networks were trained using different sets. They considered several design variables and found that ANN models were capable of providing accurate estimates of surface roughness values in an affordable way.

The turning of Ti-6Al-4V titanium alloy was investigated in the work of Sangwan et al. [7] to minimize surface roughness using ANN-GA approach. A feed forward neural network was proposed for training and testing of the neural network model. The predicted results were found to be in good agreement with the obtained experimental results.

A comparison between linear regression models and ANN approach has been studied in the work of Acayaba and Escalona [8]. A target of saving cost, effort, and machining time leads to the necessity of predicting surface roughness prior to performing machining operations. They used experimental data to validate their claim and found that using ANN outperforms linear modeling. Instead of using GA like other researchers previously listed, this research employed a simulated annealing (SA) optimization algorithm to optimize cutting parameters for minimizing surface roughness. Results show similar findings as reported previously with no major significant improvement.

A more concise investigation focusing only on the three major cutting parameters influencing the surface roughness was presented in the work of Bajić et al. [9]. Cutting speed, feed rate, and depth of cut are optimized using regression analysis and ANN. Results obtained show no superiority of one approach over the other, and both gave a good prediction of the surface roughness.

A new approach that integrates artificial intelligence (AI) with ANN and GA has been introduced by Gupta et al. [10], and the paper illustrates the impact of using AI on the quality and type of results obtained for the surface roughness prediction. They analyzed the experimental data using support vector regression (SVR) defining the tool wear and power required as output parameters.

Grade-H high-strength steel had its share in the investigation for better surface quality studied by Abbas et al. [11] in their work. They emphasized that the key factors for the manufacturing of parts produced using Grade-H high-strength steel are parts accuracy and surface roughness MRR. Identifying the final surface roughness of produced parts prior to machining is crucial to ensure that those parts will not be rejected. The rejection of these parts at any processing stage will represent huge problems to any factory because the processing and raw material of these parts are very expensive. ANN was used in this work to determine the optimized cutting parameters to ensure minimum surface roughness during the turning operations.

As a continuation of their work, Abbas et al. [12] investigated the turning of high-strength steel focusing on three main cutting parameters: cutting speed, feed rate, and depth of cut. Their results included a Pareto frontier between surface roughness and machining time of finished components made from high-strength steel using the ANN model that was later used to determine the optimum cutting conditions. This study showed the feasibility of integrating optimization algorithms with computer-aided manufacturing CAM systems using Matlab.

A quantitative approach to evaluate the cutting process and its stability was demonstrated in the work of Yamane et al. [13], and they used the turning operation as a base for their study aiming at identifying the machining system deviation from a perfect process. Such a deviation can be identified by monitoring the machined surface and comparing it with the cutter profile. Adhesion and builtup edge produced during machining operation can then be easily noticed and monitored. Excessive vibration and the accuracy of spindle rotation can also be recorded and is a good indication of system instability and related directly to the quality of parts produced. Their conclusion was that the proposed method can be successfully implemented to evaluate turning operations.

The influence of the type of inserts used in the machining process on the quality of the surface produced was investigated in the work of D'Addona and Raykar [14]. They compared wiper inserts to conventional ones in the turning operation of oil hardened nonshrinking steel used in the manufacture of strain gauges and measuring instruments. Surface roughness was a major factor in this study as it is a very important aspect in the performance of those devices.

They used analysis of variance (ANOVA) and analysis of means (AOM) plots to evaluate their results.

Pu et al. [15] reported that magnesium alloys are gaining a lot of intension from researchers in the literature due to their advanced properties over conventional materials used in the automotive industry as well as medical applications such as biodegradable implants. One of the major factors looked at in machining of this alloy is the surface integrity. Pu et al. [15] investigated the effect of machining AZ31B Mg alloy under dry conditions as well as using liquid nitrogen as a lubricant. They concluded that using cryogenic machining with large nose radii improved several material performance criteria such as surface finish and grain size refinement.

Increasing productivity and maximizing material removal rate (MMR) have been also investigated for the machining of magnesium alloys. Using very high cutting speeds has its drawbacks and has been analyzed by Tomac et al. [16] in their work. They concluded that using speeds in excess of 600 m/min will result in buildup edge on the flank face of the cutting tool. They supported their argument with microstructure pictures of tool inserts as well as the machined surface of three different Mg-Al-Zn alloys.

Different coating materials have been used in industry to reduce the builtup edge effect appearing on the tool flank face during turning operations of magnesium alloys at high speeds. Tönshoff and Winkler [17] reported the different interactions happening between the cutting tool inserts, the coating, and the workpiece materials in turning of AZ91 HP at very high speeds ranging between 900 and 2400 m/min. They concluded that cutting tools with polycrystalline diamond (PCD) inserts can significantly reduce the cutting forces and hence the frictions at the tool-workpiece interface.

The optimization of cutting parameters is another venue pursued by researchers to improve surface quality of machined magnesium alloys. Wojtowicz et al. [18] studied the effect of changing cutting parameters on the turning of AZ91 HP. Parameters explored include cutting speed, feed rate, depth of cut, and tool nose radius. Surface integrity and increasing fatigue life were the major optimized parameters of the machined components, and other reported parameters include microstructure, grain size and residual stresses improve fatigue life. They also supported the argument provided by Tönshoff and Winkler [17] regarding the superior performance of PCD coating tool inserts at high cutting speeds and feed rates.

In this paper, an ANN model has been employed to estimate and optimize the produced surface roughness of AZ61 material during turning operations. This method proves to be more efficient and provided the manufacturing engineers with a good tool to be utilized to effectively predict the quality of the surface produced in an economical and time saving manner. Thus this eliminates the possibility of part rejection due to manufacturing process errors that costs the factory time and money and wasted raw material.

2. Materials and Methods

Table 1 presents the chemical composition of magnesium alloy AZ61 which contains zinc and aluminum with 1 and 6

percent content, respectively. The microstructure of the composition is analyzed at the previously mentioned aluminum concentration, and the phase diagram shows a magnesium-rich phase interacting with an $Al_{12}Mg_{17}$ composite. Zinc along with other traces is found to have no effect on the alloy microstructure.

The machining of test specimens is done using Emco mill concept 45 CNC turning machine equipped with Sinumeric 840-D. The diameter of the workpiece is equal to 40 mm with a length of 100 mm. Tool holder specification is SVJCL2020K16, while the insert is VCGT160404 FN-ALU. The cutting edge angle, nose radius, and clearance angle are set at 35°, 0.4 mm, and 5°, respectively. All experiments were conducted in wet conditions while the cutting parameters are controlled via CNC part program. The surface roughness tester TESA Rugosurf 90-G is used to evaluate the produced surface roughness. A sketch of the test specimen is shown in Figure 1. The test plan was implemented through 64 turning runs. These runs were divided into 16 groups. Each of four groups was subjected to one common cutting speed (125, 150, 175, and 200 m/min). Each group was machined using four levels of cutting depth (0.30, 0.60, 0.90, and 1.12 mm). Each depth was processed using feed rate having four levels (0.05, 0.010, 0.15, and 0.20 mm/rev). Full listing of all samples and the resulting measured surface roughness are provided in Appendix A.

Multivariable regression analysis was used to build a mathematical model relating the process outcome (surface roughness R_a) with the three studied input parameters (cutting speed (V), depth of cut (d), and feed rate (fr)). 56 experiments were conducted that cover the input parameter range described previously. Eight extra experimental runs were carried out to be used in testing both the regression and ANN models.

Regression was conducted using Minitab 17 software with stepwise technique to eliminate the insignificant terms from the model. The model was fitted in the form given by the following equation [19]:

$$
\begin{aligned}
R_a = \beta_o &+ \sum_{i=1}^{k}\beta_i X_i + \sum_{i=1}^{k}\beta_{ii}X_i^2 + \sum\sum_{i<j}\beta_{ij}X_iX_j \\
&+ \sum\sum_{i<j<k}\beta_{ijk}X_iX_jX_k + \sum\sum_{i\neq j}\beta_{iij}X_i^2X_j + \varepsilon_i,
\end{aligned}
\tag{1}
$$

where β_o is the constant term, β_i represents the linear effects, β_{ii} represents the pure quadratic effects, β_{ij} represents the second level interaction effects, β_{ijk} the third level interaction effects, β_{iij} represents the effect of interaction between linear and quadratic terms, and ε_i represents the error in predicting experimental surface roughness. The material removal rate (MRR) was calculated using (2) for each run. Desirability function approach was used to maximize MRR maintaining R_a below 0.4 as a maximum limit for the surface roughness value:

$$
MRR = 1000\, V * \text{fr} * d,
\tag{2}
$$

where MRR is the volume removed per unit time (mm³/min.), V is the cutting speed (m/min.), fr is the feed

TABLE 1: AZ61 magnesium alloy chemical composition.

Element	Aluminium	Zinc	Copper	Silicon	Iron	Nickel	Magnesium
% by mass	5.95	0.95	<0.03	<0.01	<0.01	<0.005	Balance

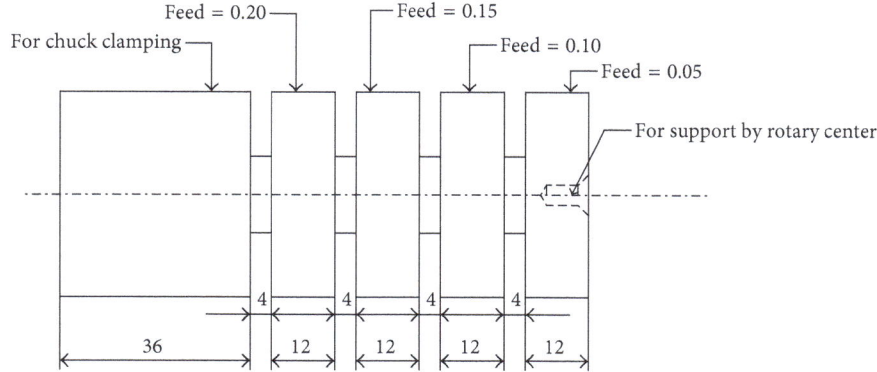

FIGURE 1: Working drawing of the test workpiece.

rate (mm/rev), and d is the depth of cut (mm). Multivariable regression analysis was used to build a mathematical model relating the process outcome (surface roughness R_a) with the three studied input parameters (cutting speed (V), depth of cut (d), and feed rate (fr)). 56 experiments were conducted that cover the input parameters range described previously. Eight extra experimental runs were carried out to be used in testing both the regression and ANN models.

3. Results and Discussions

The regression-fitted mathematical model is given by (3). Box-Cox transformation was used to normalize the residuals with $\lambda = 0$ (natural log) for R_a. Anderson–Darling test was conducted to check the normality of residuals with a result of p value $= 0.885 > 0.05$. The null hypothesis of such a test is that the data are normal, and a p value < 0.05 proves nonnormality.

Determination coefficient (R^2), mean square error (MSE), and mean absolute error (MAE) were calculated to be 0.975, 0.02, and 0.12, respectively. Figure 2 shows a scatter plot for the predicted R_a versus measured R_a. It is clear from the figure that the relation between them is very close to linear with calculated R^2 equals 0.97.

$$\ln(R_a) = -2.6420 + 0.1874\ d + 16.516\ \text{fr}. \qquad (3)$$

Desirability function approach was used to estimate the values of studied process parameters that maximize the MRR keeping R_a at levels not exceeding a practical value of 0.4 μm. The optimization plot, illustrated in Figure 3, shows that an optimum MRR of 13,928 mm^3/min with $R_a = 0.4\ \mu$m is obtained at cutting speed 200 m/min., depth of cut 1.12 mm, and feed rate 0.09 mm/rev.

ANN modeling was used to predict the surface roughness of AZ61 magnesium alloy. The three input parameters cutting speed (V), depth of cut (d), and feed rate (fr) were

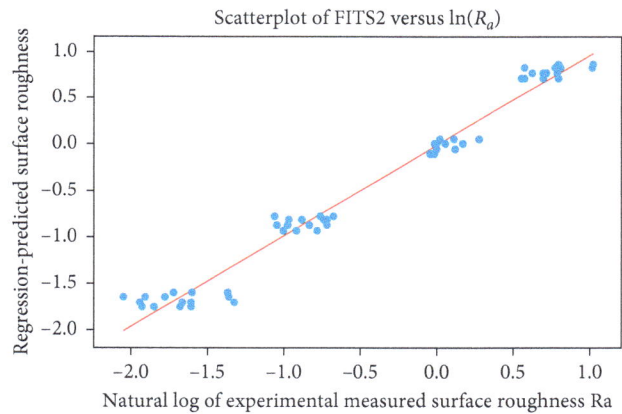

FIGURE 2: Experimental readings versus regression model predictions.

taken into account to predict the surface roughness as an output parameter. The data were taken from the experimental result. The transfer function was selected as a TanhAxon. The data were divided in two parts as training and test data. 80% of data were used for training stage while the left 20% of data were used for test stage to understand the performance of the developed ANN model. The best ANN model predicting the R_a value between developed trial models was obtained according to the values of R^2 and MSE. Figure 4 gives the experimental and ANN-predicted results. It can be seen from the figure that a good agreement was obtained between experimental and ANN-predicted results. It is also seen from Figure 5 that the value of R^2 is 0.9629 for experimental surface R_a and ANN-predicted surface R_a, while R^2 is 0.9931 for experimental MRR and ANN-predicted MRR in Figure 6.

Optimal		S	d	fr
D: 0.7016	High	200.0	1.120	0.20
	Cur	[200.0]	[1.120]	[0.0902]
Predict	Low	125.0	0.30	0.050

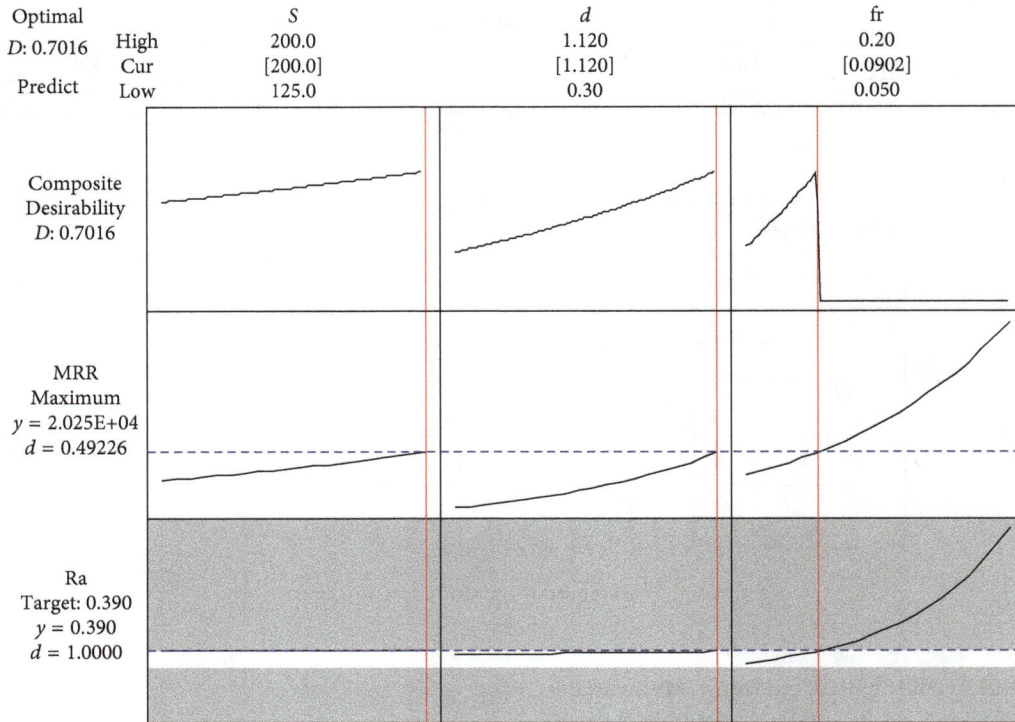

FIGURE 3: Optimization plot for R_a and MRR.

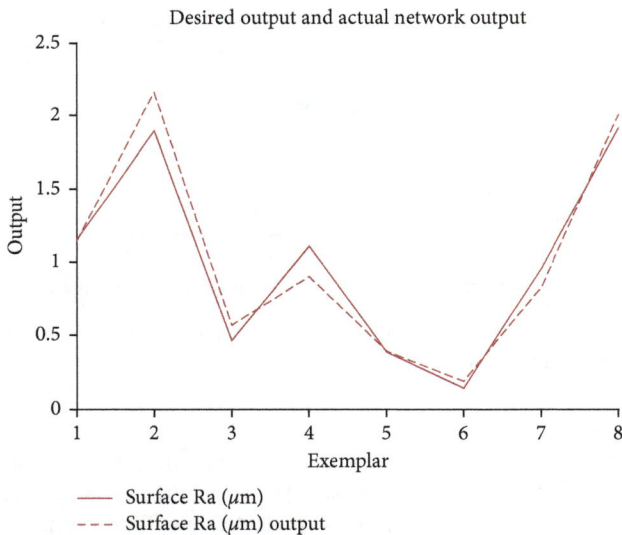

FIGURE 4: Desired output and actual network output.

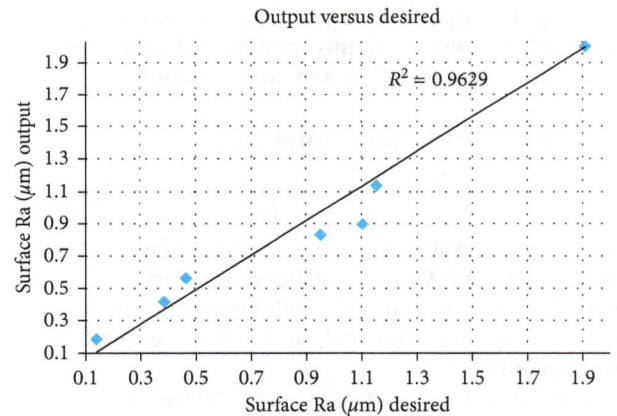

FIGURE 5: Comparison between ANN predicted and desired output for R_a.

Predictability of both regression and ANN models were compared using eight extra experimental cases that were not included in the modeling phase. Table 2 illustrates the results of this comparison. Figures 4–6 and Table 2 clearly show the good agreement between the results. From the table, both ANN and regression-predicted R_a and MRR values can be acceptable when they are compared with experimental results.

Feed rate has a direct impact on product surface finish. Although increasing feed rate can result in a significant increase in the material removal rate (MMR) and increase productivity, it will also increase cutting forces resulting in higher tool-workpiece friction associated with poor surface finish. Horizontal markings as well as vertical ones in surface roughness profile can be detected when examining the surface profile produced using high feed rates. MRR and surface roughness are two contradicting objectives in determining an optimized value for the feed rate. Optimization algorithms are often used to come up with optimized cutting parameters for different machining processes. Figure 7 represents an optical microscopy view of machined surface, while Figure 8 shows the surface roughness profile produced by

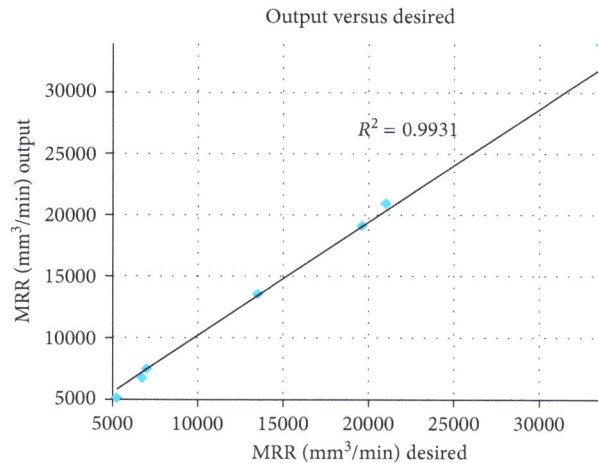

FIGURE 6: Comparison between ANN predicted and desired output for MRR.

TABLE 2: Comparison between the regression model and ANN predictions.

Machining parameters			Measured R_a (μm)	Regression model		ANN	
Speed	Depth	Feed		Predicted R_a	Residual	Predicted R_a	Residual
125	1.12	0.15	1.154	1.05	0.11	1.14	0.01
200	1.12	0.2	1.896	2.39	−0.49	2.16	−0.26
175	1.12	0.1	0.464	0.46	0.01	0.57	−0.11
150	0.6	0.15	1.1058	0.95	0.16	0.90	0.21
175	0.3	0.1	0.386	0.39	−0.01	0.39	−0.01
125	1.12	0.05	0.144	0.20	−0.06	0.19	−0.05
150	0.3	0.15	0.954	0.90	0.06	0.83	0.12
150	1.12	0.2	1.911	2.39	−0.48	2.00	−0.09

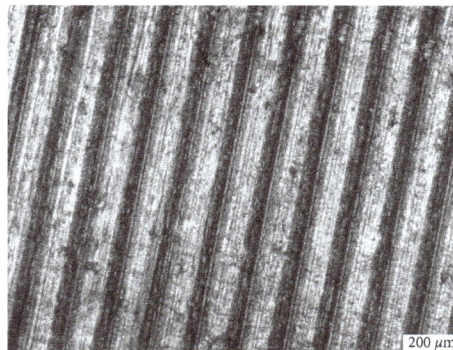

FIGURE 7: Optical microscopy for machined surface with $V = 125$ m/min, $d = 0.30$ mm, and fr $= 0.20$ mm/rev.

the surface roughness tester under the following cutting conditions: a speed of 125 m/min, a depth of cut of 0.3 mm, and a feed rate of 0.20 mm/rev. Another set of views and graph is also provided in Figures 9 and 10 for different cutting conditions summarized as follows: a cutting speed of 125 m/min, a depth of cut of 0.3 mm, and a feed rate of 0.05 mm/rev. From those results, we can conclude that reducing feed rate will produce thinner surface roughness markings, and increasing the feed rate is associated with the presence of distant thick surface roughness markings.

4. Conclusions

Optimization and estimation of R_a and MRR in cutting conditions of AZ61 magnesium alloy were realized by ANN modeling and regression analysis. The results of the developed ANN-predicted model and regression model were

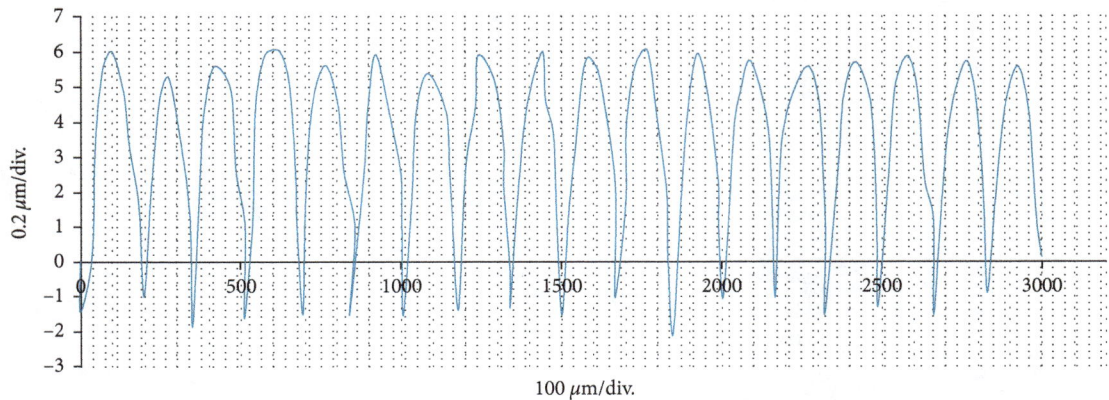

FIGURE 8: Profile of surface roughness graph by the surface roughness tester for machined surface with $V = 125$ m/min, $d = 0.30$ mm, and fr = 0.20 mm/rev.

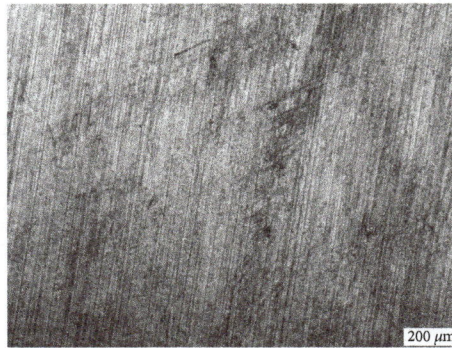

FIGURE 9: Optical microscopy for machined surface with $V = 125$ m/min, $d = 0.30$ mm, and fr = 0.05 mm/rev.

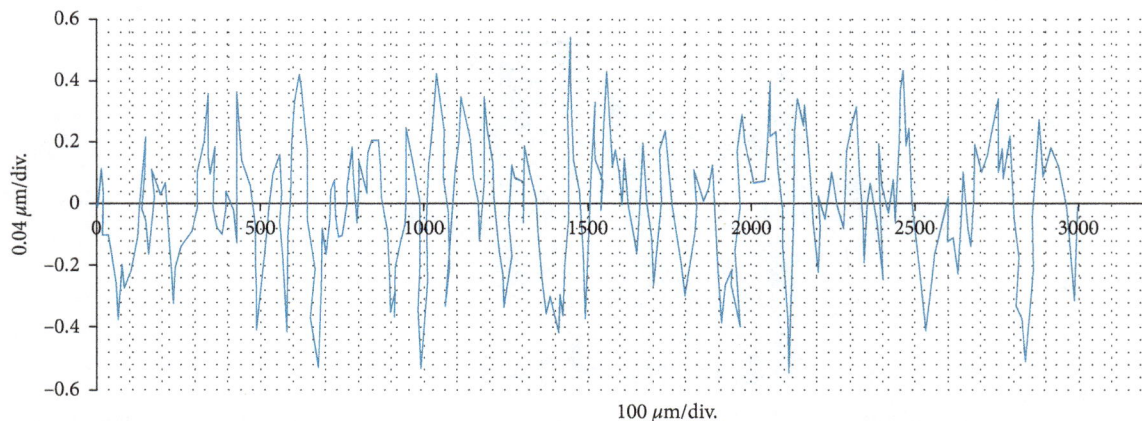

FIGURE 10: Profile of surface roughness graph by the surface roughness tester for machined surface with $V = 125$ m/min, $d = 0.30$ mm, and fr = 0.05 mm/rev.

compared with experimental results. The results showed that a good agreement was obtained for both developed ANN-predicted and regression analysis results. In addition, the compatibility between the developed ANN model and experimental results showed that ANN approach is an accurate method to estimate surface R_a and MRR. Optical microscopy views and the corresponding surface roughness profile for different sets of cutting parameters were utilized on two machined surfaces to showcase the direct effect of increasing the feed rate on surface finish. A hypothetical analysis relating the higher surface roughness values associated with the increase in feed rate is reported.

Appendix

The listing of all samples and the resulting measured surface roughness are provided in Table 3.

TABLE 3: Listing of four-level full factorial samples.

Sample ID	Group	Cutting speed (m/min)	Depth of cut (mm)	Feed (mm/rev)	Surface R_a (μm)	MRR (mm^3/min)
1		125	0.30	0.05	0.185	1875
2	1	125	0.30	0.10	0.457	3750
3		125	0.30	0.15	0.972	5625
4		125	0.30	0.20	1.779	7500
5		125	0.60	0.05	0.188	3750
6	2	125	0.60	0.10	0.375	7500
7		125	0.60	0.15	0.998	11250
8		125	0.60	0.20	2.205	15000
9		125	0.90	0.05	0.168	5625
10	3	125	0.90	0.10	0.378	11250
11		125	0.90	0.15	1.059	16875
12		125	0.90	0.20	2.254	22500
13		125	1.12	0.05	0.144	7000
14	4	125	1.12	0.10	0.466	14000
15		125	1.12	0.15	1.154	21000
16		125	1.12	0.20	2.786	28000
17		150	0.30	0.05	0.156	2250
18	5	150	0.30	0.10	0.365	4500
19		150	0.30	0.15	0.954	6750
20		150	0.30	0.20	1.737	9000
21		150	0.60	0.05	0.143	4500
22	6	150	0.60	0.10	0.349	9000
23		150	0.60	0.15	1.1058	13500
24		150	0.60	0.20	1.866	18000
25		150	0.90	0.05	0.128	6750
26	7	150	0.90	0.10	0.413	13500
27		150	0.90	0.15	0.999	20250
28		150	0.90	0.20	1.782	27000
29		150	1.12	0.05	0.201	8400
30	8	150	1.12	0.10	0.508	16800
31		150	1.12	0.15	1.020	25200
32		150	1.12	0.20	1.911	33600
33		175	0.30	0.05	0.199	2625
34	9	175	0.30	0.10	0.386	5250
35		175	0.30	0.15	0.982	7875
36		175	0.30	0.20	2.005	10500
37		175	0.60	0.05	0.199	5250
38	10	175	0.60	0.10	0.432	10500
39		175	0.60	0.15	1.128	15750
40		175	0.60	0.20	2.054	21000
41		175	0.90	0.05	0.256	7875
42	11	175	0.90	0.10	0.486	15750
43		175	0.90	0.15	1.187	23625
44		175	0.90	0.20	2.759	31500
45		175	1.12	0.05	0.178	9800
46	12	175	1.12	0.10	0.464	19600
47		175	1.12	0.15	1.316	29400
48		175	1.12	0.20	2.214	39200

Table 3: Continued.

Sample ID	Group	Cutting speed (m/min)	Depth of cut (mm)	Feed (mm/rev)	Surface R_a (μm)	MRR (mm^3/min)
49		200	0.30	0.05	0.145	3000
50	13	200	0.30	0.10	0.398	6000
51		200	0.30	0.15	0.955	9000
52		200	0.30	0.20	2.211	12000
53		200	0.60	0.05	0.265	6000
54	14	200	0.60	0.10	0.487	12000
55		200	0.60	0.15	0.999	18000
56		200	0.60	0.20	1.999	24000
57		200	0.90	0.05	0.148	9000
58	15	200	0.90	0.10	0.477	18000
59		200	0.90	0.15	0.987	27000
60		200	0.90	0.20	2.165	36000
61		200	1.12	0.05	0.254	11200
62	16	200	1.12	0.10	0.345	22400
63		200	1.12	0.15	1.115	33600
64		200	1.12	0.20	1.896	44800

Conflicts of Interest

The authors declare that they have no conflicts of interest.

Acknowledgments

This project was supported by King Saud University, Deanship of Scientific Research, College of Engineering Research Center.

References

[1] I. Asiltürk, "Predicting surface roughness of hardened AISI 1040 based on cutting parameters using neural networks and multiple regression," *International Journal of Advanced Manufacturing Technology*, vol. 63, no. 1–4, pp. 249–257, 2012.

[2] R. Mokhtari Homami, A. Fadaei Tehrani, H. Mirzadeh, B. Movahedi, and F. Azimifar, "Optimization of turning process using artificial intelligence technology," *International Journal of Advanced Manufacturing Technology*, vol. 70, no. 5–8, pp. 1205–1217, 2014.

[3] F. Jafarian, M. Taghipour, and H. Amirabadi, "Application of artificial neural network and optimization algorithms for optimizing surface roughness, tool life and cutting forces in turning operation," *Journal of Mechanical Science and Technology*, vol. 27, no. 5, pp. 1469–1477, 2013.

[4] Y. Karpat and T. Özel, "Multi-objective optimization for turning processes using neural network modeling and dynamic-neighborhood particle swarm optimization," *International Journal of Advanced Manufacturing Technology*, vol. 35, no. 3-4, pp. 234–247, 2007.

[5] C. Natarajan, S. Muthu, and P. Karuppuswamy, "Prediction and analysis of surface roughness characteristics of a non-ferrous material using ANN in CNC turning," *International Journal of Advanced Manufacturing Technology*, vol. 57, no. 9–12, pp. 1043–1051, 2011.

[6] F. J. Pontes, A. P. D. Paiva, P. P. Balestrassi, J. R. Ferreira, and M. B. D. Silva, "Optimization of Radial Basis Function neural network employed for prediction of surface roughness in hard turning process using Taguchi's orthogonal arrays," *Expert Systems with Applications*, vol. 39, no. 9, pp. 7776–7787, 2012.

[7] K. S. Sangwan, S. Saxena, and G. Kant, "Optimization of machining parameters to minimize surface roughness using integrated ANN-GA approach," *Procedia CIRP*, vol. 29, pp. 305–310, 2015.

[8] G. M. A. Acayaba and P. M. D. Escalona, "Prediction of surface roughness in low speed turning of AISI316 austenitic stainless steel," *CIRP Journal of Manufacturing Science and Technology*, vol. 11, pp. 62–67, 2015.

[9] D. Bajić, B. Lela, and G. Cukor, "Examination and modelling of the influence of cutting parameters on the cutting force and the surface roughness in longitudinal turning," *Strojniski Vestnik/Journal of Mechanical Engineering*, vol. 54, no. 5, pp. 322–333, 2008.

[10] A. K. Gupta, S. C. Guntuku, R. K. Desu, and A. Balu, "Optimisation of turning parameters by integrating genetic algorithm with support vector regression and artificial neural networks," *International Journal of Advanced Manufacturing Technology*, vol. 77, no. 1–4, pp. 331–339, 2015.

[11] A. T. Abbas, M. Alata, A. E. Ragab, M. M. El Rayes, and E. A. El Danaf, "Prediction model of cutting parameters for turning high strength steel grade-H: comparative study of regression model versus ANFIS," *Advances in Materials Science and Engineering*, vol. 2017, Article ID 2759020, 12 pages, 2017.

[12] A. T. Abbas, D. Y. Pimenov, I. N. Erdakov, T. Mikolajczyk, E. A. El Danaf, and M. A. Taha, "Minimization of turning time for high-strength steel with a given surface roughness using the Edgeworth–Pareto optimization method," *International Journal of Advanced Manufacturing Technology*, vol. 93, pp. 2375–2392, 2017.

[13] Y. Yamane, T. Ryutaro, S. Tadanori, I. Martinez Ramirez, and Y. Keijiaa Hiroshim, "A new quantitative evaluation for characteristic of surface roughness in turningYasuo," *Precision Engineering*, vol. 50, pp. 20–26, 2017.

[14] D. M. D'Addona and S. J. Raykar, "Analysis of surface roughness in hard turning using wiper insert geometry," *Procedia CIRP*, vol. 41, pp. 841–846, 2016.

[15] Z. Pu, J. C. Outeiro, A. C. Batista, O. W. Dillon, D. A. Puleo, and I. S. Jawahir, "Enhanced surface integrity of AZ31B Mg alloy by cryogenic machining towards improved functional

performance of machined components," *International Journal of Machine Tools and Manufacture*, vol. 56, pp. 17–27, 2012.

[16] N. Tomac, K. Tonnessen, and F. O. Rasch, "Formation of flank build-up in cutting magnesium alloys," *CIRP Annals-Manufacturing Technology*, vol. 40, no. 1, pp. 79–82, 1991.

[17] H. K. Tönshoff and J. Winkler, "The influence of tool coatings in machining of magnesium," *Surface and Coatings Technology*, vol. 94-95, pp. 610–616, 1997.

[18] N. Wojtowicz, I. Danis, F. Monies, P. Lamesle, and R. Chieragati, "The influence of cutting conditions on surface integrity of a wrought magnesium alloy," *Procedia Engineering*, vol. 63, pp. 20–28, 2013.

[19] E. A. Al Bahkali, A. E. Ragab, E. A. El Danaf, and A. T. Abbas, "An investigation of optimum cutting conditions in turning nodular cast iron using carbide inserts with different nose radius," *Proceedings of the Institution of Mechanical Engineers, Part B: Journal of Engineering Manufacture*, vol. 230, no. 9, pp. 1584–1591, 2016.

17

Effects of Magnesium Content on Structure and Electrochemical Properties of La-Mg-Pr-Al-Mn-Co-Ni Hydrogen Storage Alloys

L. M. C. Zarpelon, E. P. Banczek, L. G. Martinez, N. B. Lima, I. Costa, and R. N. Faria

Center of Science and Materials Technology, Nuclear and Energy Research Institute (IPEN-CNEN/SP), São Paulo, SP, Brazil

Correspondence should be addressed to R. N. Faria; rfaria@ipen.br

Academic Editor: Jörg M. K. Wiezorek

The discharge capacity, microstructures, and corrosion resistance of some as-cast alloys represented by the formula $La_{0.7-x}Mg_xPr_{0.3}Al_{0.3}Mn_{0.4}Co_{0.5}Ni_{3.8}$, where $x = 0.0$, 0.1, 0.3, 0.5, and 0.7, were investigated by SEM/EDX, XRD, and electrochemical measurements. The partial substitution of La by Mg refined the grain structure while the total substitution changed it from equiaxed to columnar. Three phases were detected: a major phase (M), a grey phase (G), and a dark phase (D). The compositions analyzed by EDX suggested that the M phase was close to a $LaNi_5$ phase. With the increase of the Mg content, the analyses revealed a G phase with composition close to a RMg_2Ni_9 (R = La,Pr) and a D phase close to a $MgNi_2$ phase. The XRD analysis and Rietveld refinement corroborated the EDX results. The corrosion resistance of the alloys was evaluated in $6.0 \, mol \cdot L^{-1}$ KOH solution, and the results showed that the substitution of La by Mg was beneficial for this alloy property. Nevertheless, Mg addition was deleterious to the discharge capacity of the electrodes.

1. Introduction

Over the past years, extensive research has been concentrated on the study of hydrogen storage alloys as a negative electrode of nickel-metal hydride (Ni/MH) secondary battery, as shown in recent reviews on the subject [1–5]. Commercial alloy systems for Ni/MH batteries are rare earth-based AB_5-type alloys, Ti- and Zr-based AB_2-type alloys, and, recently, A_2B_7- and AB_3-type RE-Mg-Ni-based superlattice alloys. Although representing the first generation of negative electrodes, the rare earth-based AB_5-type alloys (with discharge capacity limited to 320 mAh/g) are very popular in use on commercial Ni/MH batteries [1]. Nickel substitution in La-Ni-type electrodes of AB_5 system multicomponent metal hydride alloys has been widely reported, and the purpose of alloy modification is to improve the electrode performance. Aluminum, manganese, and/or cobalt are always present in Ni/MH electrodes. Lanthanum is frequently substituted for cerium, neodymium, and/or praseodymium. Magnesium has also been included in this type of hydrogen storage alloys as an element that can increase the number of hydrogen atoms stored per metal atom [6]. Mg additions vary from impurity levels to considerable atomic concentrations aiming to reduce costs (by reducing the amount of Co) and improving cyclic stability or durability of the Ni/MH batteries [7–11].

Reported microstructural investigations and chemical analyses for as-cast hydrogen storage alloys with Mg addition are very scarce. This paper addresses this aspect and reports the results of a study with partial and total substitution of La by Mg, on hydrogen storage $La_{0.7-x}Mg_xPr_{0.3}Al_{0.3}Mn_{0.4}Co_{0.5}Ni_{3.8}$ as-cast alloys ($x = 0.0$, 0.1, 0.3, 0.5 and 0.7). A thorough investigation of the microstructures of these alloys and the phases present has been carried out using SEM/EDX and XRD. Moreover, it is well known that the Ni/MH batteries work in a strong oxidizing medium composed of high-concentration alkaline electrolyte. Therefore, among the desired properties of negative electrodes alloys, high corrosion resistance is essential for long cycle lifetime [6–13]. In the present study, the corrosion resistance of the La-Mg-Pr-Al-Mn-Co-Ni alloys in $6.0 \, mol \cdot L^{-1} \cdot KOH$ solution was also investigated.

2. Experimental Procedures

The alloys investigated in this study were commercially prepared in 5 kg batches melted in induction heating vacuum equipment and cast in a cooled mold. The chemical

TABLE 1: Composition of the as-cast $La_{0.7-x}Mg_xPr_{0.3}Al_{0.3}Mn_{0.4}Co_{0.5}Ni_{3.8}$ alloys.

Nominal composition	x	Specified composition (at.%)						
		La	Pr	Mg	Al	Mn	Co	Ni
$La_{0.7}Pr_{0.3}Al_{0.3}Mn_{0.4}Co_{0.5}Ni_{3.8}$	0.0	11.67	5.00	—	5.00	6.67	8.33	63.33
$La_{0.6}Mg_{0.1}Pr_{0.3}Al_{0.3}Mn_{0.4}Co_{0.5}Ni_{3.8}$	0.1	10.00	5.00	1.67	5.00	6.67	8.33	63.33
$La_{0.4}Mg_{0.3}Pr_{0.3}Al_{0.3}Mn_{0.4}Co_{0.5}Ni_{3.8}$	0.3	6.67	5.00	5.00	5.00	6.67	8.33	63.33
$La_{0.2}Mg_{0.5}Pr_{0.3}Al_{0.3}Mn_{0.4}Co_{0.5}Ni_{3.8}$	0.5	3.34	5.00	8.33	5.00	6.67	8.33	63.33
$Mg_{0.7}Pr_{0.3}Al_{0.3}Mn_{0.4}Co_{0.5}Ni_{3.8}$	0.7	—	5.00	11.67	5.00	6.67	8.33	63.33

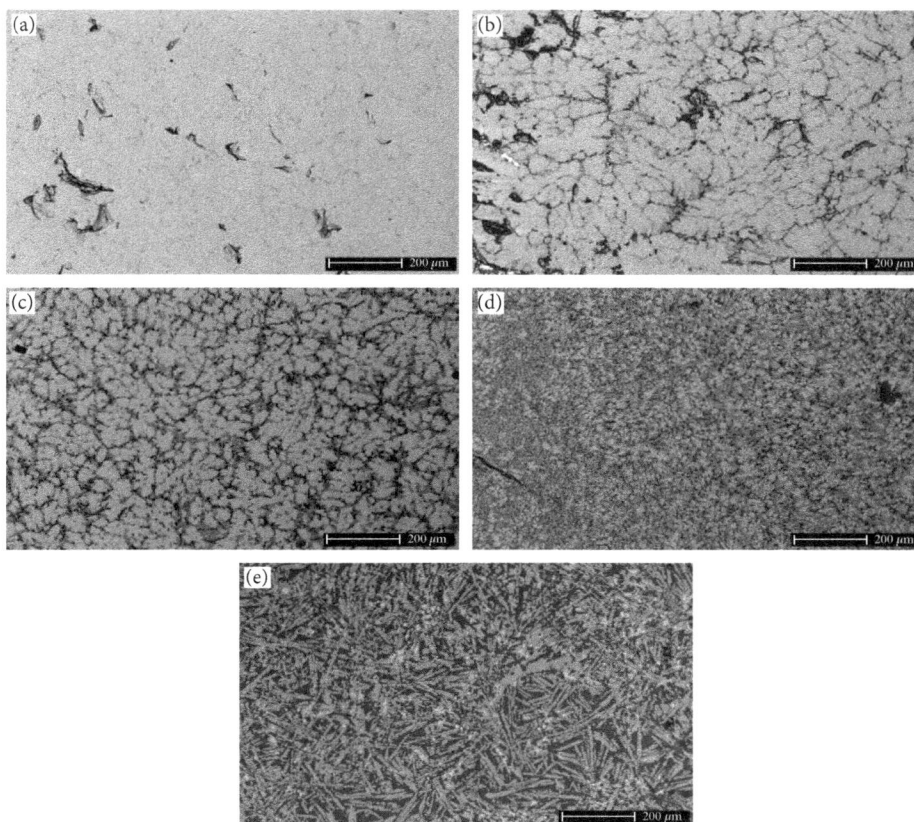

FIGURE 1: SEM micrographs showing a general view of the alloys investigated: (a) $La_{0.7}Pr_{0.3}Al_{0.3}Mn_{0.4}Co_{0.5}Ni_{3.8}$; (b) $La_{0.6}Mg_{0.1}Pr_{0.3}Al_{0.3}Mn_{0.4}Co_{0.5}Ni_{3.8}$; (c) $La_{0.4}Mg_{0.3}Pr_{0.3}Al_{0.3}Mn_{0.4}Co_{0.5}Ni_{3.8}$; (d) $La_{0.2}Mg_{0.5}Pr_{0.3}Al_{0.3}Mn_{0.4}Co_{0.5}Ni_{3.8}$; (e) $Mg_{0.7}Pr_{0.3}Al_{0.3}Mn_{0.4}Co_{0.5}Ni_{3.8}$.

analyses of the as-cast alloys are given in Table 1. For comparison convenience, a conversion to the substitution composition (atomic %) is also provided.

Agreement has been found between the composition specified values and that determined by analyses in the alloys. As per the supplier's analysis, the alloys contained sulfur, oxygen, carbon, and nitrogen as impurities (≤100 ppm). Specimens for microstructure investigations were prepared using standard metallographic methods. The microstructures of the samples were examined using a scanning electron microscopy (SEM) with energy dispersive X-ray (EDX) analysis facility (Philips XL30). Average data were obtained from various independent measurements from each phase. The crystal structure of the alloys was identified by X-ray diffraction (XRD) with a Cu K_α radiation (Rigaku DMAX-2000). The phase abundance, lattice parameters, and cell volume of the

alloy phases were obtained from Rietveld refinement method using GSAS (General Structure Analysis System) program.

The corrosion studies of the alloys was evaluated by electrochemical methods, specifically electrochemical impedance spectroscopy (EIS) and potentiodynamic polarization curves (anodic and cathodic, separately). Electrochemical measurements have been used in previous papers [14, 15]. A three-electrode cell was employed in this investigation with a Pt counter electrode and a mercurous oxide (Hg/HgO/6.0 mol·L⁻¹·KOH solution) as a standard reference electrode. The working electrode was produced using cold epoxy resin mounting with a contact wire for electric connection. The surface for exposure to the electrolyte was polished prior testing. The corrosion behavior was investigated with a Gamry frequency response analyzer (EIS 300) linked to a potentiostat (PCI4/300). The corrosion tests were performed in a 6.0 mol·L⁻¹·KOH solution at room

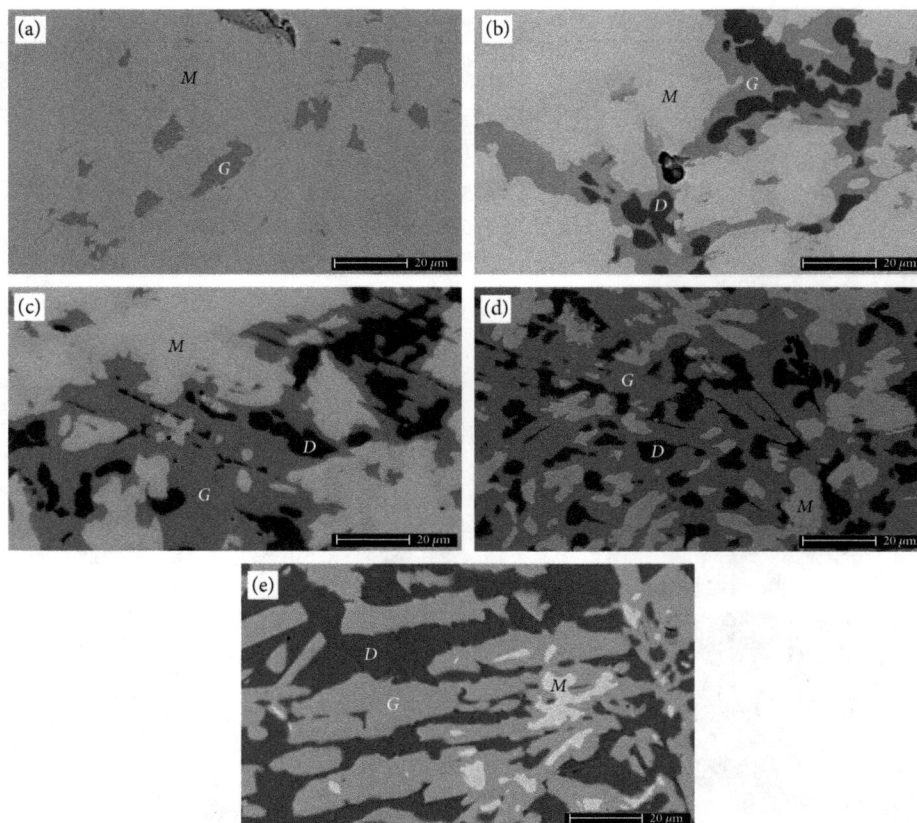

FIGURE 2: SEM micrographs showing the M, G, and D phases of the alloys investigated: (a) $La_{0.7}Pr_{0.3}Al_{0.3}Mn_{0.4}Co_{0.5}Ni_{3.8}$; (b) $La_{0.6}Mg_{0.1}Pr_{0.3}Al_{0.3}Mn_{0.4}Co_{0.5}Ni_{3.8}$; (c) $La_{0.4}Mg_{0.3}Pr_{0.3}Al_{0.3}Mn_{0.4}Co_{0.5}Ni_{3.8}$; (d) $La_{0.2}Mg_{0.5}Pr_{0.3}Al_{0.3}Mn_{0.4}Co_{0.5}Ni_{3.8}$; (e) $Mg_{0.7}Pr_{0.3}Al_{0.3}Mn_{0.4}Co_{0.5}Ni_{3.8}$.

temperature $(20 \pm 2)°C$. All the reagents employed for the test solution were per analytical (p.a.) grade.

The specimens were immersed in the test solution, and the open circuit potential (OCP) was determined as a function of time. EIS tests were performed employing perturbation voltage amplitude of ± 10 mV, relatively to the OCP, from 10 kHz to 10 mHz, with an acquisition rate of 10 points/decade. Soon after the EIS tests, the OCP was measured to test the stability of the potential. Potentiodynamic polarization curves were obtained employing a scan rate of 30 mV/s. Polarization curves with slower scan rates were unsuccessful due to the highly passive character of the alloys in the electrolyte employed (resulting only in scattered data). Four specimens of each commercial alloy were examined in this investigation.

Discharge capacity test electrodes were produced by pressing a mixture of 100 mg of the commercial alloy (powdered by manual crushing to a 270 mesh sieve) and teflonized carbon (Vulcan XC-72R with 33 wt.% polytetrafluoroethylene) in a weight ratio of 1 : 1 on both sides of a nickel mesh (10 mm diameter). The charge/discharge cycling tests were carried out at 25°C employing a commercial battery analyzer (Arbin BT-4). Discharge capacity assessments were performed in a trielectrode cell consisting of the prepared working electrode, a coiled Pt wire (0.5 mm in diameter and 300 mm long) counter electrode, and a $Hg/HgO/6.0$ mol·L^{-1}·KOH solution reference electrode in a 6.0 mol·L^{-1}·KOH electrolyte solution. The working electrode was charged at 100 mA/g for 5 h and subsequently discharged at 50 mA/g to the cutoff potential of −0.6 V versus the Hg/HgO reference electrode.

3. Results and Discussion

3.1. SEM and EDX Characterization. Backscattered electron micrographs showing a general view and details of the as-cast alloys are shown in Figures 1 and 2, respectively. The $La_{0.7}Pr_{0.3}Al_{0.3}Mn_{0.4}Co_{0.5}Ni_{3.8}$ alloy (Figure 2(a)) is composed of a major phase (M) and a grey phase (G) in the grain boundaries. The other alloys are composed of three phases: M, G, and a dark phase (D). The proportion of these three phases in the alloys changed as Mg was included and also as its content increased. Total substitution of La by Mg ($x = 0.7$) makes the M phase the minor phase in the $Mg_{0.7}Pr_{0.3}Al_{0.3}Mn_{0.4}Co_{0.5}Ni_{3.8}$ alloy (Figure 2(e)). In this case, the M phase is inside the grey phase. The morphologies of the as-cast alloys also reveal that substitution of lanthanum by magnesium results in a marked refinement in the grain structure of the alloys (Figure 1). Clearly, the magnesium-free alloy shows a coarse equiaxed grain structure, whereas the lanthanum-free alloy reveals a fine structure with columnar grain. Previous study has shown that the total substitution of lanthanum by praseodymium in the alloy changed the grain structure from equiaxed to columnar [16]. This result and the present observation

TABLE 2: Composition determined using EDX at the centers of the M phase in the as-cast $La_{0.7-x}Mg_xPr_{0.3}Al_{0.3}Mn_{0.4}Co_{0.5}Ni_{3.8}$ alloys.

x (N^*)	Analyzed composition (at.%)							Ratio**
	La	Pr	Mg	Al	Mn	Co	Ni	
0.0 (5)	12.0 ± 0.2	4.7 ± 0.4	—	5.5 ± 0.5	5.8 ± 1.4	8.6 ± 0.2	63.4 ± 1.3	5.0
0.1 (2)	11.0 ± 0.5	5.3 ± 0.1	—	3.6 ± 1.1	3.1 ± 1.1	7.7 ± 0.2	69.4 ± 2.3	5.1
0.3 (11)	8.5 ± 0.1	7.0 ± 0.4	—	4.2 ± 0.6	3.1 ± 0.9	8.2 ± 0.4	68.0 ± 1.5	5.4
0.5 (11)	5.9 ± 0.2	9.7 ± 0.3	—	3.6 ± 0.4	3.1 ± 0.5	8.0 ± 0.4	70.0 ± 0.7	5.4
0.7 (5)	—	14.0 ± 0.2	—	3.4 ± 0.4	2.8 ± 0.3	7.6 ± 0.3	72.0 ± 0.8	6.1

*Number of independent measurements from the M phase. **(La,Pr)/(Al,Mn,Co,Ni).

TABLE 3: Composition determined using EDX at the centers of the G phase in the as-cast $La_{0.7-x}Mg_xPr_{0.3}Al_{0.3}Mn_{0.4}Co_{0.5}Ni_{3.8}$ alloys.

x (N^*)	Analyzed composition (at.%)							Ratio**
	La	Pr	Mg	Al	Mn	Co	Ni	
0.0 (8)	8.1 ± 0.4	2.3 ± 0.4	—	4.0 ± 0.3	21.8 ± 1.4	8.5 ± 0.8	55.3 ± 1.3	—
0.1 (4)	6.4 ± 0.3	2.9 ± 0.2	10.9 ± 0.4	2.9 ± 0.2	10.3 ± 0.4	7.9 ± 0.3	58.7 ± 0.5	1:2.3:7.5
0.3 (7)	5.0 ± 0.3	4.3 ± 0.2	12.0 ± 0.6	3.7 ± 0.1	6.5 ± 0.5	7.4 ± 0.2	61.0 ± 0.3	1:2.0:8.0
0.5 (2)	3.5 ± 0.2	5.9 ± 0.3	12.4 ± 0.4	2.8 ± 0.1	4.8 ± 0.5	6.8 ± 0.6	63.8 ± 0.2	1:1.8:7.8
0.7 (6)	—	9.1 ± 0.3	12.5 ± 0.5	1.7 ± 0.2	3.8 ± 0.3	5.8 ± 0.5	67.0 ± 0.7	1:1.8:8.2

*Number of independent measurements from the G phase. **(La,Pr)/(Mg)/(Al,Mn,Co,Ni).

TABLE 4: Composition determined using EDX at the centers of the D phase in the as-cast $La_{0.7-x}Mg_xPr_{0.3}Al_{0.3}Mn_{0.4}Co_{0.5}Ni_{3.8}$ alloys.

x (N^*)	Analyzed composition (at.%)							Ratio**
	La	Pr	Mg	Al	Mn	Co	Ni	
0.1 (4)	<1	<1	<1	16–17	23–24	7–8	49–50	1.4
0.3 (6)	<1	<1	1–8	10–17	15–16	15–16	56–57	2.8
0.5 (8)	<1	<1	1–18	3–11	12–13	15–16	60–63	3.1
0.7 (5)	—	<1	21–23	1–2	7–9	5–13	61–65	2.0

*Number of independent measurements from the D phase. **(Pr,Mg,Al,Mn)/(Co,Ni).

indicate that the presence of lanthanum in the alloys favors an equiaxed grain structure.

The chemical composition of the major phase in the different alloys, as determined by EDX, is presented in Table 2. The M phase in the Mg-free alloy revealed a (La,Pr) : (Al, Mn,Co,Ni) atomic ratio of 5.0, indicating it to be a 1 : 5 phase (similar to $LaNi_5$). This result would be in agreement with the $LaNi_5$ matrix phase observed in previous studies [17–20]. The subsequent alloys ($x > 0.0$) showed that this (La,Pr) : (Al, Mn,Co,Ni) atomic ratio increases due to the substitution of La by Mg. Despite the increase of the praseodymium content, the total rare earth content in the M phase decreased from 16.7 at.% to 14.0 at.% with the total substitution of La by Mg. Conversely, the nickel content in the M phase of these alloys increased from 63.4 at.% to 72.0 at.% with this substitution. The amount of aluminum and manganese in this phase decreased somewhat with increasing magnesium content in the alloys. The cobalt content showed an average close to 8 at.%.

The chemical composition of the grey phase observed in all alloys, also determined using EDX, is shown in Table 3. As in the previous case, in all alloys the La content showed a decrease, whereas the Pr content increases in this phase. The measured value of the rare earth (9.1~10.4 at.%) was

more than half of that found in the M phase (14.0~16.7 at.%). The magnesium content in this grey phase was as high as 12.5 at.%, which indicates the preferential concentration of this element in this phase. Similar to that observed in the matrix phase, the amount of Al and Mn in the grey phase diminished with the substitution of La by Mg. The aluminum content in the G phase (1.7~4.0 at.%) was somewhat lower than that found in the M phase (3.4~5.5 at.%). It was also detected a manganese concentration peak in the Mg-free alloy (~22 at.%), which indicated the presence of a possible phase rich in Mn, in a very reduced amount, as observed in the correspondent micrograph (Figure 2(a)). The presence of this phase can be attributed to the as-cast state of the alloys. On the other hand, in the alloys with $x > 0.0$, the grey phase revealed a (La,Pr) : Mg : (Al,Mn,Co, Ni) atomic ratio close to 1 : 2 : 9. This ratio would indicate the presence of phase similar to the RMg_2Ni_9 type, in agreement with the system AB_2C_9 [21, 22]. The average cobalt content (~7 at.%) was close to that in the M phase (~8 at.%).

The chemical composition of the dark phase in the different alloys is presented in Table 4. This phase has not been found in the Mg-free alloy. Apparently, magnesium influences the microstructure and also the phases that can be found in the alloys. In this phase, the rare earth content was

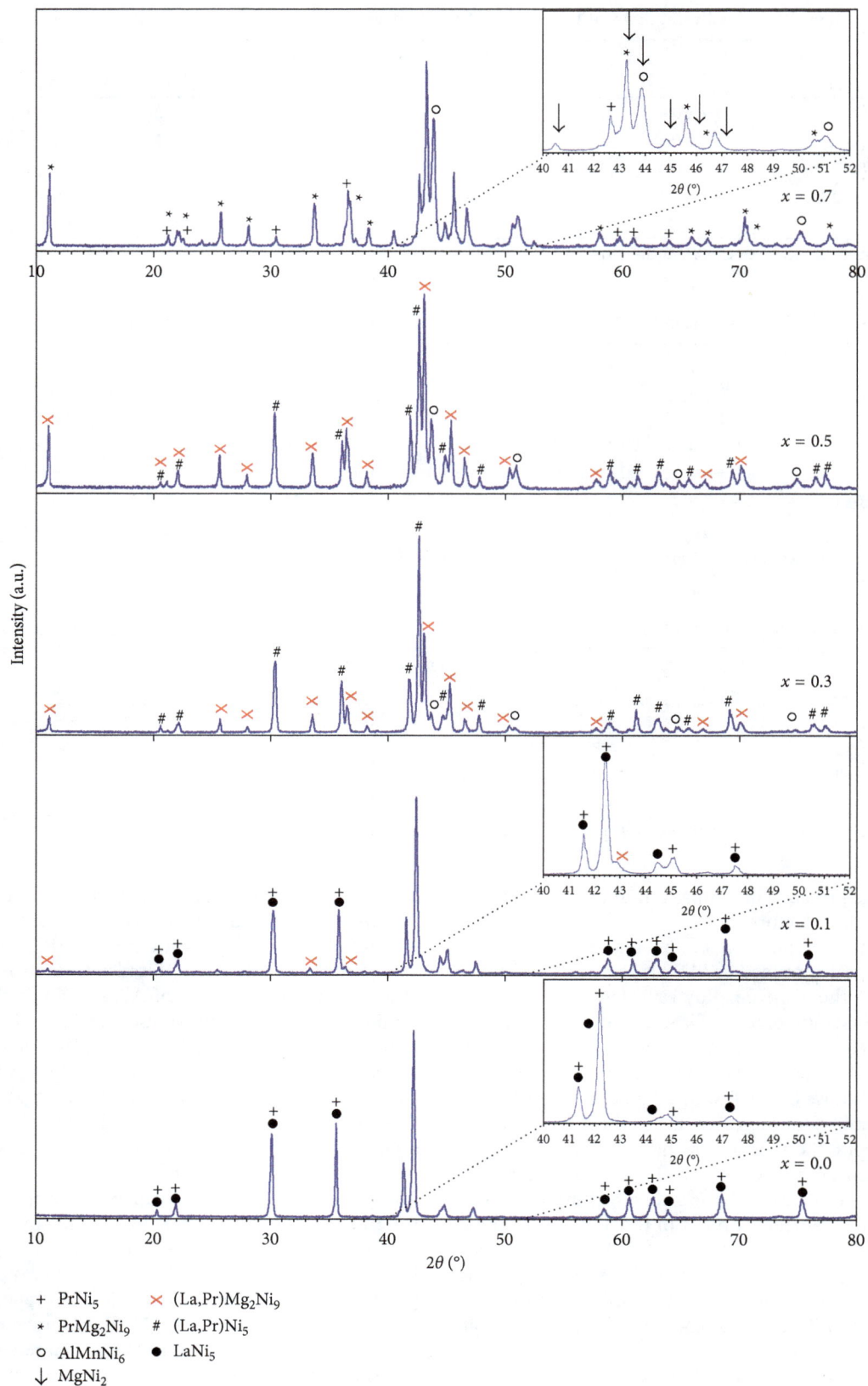

+ PrNi$_5$ × (La,Pr)Mg$_2$Ni$_9$
* PrMg$_2$Ni$_9$ # (La,Pr)Ni$_5$
○ AlMnNi$_6$ ● LaNi$_5$
↓ MgNi$_2$

FIGURE 3: XRD profiles of the La$_{0.7-x}$Mg$_x$Pr$_{0.3}$Al$_{0.3}$Mn$_{0.4}$Co$_{0.5}$Ni$_{3.8}$ (x = 0.0–0.7) alloys.

TABLE 5: The characteristics of alloy phases in the $La_{0.7-x}Mg_xPr_{0.3}Al_{0.3}Mn_{0.4}Co_{0.5}Ni_{3.8}$ ($x = 0.0$–0.7) alloys.

Samples	Phases	Space group	Parameters of fit[a]	Phase abundance (wt.%)	Lattice parameter (Å)		Cell volume (Å^3)
					a	c	
$x = 0.0$	LaNi$_5$	P6/mmm	$R_p = 0.2660$	74.2	5.047	4.063	89.6 ± 0.5
	PrNi$_5$	P6/mmm	$R_{wp} = 0.4664$, $\chi^2 = 2.757$	25.8	5.034	4.034	88.5 ± 0.5
$x = 0.1$	LaNi$_5$	P6/mmm	$R_p = 0.2416$	43.0	5.017	4.062	88.5 ± 0.5
	PrNi$_5$	P6/mmm	$R_{wp} = 0.4440$, $\chi^2 = 2.467$	45.5	5.020	4.019	87.7 ± 0.5
	(La,Pr)Mg$_2$Ni$_9$	R-3m		11.5	4.950	24.299	516 ± 1
$x = 0.3$	(La,Pr)Ni$_5$	P6/mmm	$R_p = 0.2416$	60.7	4.996	4.051	87.6 ± 0.5
	(La,Pr)Mg$_2$Ni$_9$	R-3m	$R_{wp} = 0.4440$, $\chi^2 = 2.467$	29.3	4.936	24.104	509 ± 1
	AlMnNi$_6$	Pm3m		10.0	3.603	3.603	46.8 ± 0.5
$x = 0.5$	(La,Pr)Ni$_5$	P6/mmm	$R_p = 0.2469$	32.6	4.981	4.041	86.8 ± 0.5
	(La,Pr)Mg$_2$Ni$_9$	R-3m	$R_{wp} = 0.6088$, $\chi^2 = 2.148$	42.6	4.932	23.993	505 ± 1
	AlMnNi$_6$	Pm3m		24.8	3.590	3.590	46.3 ± 0.5
$x = 0.7$	PrNi$_5$	P6/mmm	$R_p = 0.2056$	5.8	4.943	4.028	85.2 ± 0.5
	PrMg$_2$Ni$_9$	R-3m	$R_{wp} = 0.3734$, $\chi^2 = 2.249$	46.4	4.918	23.906	501 ± 1
	AlMnNi$_6$	Pm3m		33.5	3.583	3.583	46.0 ± 0.5
	MgNi$_2$	P6$_3$/mmc		14.3	4.834	15.824	320 ± 1

[a]R_p = pattern factor; R_{wp} = weighted pattern factor; χ^2 = goodness of fit.

close or below the EDX detection limit, indicating the preferential concentration of these elements in the other detected phases. For $x = 0.1$, the dark phase was richer in manganese and aluminum in contrast to the other constituents (except Ni). Only in the La-free alloy ($x = 0.7$) was observed a magnesium concentration peak. The approximated atomic relation between Mg and Ni in this alloy suggests the presence of similar phase to the MgNi$_2$ phase. This observation would be in agreement upon observations in the synthesis of ternary RMg$_2$Ni$_9$ alloys by sintering of powders of MgNi$_2$ and RNi$_5$ (R = La, Ce, Pr, Nd, Sm, and Gd) [22]. Except for $x = 0.7$, the values measured showed a preferential presence of aluminum in the D phase in comparison with the concentrations found in the M and G phases. In the case of manganese, after the peak of concentration in $x = 0.1$, it was observed a decrease in the other compositions. For the cobalt content, some variations with higher values of $x = 0.3$ and 0.5 were observed. It must be kept in mind that the characterization of the alloys was carried out in the as-cast state, and deviation in the results can be attributed to the alloys' heterogeneous condition. This is consistent with previous studies [18, 23], where it was shown that, in a La$_{0.7}$Mg$_{0.3}$Al$_{0.2}$Mn$_{0.1}$Co$_{0.75}$Ni$_{2.45}$ alloy, annealing at high temperature was essential to achieve a homogeneous composition. EDX determinations indicate phase compositions close to LaNi$_5$ in the phase matrix, RMg$_2$Ni$_9$ (R = La, Pr) in the grey phase, and MgNi$_2$ in the dark phase.

3.2. XRD Characterization.
The XRD profiles of the La$_{0.7-x}$Mg$_x$Pr$_{0.3}$Al$_{0.3}$Mn$_{0.4}$Co$_{0.5}$Ni$_{3.8}$ ($x = 0.0, 0.1, 0.3, 0.5, 0.7$) as-cast alloys are shown in Figure 3. It can be observed that the diffraction peaks were shifted to higher angles due to the La substitution (atomic radius 2.74 Å) by Mg (atomic radius 1.72 Å), also noting the constant presence of Pr (atomic radius 2.67 Å) in all compositions. The identification of phases by XRD was based on the results obtained by SEM/EDX. The identified phases, the lattice parameters, unit cell volumes, and Rietveld phase abundance are listed in Table 5 and are also shown in Figures 4 and 5. The substitution led to a decrease in the lattice parameters and unit cells of the phases present in the alloys. There have been identified phases with similar crystalline symmetry and spatial group as reference phases LaNi$_5$, PrNi$_5$, LaMg$_2$Ni$_9$, PrMg$_2$Ni$_9$, AlMnNi$_6$, and MgNi$_2$. However, for the a and c parameters, there have been observed variations when compared to these reference phases. There have been attributed to possible substitutions in the crystalline lattice in the sites of La (La,Pr) and of Ni (Al,Mn,Co,Ni) as a consequence of the as-cast state of the alloys. Both, this condition and the increased substitution of La by Mg, influenced the phase's relative abundance, as shown in Figure 5. Also shown in this figure are the correspondence with the matrix phase (M), grey phase (G), and dark phase (D), all determined by SEM/EDX analyses for the Mg alloys.

The phase abundance of the phase similar to LaNi$_5$ for $x = 0.0$ and 0.1 diminished from 74.2% to 43.0%, whereas for the phase similar to PrNi$_5$ increased from 25.8% to 45.5%. It was also detected for $x = 0.1$ a phase similar to (La,Pr)Mg$_2$Ni$_9$ (11.5%). The abundance of this phase showed a steady increase with the amount of Mg in the alloy ($x = 0.1$ to 0.5) and the detection of a PrMg$_2$Ni$_9$-type phase on the total substitution of La by Mg (Mg$_{0.7}$Pr$_{0.3}$Al$_{0.3}$Mn$_{0.4}$Co$_{0.5}$Ni$_{3.8}$).

(La,Pr)Ni$_5$ shifted from major abundance (60.7%) for $x = 0.3$ to minor abundance (5.8%) for $x = 0.7$. It has also been identified from $x = 0.3$ a phase similar to the AlMnNi$_6$ phase in increasing abundance that reached 33.5% for $x = 0.7$. In this condition, that is, total substitution of La by Mg, it has been detected a phase similar to the MgNi$_2$ phase (14.3%). The identification of phases present in the alloys by

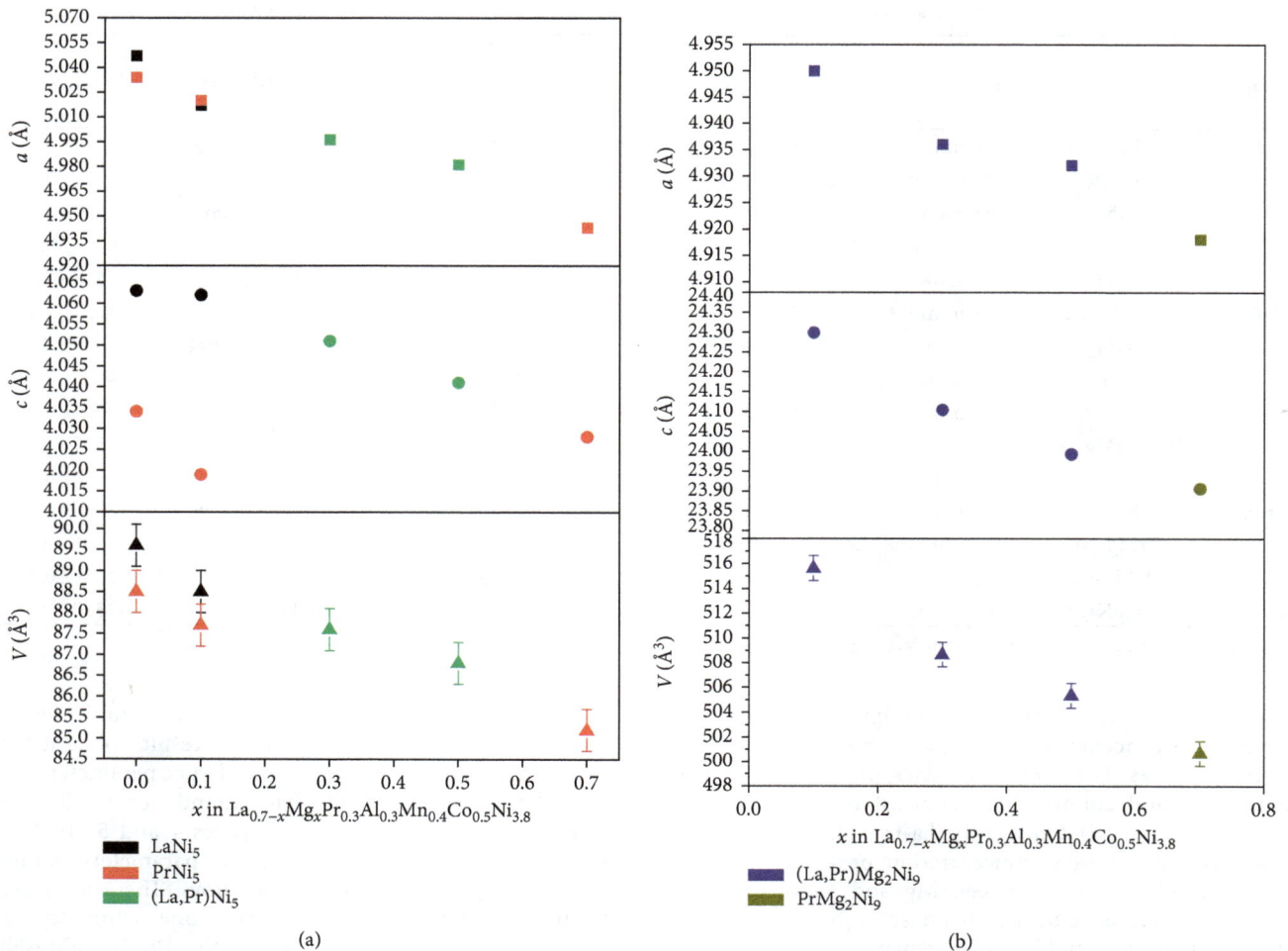

FIGURE 4: The lattice parameters and cell volumes of the $La_{0.7-x}Mg_xPr_{0.3}Al_{0.3}Mn_{0.4}Co_{0.5}Ni_{3.8}$ ($x = 0.0–0.7$) alloys: (a) $LaNi_5$, $PrNi_5$, and (La, Pr)Ni_5 phases; (b) (La,Pr)Mg_2Ni_9 and $PrMg_2Ni_9$ phases.

XRD and SEM/EDX led to the following correspondences: M to $LaNi_5$ and $PrNi_5$-type phases, G to $LaMg_2Ni_9$ and $PrMg_2Ni_9$-type phases, and D to $MgNi_2$ and $AlMnNi_6$-type phases.

3.3. Electrochemical Characterization

3.3.1. Corrosion. Figure 6 shows the anodic potentiodynamic polarization curves for the various tested alloys. At the corrosion potential, most alloys presented very low current densities (i), in the order of 10^{-5} to 10^{-6} A/cm², typical of passive materials, but the lowest i value was related to the alloy with the highest Mg-content ($x = 0.7$). For the alloy without Mg ($x = 0$), the current density largely increased with overpotential showing that the passive film on this alloy had the lowest resistance. The passive films on the alloys with 0.6 and 0.4 at.% La content showed intermediate resistances, between that without Mg and that without La. These results show that electrochemical behavior of the alloys was highly dependent on their microstructures. The substitution of La by Mg that led to refinement in the grain structure also had a beneficial effect on their corrosion resistance.

The polarization curves show that from the corrosion potential up to 0.7 V the effect of the alloy composition and consequently its microstructure on the electrochemical behavior is clearly seen with the lowest current densities associated with the alloys with lower La contents, that is, increasing Mg concentration. At potentials of nearly 0.7 V, all alloys showed a current density increase, and from 0.8 V upwards the electrochemical behavior of the alloys with Mg-contents of 0.0, 0.1, 0.3, and 0.5 at.% was similar. On the other hand, for the alloy with total substitution of La by Mg, the current density values were lower than the other alloys in the whole potential range.

The Nyquist diagrams for the various alloys studied are shown in Figure 7. The EIS results showed higher impedances associated with the samples with the higher Mg-contents ($x = 0.5$ and $x = 0.7$), indicating a passive behavior for these alloys. The lowest impedances were associated with the alloy without Mg and supported the polarization results that indicated that this alloy presented the lowest corrosion resistance.

The Bode phase angle diagrams in Figure 8 show two time constants for all studied alloys with lower phase angles

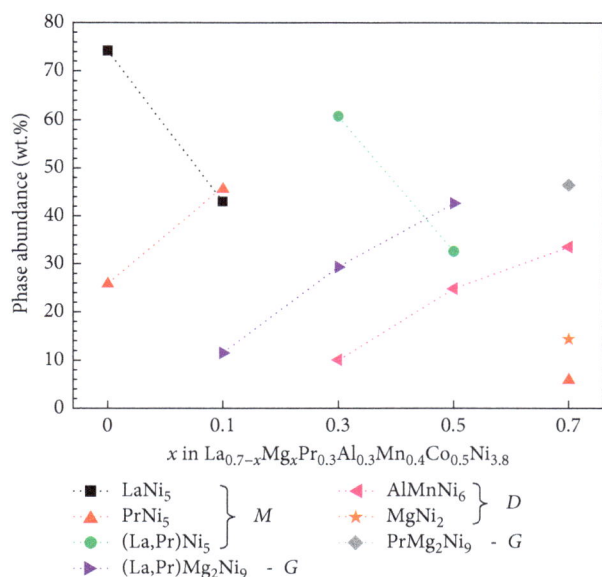

FIGURE 5: The phase abundance of the $La_{0.7-x}Mg_xPr_{0.3}Al_{0.3}Mn_{0.4}Co_{0.5}Ni_{3.8}$ ($x = 0.0$–0.7) alloys.

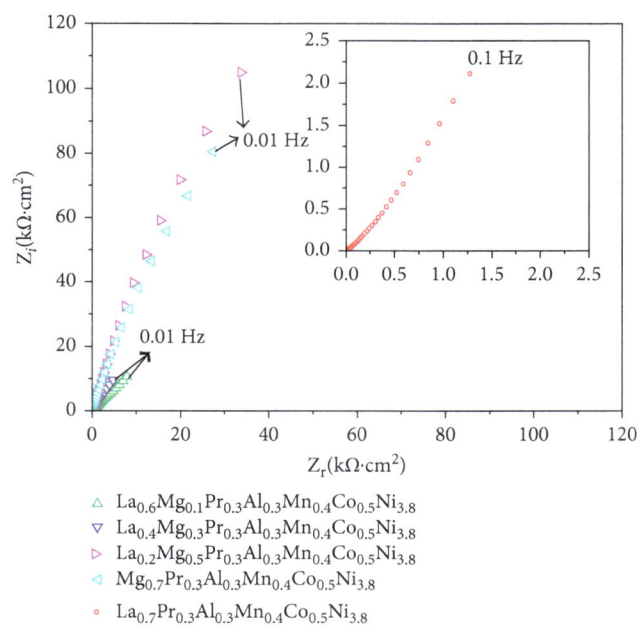

FIGURE 7: Nyquist diagrams of the $La_{0.7-x}Mg_xPr_{0.3}Al_{0.3}Mn_{0.4}Co_{0.5}Ni_{3.8}$ ($x = 0$, 0.1, 0.3, 0.5, and 0.7) alloys tested in 6.0 mol·L^{-1}·KOH solution.

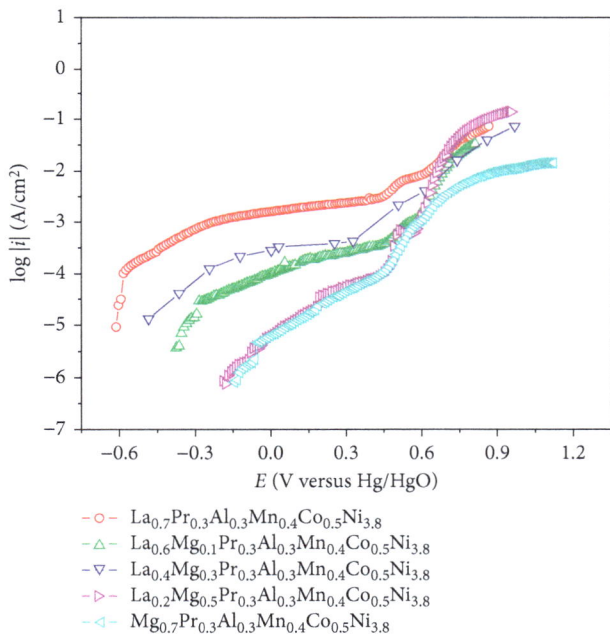

FIGURE 6: Anodic polarization curves of the $La_{0.7-x}Mg_xPr_{0.3}Al_{0.3}Mn_{0.4}Co_{0.5}Ni_{3.8}$ ($x = 0$, 0.1, 0.3, 0.5, and 0.7) alloys in 6.0 mol·L^{-1}·KOH solution.

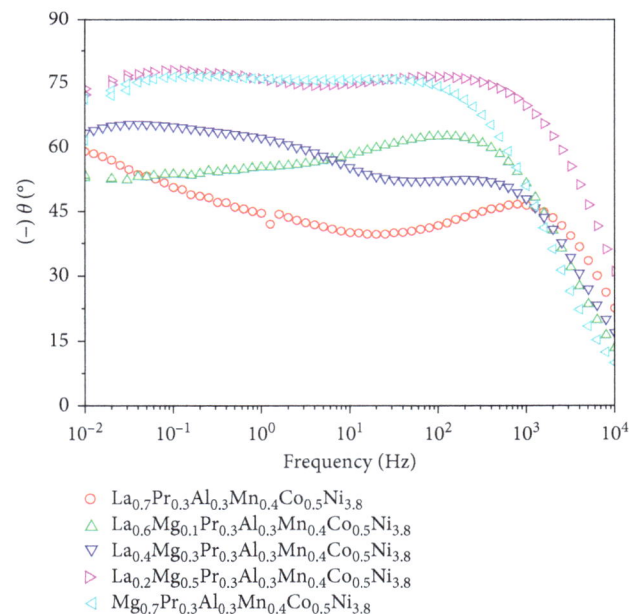

FIGURE 8: Bode phase angle diagrams of the $La_{0.7-x}Mg_xPr_{0.3}Al_{0.3}Mn_{0.4}Co_{0.5}Ni_{3.8}$ ($x = 0$, 0.1, 0.3, 0.5 and 0.7) alloys tested in 6.0 mol·L^{-1}·KOH solution.

related to the alloys with low Mg contents (0.0, 0.1, and 0.3 at.%). For the alloys with 0.5 or 0.7 at.% Mg (without La), the time constant at high frequencies is seen at approximately 1 kHz, whereas for the other alloys, this time constant occurs at nearly 0.1 kHz. For all alloys, the second time constant is seen at approximately the same frequency, 0.1 Hz. The high frequency time constant is related to the surface oxide/hydroxide film, whereas the one at lower frequencies

is associated with charge transfer processes. These results indicate that more protective surface films are formed on the alloys with the high Mg contents. This is most likely related to the refined grain microstructure associated with these alloys and shows that the substitution of La by Mg increases the alloy corrosion resistance mainly due to the formation of a more protective oxide film on their surface comparatively

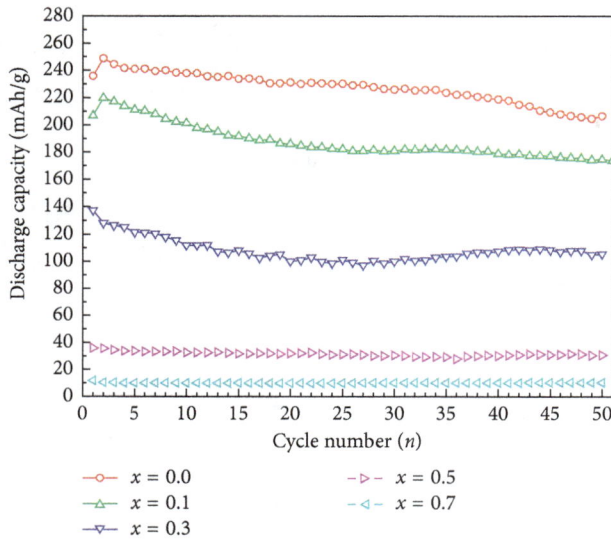

FIGURE 9: Discharge capacity of the $La_{0.7-x}Mg_xPr_{0.3}Al_{0.3}Mn_{0.4}Co_{0.5}Ni_{3.8}$ alloys versus cycle number.

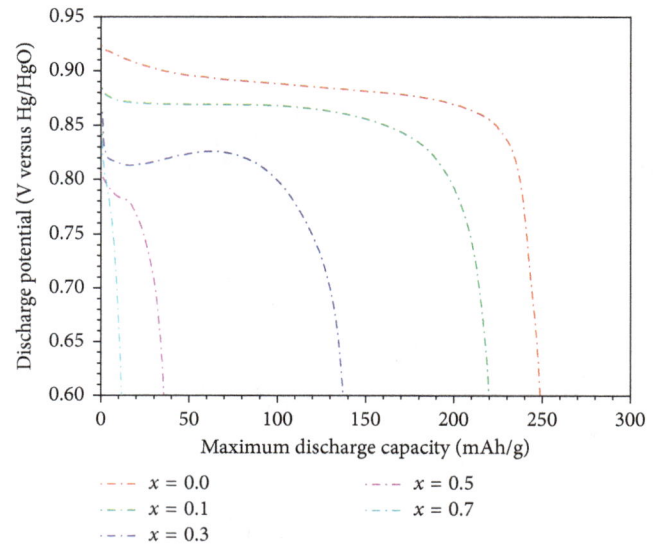

FIGURE 10: The discharge potential curves of the $La_{0.7-x}Mg_xPr_{0.3}Al_{0.3}Mn_{0.4}Co_{0.5}Ni_{3.8}$ ($x = 0.0-0.7$) alloys.

to that formed on the alloys with larger La contents. Passive surface films such as $La(OH)_3$ and $Mg(OH)_2$ have been reported in literature for La-Mg-Ni ternary alloy systems [24, 25]. Due to the very low reduction potentials associated with Mg and rare earth elements, such as La and Pr, in alkaline environments, these types of passive films are fairly stable [26].

3.3.2. Discharge Capacity.

Figure 9 shows the discharge capacity characteristics of the alloy electrodes with cycling. The range of measured discharge capacity values, limited to 250 mAh/g, could be attributed to superficial oxide formation during the preparation of the electrodes in the air. It can be seen that the electrodes can be easily activated within 2 cycles. Higher discharge capacity values were obtained for the Mg-free alloy with a maximum of 250 mAh/g and for $x = 0.1$ (maximum of 220 mAh/g). A significant decrease was observed for $x = 0.3$ (maximum ~140 mAh/g), and a dramatic reduction in this property with higher amount of Mg was observed. Clearly, increasing the magnesium content on the alloys decreased the discharge capacity of the negative electrodes. From the structural viewpoint, the increasing amount of Mg decreased the presence of the phase similar to the $LaNi_5$-type phase in the alloys, and the consequent increase of the abundance of the others observed phases (phases similar to the $(La,Pr)Mg_2Ni_9$ and $MgNi_2$ phases). Notwithstanding the limitation of the discharge capacity showed experimentally by the $LaNi_5$ phase (hexagonal, type $CaCu_5$) [5, 27], the performance of this phase still overcame the $(La,Pr)Mg_2Ni_9$ and $MgNi_2$ phases. The $(La,Pr)Mg_2Ni_9$ phases follow the structural model formed by the relation between the subunities $(MgNi_2)/(LaNi_5)$ equal to 2. The subunity $MgNi_2$ present (Laves phases, hexagonal type $MgNi_2$) shows some slow performance for hydrogen storage very similar to that individually presented by the $MgNi_2$ alloy ($H/M{\sim}0.33\%$ in mass) [21]. Conversely, the hydride

corresponding to the $LaNi_5$ phase shows a hydrogen storage capacity H/M of 1.4% in mass [28]. Thus, comparatively, the higher abundance of the phase similar to the $LaNi_5$ phase demonstrated to be beneficial to a higher discharge capacity of the electrodes. Furthermore, the refining of the alloys' microstructure that followed the increasing substitution of La by Mg showed unfavorable to the discharge capacity. Counteracting this effect, the refining proved to be beneficial to the corrosion resistance, as indicated by the experimental results. Thus, magnesium addition proved to be favorable to the corrosion properties, but high amounts have shown to be deleterious to the discharge capacity of the alloy electrodes tested. The overall discharge capacities of the present as-cast $La_{0.7-x}Mg_xPr_{0.3}Al_{0.3}Mn_{0.4}Co_{0.5}Ni_{3.8}$ alloys were somewhat superior to those obtained using mechanical alloying with a composition of $La_{1-x}Mg_xAl_{0.4}Mn_{0.3}Co_{0.3}Ni_{3.8}$ ($x = 0-0.2$) [7]. On the other hand, better discharge capacities were obtained with La-rich mischmetal (MI) electrodes of $MI_{1-x}Mg_xAl_{0.10}Mn_{0.10}Co_{0.55}Ni_{3.0}$ ($x = 0.05-0.30$) alloys [8]. These discrepancies can be attributed to the alloy preparation method, to testing conditions, and to the slight differences in the alloys compositions.

Figure 10 shows the discharge potential curves of the as-cast $La_{0.7-x}Mg_xPr_{0.3}Al_{0.3}Mn_{0.4}Co_{0.5}Ni_{3.8}$ ($x = 0.0-0.7$) alloys. It is well known that the principle of operation of a Ni/MH battery is based on the ability of certain metals, alloys, or intermetallics to absorb hydrogen in a reversible way. The hydride formation/decomposition process occurs via electrochemical charge transfer reaction [5, 29]. It can be seen in Figure 10 that for the first three conditions of substitution of La by Mg ($x = 0.0$, $x = 0.1$, and $x = 0.3$) it is possible to identify two regions that are commonly present in the curves of discharge potentials. The relatively flat part is called potential plateau and corresponds to the discharge process controlled by charge transfer. On the other hand, the final part of the curve represents the striking decrease of the potential due to the depletion of hydrogen atoms from the electrode surface. Additionally, the discharge potential also decreases with the increase of the internal resistance of the alloy. The internal resistance of alloy generally includes contact resistance, charge transfer resistance, and diffusion resistance [30]. The $La_{0.7}Pr_{0.3}Al_{0.3}Mn_{0.4}Co_{0.5}Ni_{3.8}$ $(x = 0.0)$ electrode alloy presented the longest potential plateau, indicating more effective charge transfer process and lower internal resistance compared to the other studied alloys. Again, the substitution of La by Mg in the alloys studied, considering the structural changes that resulted, proved to be deleterious to the discharge capacity of the electrodes.

4. Conclusions

From the structures and electrochemical properties of the as-cast $La_{0.7-x}Mg_xPr_{0.3}Al_{0.3}Mn_{0.4}Co_{0.5}Ni_{3.8}$ $(x = 0.0-0.7)$ alloys investigated in this study, some conclusion can be summarized:

(1) The substitution of La by Mg in the hydrogen storage alloys refined the grain microstructure. Total substitution changed the structure of the grains from equiaxed to columnar. The investigated alloys consisted of $LaNi_5$-, $PrNi_5$-, $LaMg_2Ni_9$-, and $PrMg_2Ni_9$-type phases, and the lattice parameters and cell volumes of the phases decreased with increasing Mg content.

(2) The electrochemical results showed that the Mg addition leads to corrosion improvement likely due to a more protective oxide film formed on the refined grain microstructure of the substrate. The results also indicated that at nonpolarized conditions all alloys showed very low current densities, indicating a passive behavior. Hence, the corrosion resistance of the studied alloys is not a limiting property for their use as negative electrodes in Ni/MH batteries.

(3) Magnesium inclusion in the alloys studied proved to be deleterious to the discharge capacity, with the maximum discharge capacity decreased with the amount of Mg substitution, from 250 mAh/g $(x = 0.0)$ to 12 mAh/g $(x = 0.7)$, which corresponds to a fact that the abundance of a similar $LaNi_5$ reference phase decrease with increasing Mg content.

(4) Based on the effect of substitution of Mg by La on the characteristics of the alloys and performance of the electrodes, it has been found that the $La_{0.7}Pr_{0.3}Al_{0.3}Mn_{0.4}Co_{0.5}Ni_{3.8}$ alloy could work as a base alloy for nonstoichiometry in the system AB_5 together with the addition of small amounts of Mg aiming the improvement of Ni/MH battery electrodes.

Conflicts of Interest

The authors declare that they have no conflicts of interest.

Acknowledgments

The authors wish to thank FAPESP and IPEN-CNEN/SP for the financial support and infrastructure made available to carry out this investigation.

References

[1] L. Ouyang, J. Huang, H. Wang, J. Liu, and M. Zhu, "Progress of hydrogen storage alloys for Ni-MH rechargeable power batteries in electric vehicles: a review," *Materials Chemistry and Physics*, vol. 200, pp. 164–178, 2017.

[2] H. Zhang, X. Zheng, X. Tian, Y. Liu, and X. Li, "New approaches for rare earth-magnesium based hydrogen storage alloys," *Progress in Natural Science: Materials International*, vol. 27, no. 1, pp. 50–57, 2017.

[3] K. Young and J. Nei, "The current status of hydrogen storage alloy development for electrochemical applications," *Materials*, vol. 6, no. 10, pp. 4574–4608, 2013.

[4] Y. Liu, H. Pan, M. Gao, and Q. Wang, "Advanced hydrogen storage alloys for Ni/MH rechargeable batteries," *Journal of Materials Chemistry*, vol. 21, no. 13, pp. 4743–4755, 2011.

[5] Y. Liu, Y. Cao, L. Huang, M. Gao, and H. Pan, "Rare earth-Mg-Ni-based hydrogen storage alloys as negative electrode materials for Ni/MH batteries," *Journal of Alloys and Compounds*, vol. 509, no. 3, pp. 675–686, 2011.

[6] F. Feng, M. Geng, and D. O. Northwood, "Electrochemical behaviour of intermetallic-based metal hydrides used in Ni/Metal hydride (MH) batteries: a review," *International Journal of Hydrogen Energy*, vol. 26, no. 7, pp. 725–734, 2001.

[7] D. J. Cuscueta, M. Melnichuk, H. A. Peretti, H. R. Salva, and A. A. Ghilarducci, "Magnesium influence in the electrochemical properties of La–Ni base alloy for Ni–MH batteries," *International Journal of Hydrogen Energy*, vol. 33, no. 13, pp. 3566–3570, 2008.

[8] Z. Zhang, S. Han, Y. Li, and T. Jing, "Electrochemical properties of $Ml_{1-x}Mg_xNi_{3.0}Mn_{0.10}Co_{0.55}Al_{0.10}$ $(x = 0.05-0.30)$ hydrogen storage alloys," *Journal of Alloys and Compounds*, vol. 431, no. 1-2, pp. 208–211, 2007.

[9] R. Tang, Y. Liu, C. Zhu, J. Zhu, and G. Yu, "Effect of Mg on the hydrogen storage characteristics of $Ml_{1-x}Mg_xNi_{2.4}Co_{0.6}$ $(x = 0-0.6)$ alloys," *Materials Chemistry and Physics*, vol. 95, no. 1, pp. 130–134, 2006.

[10] T. Ozaki, H. B. Yang, T. Iwaki et al., "Development of Mg-containing $MmNi_5$-based alloys for low-cost and high-power Ni–MH battery," *Journal of Alloys and Compounds*, vol. 408–412, pp. 294–300, 2006.

[11] X. Zhang, Y. Chai, W. Yin, and M. Zhao, "Crystal structure and electrochemical properties of rare earth non-stoichiometric AB_5-type alloy as negative electrode material in Ni-MH battery," *Journal of Solid State Chemistry*, vol. 177, no. 7, pp. 2373–2377, 2004.

[12] A. K. Shukla, S. Venugopalan, and B. Hariprakash, "Nickel-based rechargeable batteries," *Journal of Power Sources*, vol. 100, no. 1-2, pp. 125–148, 2001.

[13] K. Hong, "The development of hydrogen storage alloys and the progress of nickel hydride batteries," *Journal of Alloys and Compounds*, vol. 321, no. 2, pp. 307–313, 2001.

[14] E. P. Banczek, L. M. C. Zarpelon, R. N. Faria, and I. Costa, "Corrosion resistance and microstructure characterization of rare-earth-transition metal-aluminum-magnesium alloys," *Journal of Alloys and Compounds*, vol. 479, no. 1-2, pp. 342–347, 2009.

[15] L. M. C. Zarpelon and R. N. Faria, "Microstructure and electrochemical characteristics of LaPrMgAlMnCoNi hydrogen storage alloys for nickel-metal hydrides batteries," *Materials Science Forum*, vol. 802, pp. 421–426, 2014.

[16] L. M. C. Zarpelon, E. Galego, H. Takiishi, and R. N. Faria, "Microstructure and composition of rare earth-transition metal-aluminium-magnesium alloys," *Materials Research*, vol. 11, no. 1, pp. 17–21, 2008.

[17] H. Pan, X. Wu, M. Gao, N. Chen, Y. Yue, and Y. Lei, "Structure and electrochemical properties of $La_{0.7}Mg_{0.3}Ni_{2.45-x}Co_{0.75}Mn_{0.1}Al_{0.2}W_x$ ($x = 0$–0.15) hydrogen storage alloys," *International Journal of Hydrogen Energy*, vol. 31, no. 4, pp. 517–523, 2006.

[18] H. Pan, N. Chen, M. Gao, R. Li, Y. Lei, and Q. Wang, "Effects of annealing temperature on structure and the electrochemical properties of $La_{0.7}Mg_{0.3}Ni_{2.45}Co_{0.75}Mn_{0.1}Al_{0.2}$ hydrogen storage alloy," *Journal of Alloys and Compounds*, vol. 397, no. 1-2, pp. 306–312, 2005.

[19] P. Zhang, X. Wei, Y. Liu, J. Zhu, Z. Zhang, and T. Zhao, "The microstructures and electrochemical properties of non-stoichiometric low-Co AB_5 alloys containing small amounts of Mg," *Journal of Alloys and Compounds*, vol. 399, no. 1-2, pp. 270–275, 2005.

[20] Y. Liu, H. Pan, M. Gao, Y. Zhu, Y. Lei, and Q. Wang, "The effect of Mn substitution for Ni on the structural and electrochemical properties of $La_{0.7}Mg_{0.3}Ni_{2.55-x}Co_{0.45}Mn_x$ hydrogen storage electrode alloys," *International Journal of Hydrogen Energy*, vol. 29, no. 3, pp. 297–305, 2004.

[21] K. Kadir, T. Sakai, and I. Uehara, "Structural investigation and hydrogen storage capacity of $LaMg_2Ni_9$ and $(La_{0.65}Ca_{0.35})(Mg_{1.32}Ca_{0.68})Ni_9$ of the AB_2C_9 type structure," *Journal of Alloys and Compounds*, vol. 302, no. 1-2, pp. 112–117, 2000.

[22] K. Kadir, T. Sakai, and I. Uehara, "Synthesis and structure determination of a new series of hydrogen storage alloys: RMg_2Ni_9 (R = La, Ce, Pr, Nd, Sm and Gd) built from $MgNi_2$ Laves-type layers alternating with AB_5 layers," *Journal of Alloys and Compounds*, vol. 257, no. 1-2, pp. 115–121, 1997.

[23] Y. Liu, H. Pan, R. Li, and Y. Lei, "Effects of Al on cycling stability of a new rare-earth Mg-based hydrogen storage alloy," *Materials Science Forum*, vol. 475–479, pp. 2457–2462, 2005.

[24] B. Liao, Y. Q. Lei, G. L. Lu, L. X. Chen, H. G. Pan, and Q. D. Wang, "The electrochemical properties of $La_xMg_{3-x}Ni_9$ ($x = 1.0$–2.0) hydrogen storage alloys," *Journal of Alloys and Compounds*, vol. 356-357, pp. 746–749, 2003.

[25] B. Liao, Y. Q. Lei, L. X. Chen, G. L. Lu, H. G. Pan, and Q. D. Wang, "Effect of the La/Mg ratio on the structure and electrochemical properties of $La_xMg_{3-x}Ni_9$ ($x = 1.6$–2.2) hydrogen storage electrode alloys for nickel–metal hydride batteries," *Journal of Power Sources*, vol. 129, no. 2, pp. 358–367, 2004.

[26] A. J. Bard, R. Parsons, and J. Jordan, *Standard Potentials in Aqueous Solutions*, IUPAC, Marcel Dekker, New York, NY, USA, 1985.

[27] J. J. Reilly, G. D. Adzic, J. R. Johnson, T. Vogt, S. Mukerjee, and J. McBreen, "The correlation between composition and electrochemical properties of metal hydride electrodes," *Journal of Alloys and Compounds*, vol. 293-295, pp. 569–582, 1999.

[28] J. J. G. Willems, "Metal hydride electrodes stability of $LaNi_5$-related compounds," *Philips Journal of Research*, vol. 39, no. 1, pp. 1–94, 1984.

[29] P. H. L. Notten and P. Hokkeling, "Double-phase hydride forming compounds: a new class of highly electrocatalytic materials," *Journal of the Electrochemical Society*, vol. 138, no. 7, pp. 1877–1885, 1991.

[30] X. Tian, W. Wei, R. Duan et al., "Preparation and electrochemical properties of $La_{0.70}Mg_xNi_{2.45}Co_{0.75}Al_{0.30}$ ($x = 0$, 0.30, 0.33, 0.36, 0.39) hydrogen storage alloys," *Journal of Alloys and Compounds*, vol. 672, pp. 104–109, 2016.

18

Microstructure and Mechanical Behavior of Hot Pressed Cu-Sn Powder Alloys

Ahmed Nassef and Medhat El-Hadek

Department of Production & Mechanical Design, Faculty of Engineering, Port Said University, Port Fouad, Port Said 42523, Egypt

Correspondence should be addressed to Medhat El-Hadek; melhadek@eng.psu.edu.eg

Academic Editor: Angela De Bonis

Cu-Sn based alloy powders with additives of elemental Pb or C were densified by hot pressing technique. The influence of densifying on the properties of the hot pressed materials was investigated. The properties, such as the hardness, compressive strength, and wear resistance of these materials, were determined. The hot pressed Cu-Sn specimens included intermetallic/phases, which were homogeneously distributed. The presence of graphite improved the wear resistance of Cu-Sn alloys three times. Similarly, the presence of lead improved the densification parameter of Cu-Sn alloys three times. There was no significant difference in the mechanical behavior associated with the addition of Pb to the Cu-Sn alloys, although Cu-Pb alloys showed considerably higher ultimate strength and higher elongation. The Cu-Sn-C alloys had lower strength compared with those of Cu-Sn alloys. Evidence of severe melting spots was noticed in the higher magnifications of the compression fracture surface of 85% Cu-10% Sn-5% C and 80% Cu-10% Sn-10% Pb alloys. This was explained by the release of load at the final event of the fracture limited area.

1. Introduction

Metal matrix composites had attracted the interest of material scientists in the last few decades due to its unique properties [1, 2]. The motivations behind developing metal matrix composites were to fabricate structures that have superior stiffness and to simultaneously have better toughness and structural integrity [3, 4]. Copper alloy powders have been used in industrial applications for many years, as they have high corrosion resistance, high ductility, moderate to high hardness and strength behavior, high thermal and electrical conductivity, and high fatigue and abrasion resistance [5]. Copper matrix composites had been considered as one of the appealing candidates in various applications [5–7]. Tin (Sn) is a malleable, ductile, and highly crystalline silvery white metal. It has a relatively low melting point of 231° compared to copper (Cu) which has a melting temperature of 1084°C. Tin resists corrosion from water but can be affected by acids and alkalis [5, 8]. Tin could be highly polished and was used as a protective coating layer for other metals [9]. In this case, the protective oxide layer prevents further oxidation, as it acts as a catalyst when oxygen is in the solution

and helps to accelerate the chemical attack [9]. Accordingly, the composites produced from the combination of these materials are expected to possess superior properties for various industrial applications [10]. Cu-Sn alloys have been widely used as self-lubricant materials for many years [5], and powder metallurgy (PM) has been the main process to fabricate these alloys. The mechanical properties of copper represent the key factor in determining the suitability of the composite materials especially in applications where the material was subjected to high loads and frictions [11]. Copper reinforced with fine and uniform tin dispersoids has shown remarkable thermal and mechanical stability at an elevated temperature as reported by Xie et al. [12]. They reported that the monotonic shear behavior of as-reflowed Cu-Sn exhibits high strain to failure values [12]. The microstructure of the Cu-Sn alloys has shown a gradual cellular to dendritic transition which reflects stabilization in the growth [13]. This leads to an increase in the ultimate tensile strengths associated with finer eutectic cells for Cu-Sn alloys [13]. Recently, Kim et al. [14] introduced graphite to Cu-Sn alloys to improve thermal properties. These new composites have proven to be very useful in the lithium ion batteries applications [15, 16]. The graphite addition improves the wear resistance. The density, impact toughness, and hardness decrease with increasing the C content. The impact toughness has a maximum value of 11.7 J/Cm^2 at 2% C and a minimum value of 4.3 J/Cm^2 at 6% C [17]. Leaded bronzes (Cu-Sn-Pb) were mainly used for bearings due to their good wear resistance. For wide range of solidification, Cu-Sn alloys are extremely difficult to be produced by casting. However, because of the wide temperature range of crystallization and the large difference in the densities of copper and lead, intensive segregation takes place during solidification of Cu-Pb alloys [17]. Consequently, it was difficult to obtain homogeneous distribution of Pb in the microstructure by melting and casting. Therefore, powder metallurgy offers a promising processing route for producing such parts. The lead, the insoluble graphite in copper, and their additions to Cu-Sn alloys have some advantages in developing antifriction alloys especially by powder metallurgy. The powder metallurgy (PM) processing technique has definite advantages when used for consolidation of prealloyed powders.

The production of such composites via the conventional melting and casting techniques was extremely difficult as a result of the nonhomogenous distribution of the dispersoids. This leads to lowering the mechanical properties of the produced composites [18]. The suitability in controlling the homogeneous distribution of different materials in the PM depends on the technique used in preparing such composites [19]. Although the use of PM techniques proved to be the best in producing a homogeneous distribution of reinforced materials in the final product, this process alone has not given convenient results [18, 19] especially with reinforced materials having extremely fine particles. A few trials, such as mechanical alloying or rapid solidification, have been examined [19, 20] but have often shown a contamination and poor economic efficiency [18]. Generally, compaction is one of the most widely used techniques to obtain powder products. The well-known press and sintering technique was widely used to obtain dense products, but difficulties may arise especially in the sintering stages. The hot compaction technique densification has two stages. In the first stage, rearrangement of particles takes place by sliding and local plastic deformation of grain surface irregularities. At the second stage, the relative motion among particles becomes very small as the relative density increases [21, 22].

The scope of this research study was to manufacture four different Cu-Sn metal matrix composites using hot pressed powder alloying process, namely, 90% Cu-10% Sn with no additions, 85% Cu-10% Sn-5% C, 80% Cu-10% Sn-10% Pb, and 90% Cu-10% Pb weight percentages. The effect of the carbon and lead addition on the microstructure was studied. The effect of graphite or lead particles powder in addition to the Cu-Sn metal matrix composites was investigated on the hardness, mechanical, and wears properties.

2. Materials and Methods

Elemental Cu, Sn, C, and Pb possess strong ionic interatomic bonding giving rise to desirable material characteristics.

Powders with purity greater than 99% with an average particle size less than 10 μm in diameter and manufactured by Alfa Aesar, USA, were used as the starting source materials. The various powder components were mechanically mixed forming the nominal composition, namely, 90% Cu-10% Sn with no additions, 85% Cu-10% Sn-5% C, 80% Cu-10% Sn-10% Pb, and 90% Cu-10% Pb weight percentages. To ensure uniformity of the particle shapes, the Cu, Sn, C, and Pb powders were mechanically milled and mixed in an agate rock mortar with high energy boll milled for half an hour with different weight ratios according to the composition design. The mechanical milling of the powders resulted in uniform sphere-like particles for Cu-Sn alloys. Figure 1 presents the (a) scanning electron microscopy (SEM) surface microstructure of Cu-Sn alloys, as the red arrows pointing to the semisolid tin during the electroless process, (b) the graphite particle, (c) the lead particle as uniform spherical shape, (d) Cu-Sn alloys with graphite particles, and (e) Cu-Sn alloys with lead particles during the electroless process.

The process starts with preparing the plating baths that contain the tin/graphite/lead particles of known weight using the electroless mixture solution and reducing agent. A uniform copper film was formed on the tin/graphite/lead surface particles in about 10–15 min deposited from the hypophosphite based solutions from the alkaline baths. In addition, a complexing agent generally citrate and ammonium salts were also used to increase the particulate bonding. The complexing agents serve the function of preventing the precipitation of basic salts, as it also affects the deposition rate and properties. The bath ph level was usually maintained at the range between 8 and 10 using ammonium hydroxide. Lower deposition rates resulted when the ph level was adjusted with sodium hydroxide. Therefore, ammonium hydroxide was used for adjusting the ph of baths. The first bath was used to produce semibright tin/graphite/lead particles deposits containing approximately 4% phosphorus [23].

Deposition rates were increased with increased bath ph or hypophosphite concentration. The deposition rate in the first bath was increased from 5.6 μm/h to 10 μm/h at 85°C by simply adding organosulfur compound of 0.2 gm./cm^3 *thiourea*. The second bath contained less citrate than the first one which resulted in a substantially greater deposition rate. However, the resulting deposits had inferior physical properties. The last bath was a typical acid electroless tin/graphite/lead particles plating bath, using reducing agent that was incapable of yielding tin/graphite/lead particles deposits from acidic solution. Some of the reported advantages of electroless method were significant cost reduction, quality improvement of the deposited materials, and elimination of cross-contamination [8, 24–26]. The weight of copper coatings was estimated by the difference in weight between the graphite particles before and after the electroless coating process.

The sintered density was obtained by both dimensional measurements as well as Archimedes density measurement technique. To compare the densification response of various compositions, the sintered densities were normalized with respect to the theoretical density. To take into account the influence of the initial as-pressed density, the compact sinterability was also expressed in terms of densification parameter,

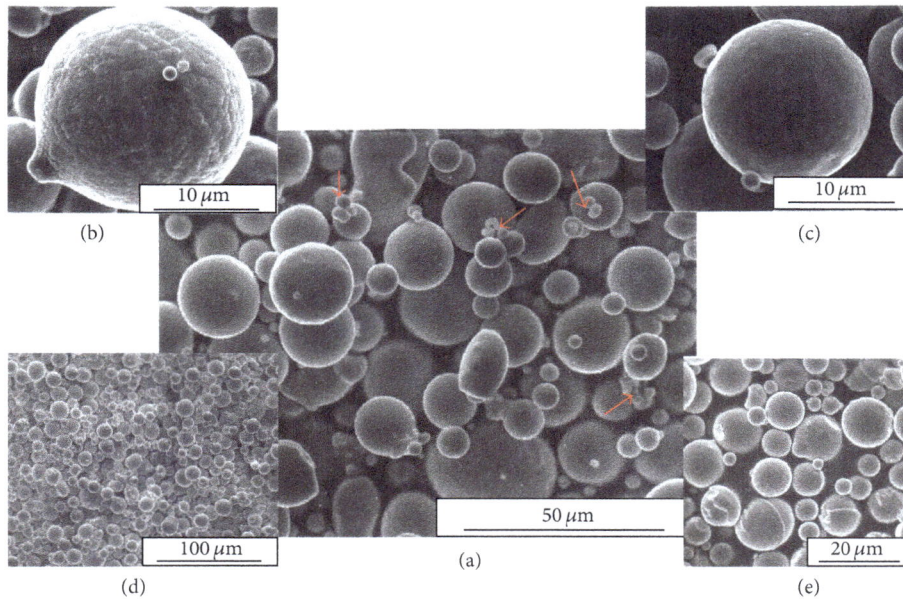

FIGURE 1: The (a) SEM surface microstructure of Cu-Sn alloys as the red arrows points to the semisolid tin during the electroless process, (b) the graphite particle, (c) the lead particle as uniform spherical shape, (d) Cu-Sn alloys with graphite particles, and (e) Cu-Sn alloys with lead particles during the electroless process.

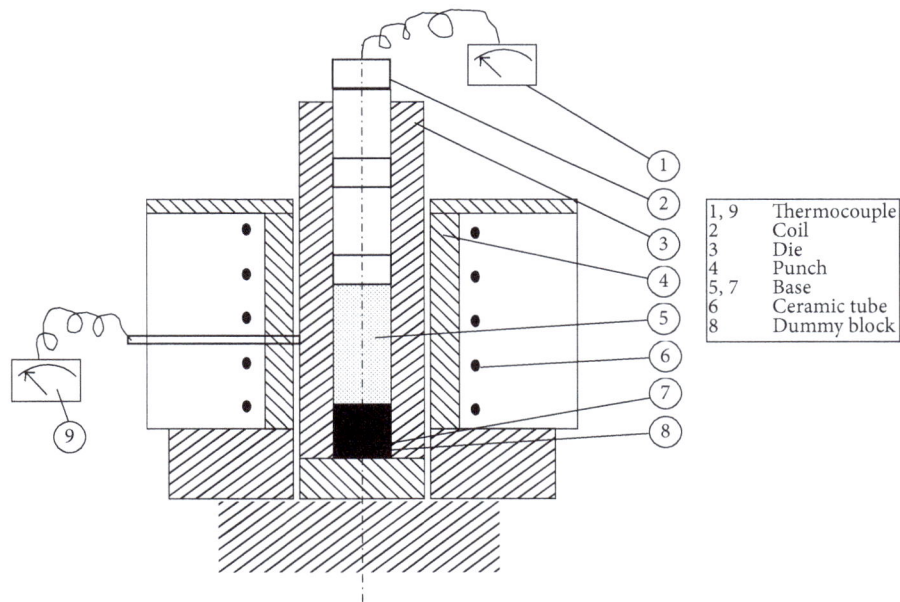

1, 9	Thermocouple
2	Coil
3	Die
4	Punch
5, 7	Base
6	Ceramic tube
8	Dummy block

FIGURE 2: The die setup of the hot pressing technique.

which was calculated as the densification parameter, ΔD = (sintered density − green density)/(theoretical density − green density), as the theoretical density (ρ_t) was calculated using the following equation [8]:

$$\rho_t = \frac{\sum_{i=1}^{n} \rho_i}{\sum_{i=1}^{n} \rho_i \cdot w_i}, \quad (1)$$

where ρ_i and w_i are the element density and weight fraction, respectively.

Hot compaction was performed in a single acting piston cylinder arrangement at room temperature in order to get 30 mm diameter and 50 mm height of the green compact, as shown in Figure 2. The die bore was smeared with intended powders reduce die wall friction, and the desired weights of mixed composites were used for each compact. A hydraulic testing machine of 200 tons capacity was used to perform the compaction of the alloy powder with constant cross head speed of 2 mm/min. The height of the green compact was

(a)

(b)

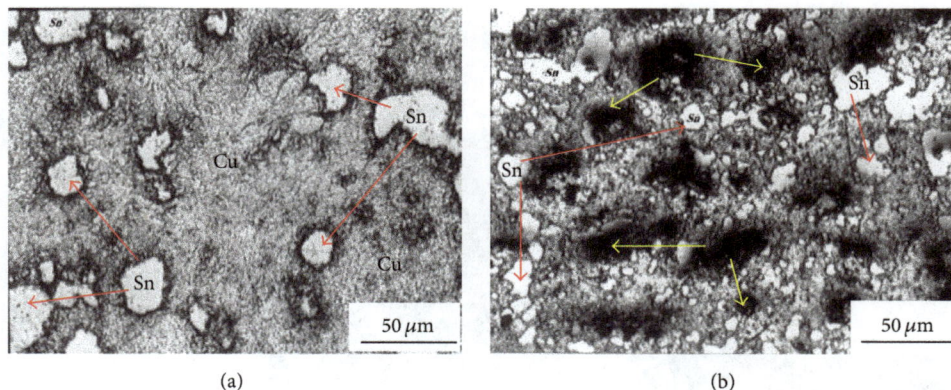

FIGURE 3: SEM micrographs of heat-treated (a) 90% Cu-10% Sn and (b) 85% Cu-10% Sn-5% C alloys.

measured directly before and after ejecting from the die. The final height was also calculated from the load displacement curve. After unloading the elastic recovery of the compacts was neglected [17]. The die temperature was measured by means of a NiCr-Ni thermocouple, which was inserted through the die and kept near its cavity. The temperature was maintained at the required level with a tolerance of ±5°C. Different mold temperatures were tested up to 550°C at constant pressure of 314.38 MPa and constant crosshead speed of 2 mm/min [11]. All hot pressed MMCs were heat-treated at about 550°C to allow the atoms to diffuse randomly into a uniform solid solution as liquid phase sintering [27, 28]. The tin melts to form a thin film surrounding the copper particles enhancing the alloying element bonding [28].

The setup was heated up to the preselected temperature which was kept constant for 30 minutes in order to homogenize the temperature throughout the powder alloy. The forming pressure was lowered for all tested hot components. After the compact operation, the samples were covered with aluminum foil and embedded in a graphite powder to protect its surface from the oxygen and nitrogen from the atmosphere during the sintering process. The specimens were sintered under liquid phase conditions at a constant heating rate of 20°C/min to a temperature of 550°C for one hour allowing tin to melt and enhance the bonding of the copper matrix. The temperature was maintained at that level with a tolerance of ±5°C.

3. Results and Discussion

3.1. Optical Investigations. The microstructure investigation on the Cu-Sn alloys was conducted using a Jeol 5400 SEM unit with a link EDS detector attachment to observe the particle morphology, particle size, particle shape, and agglomeration of particles after the fabrication process. Figure 3 shows the microstructures of the hot pressed Cu-Sn alloys with various elemental powder additions as C and Pb. The comparison between the pure Cu-Sn alloys shown in Figures 3(a) and 3(b) indicates the enhancement of diffusion and alloy formation as results of liquid phase sintering [29]. Note that the large dark particle in Figure 3(b) presents the graphite particles.

The graphite black particles were seen in Figure 3(b), as the graphite was well combined with the matrix. Two phase microstructures were presented in Figure 4, as the microstructure included α-Cu (twining structure), graphite, and precipitates around the grain boundaries. SEM micrographs and EDX analysis of these specimens were given in Figures 4(a) and 4(b), respectively.

The Cu-Sn alloy (Figure 4(a)) was composed of the major bright phase zones and the others of gray ones. The matrix of these specimens was composed of the bright areas, and point 2 was Cu-6.8% Sn which was a solid solution of Sn in Cu (α-phase) whereas the gray particles have considerably higher Sn content as point 1 that corresponds to δ-phase in the Cu-Sn system. The gray phase in Figure 4(b) as points 1 and 2 was also Sn-rich particles, containing 16.2%. This also coincides with the composition of the δ-phase.

Furthermore, the optical SEM microstructures of 80% Cu-10% Sn-10% Pb and 90% Cu-10% Pb alloys were shown in Figures 5(a) and 5(c), respectively. In Figure 5(a) the Sn-rich phase (δ) particles were observed as the addition to the Pb particles in the alloys. Separate Pb particles were defined in the microstructure as presented in higher magnification in Figure 5(b). The Pb solidified as almost pure lead forming globules at the copper grain boundaries. The structure consists of fine homogeneous Cu particles with some twining, as presented in Figure 5(c). It was noticed in higher magnifications (Figure 5(d)) for the 80% Cu-10% Sn-10% Pb alloys that the Sn particles form a uniform thin layer around the Cu particles as explained earlier during the liquid phase sintering process. A narrower Cu and Sn region, with higher inner connections between Cu and Sn particles, gives better mechanical interlock.

3.2. Vickers Microhardness Measurements. Vickers microhardness measurements were carried out for the different phase constituents of the hot pressed Cu-Sn alloys, using a load of 10 Kg for 20 sec in time, and the speed of the indenter was 100 μm/sec. To insure consistency throughout the material surface and homogeneity, a minimum of five readings were taken for each case and the average value was recorded. In all alloys the bright phase, which corresponds

General analysis, wt.%	
Cu	Sn
90.02	9.98

Spot no.	Element, wt.%	
	Cu	Sn
1	83.17	16.83
2	93.20	6.80

(a) 90% Cu-10% Sn alloy: points 1 and 2 refer to the regions analyzed by EDX

General analysis, wt.%:	
Cu	Sn
85.02	9.98

Spot no.	Element, wt.%	
	Cu	Sn
1	87.03	12.97
2	83.95	16.20

(b) 85% Cu-10% Sn-5% C alloy: points 1 and 2 refer to the regions analyzed by EDX

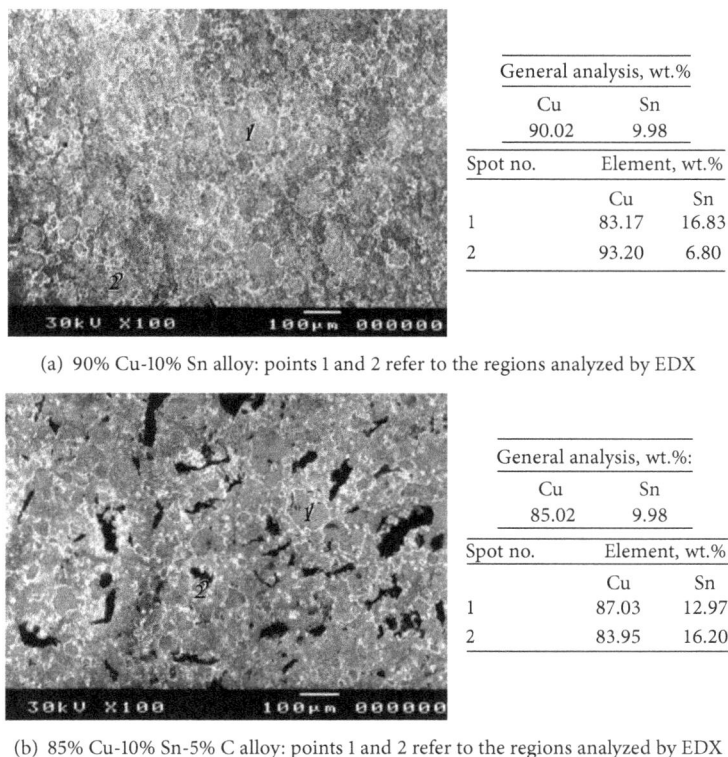

FIGURE 4: SEM micrographs and EDX analysis of heat-treated (a) 90% Cu-10% Sn and (b) 85% Cu-10% Sn-5% C alloys.

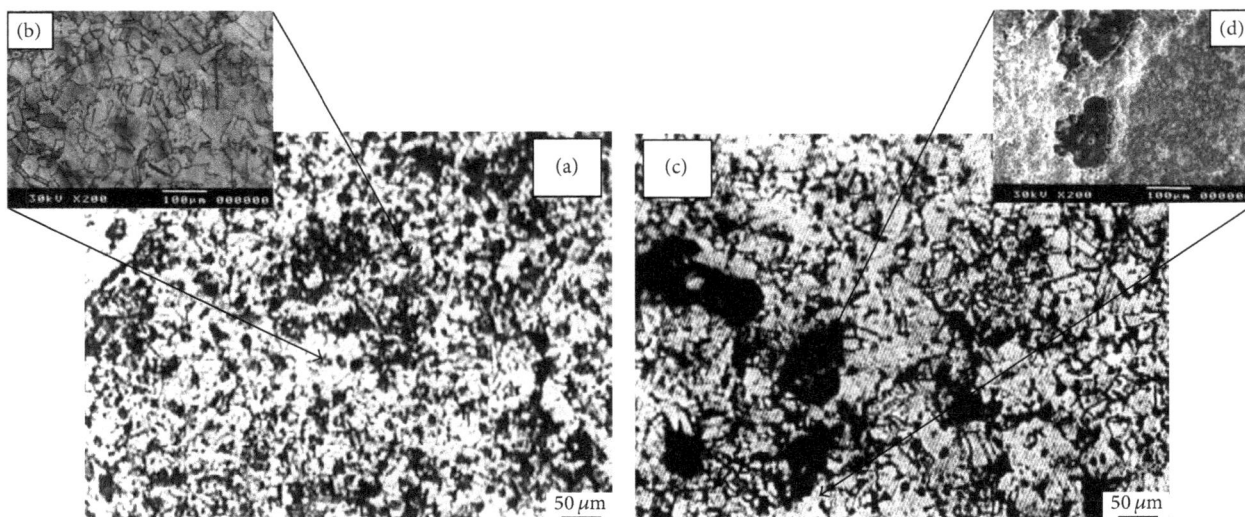

FIGURE 5: Optical SEM micrographs of heat-treated (a) 80% Cu-10% Sn-10% Pb, (b) with higher magnification, (c) 90% Cu-10% Pb alloys, and (d) with higher magnifications.

to δ-phase, had Vickers microhardness measurements about two times greater than the corresponding matrix as presented in Table 1. It was observed that the 85% Cu-10% Sn-5% C alloys exhibit lower hardness values compared to the other Cu-Sn alloys for the matrix and δ-phase. This decrease was attributed to the presence of graphite in the copper matrix.

As higher inner connections between Cu and Sn particles have the higher mechanical interlock of these alloys, the introduction of large C particles weakens the interlock

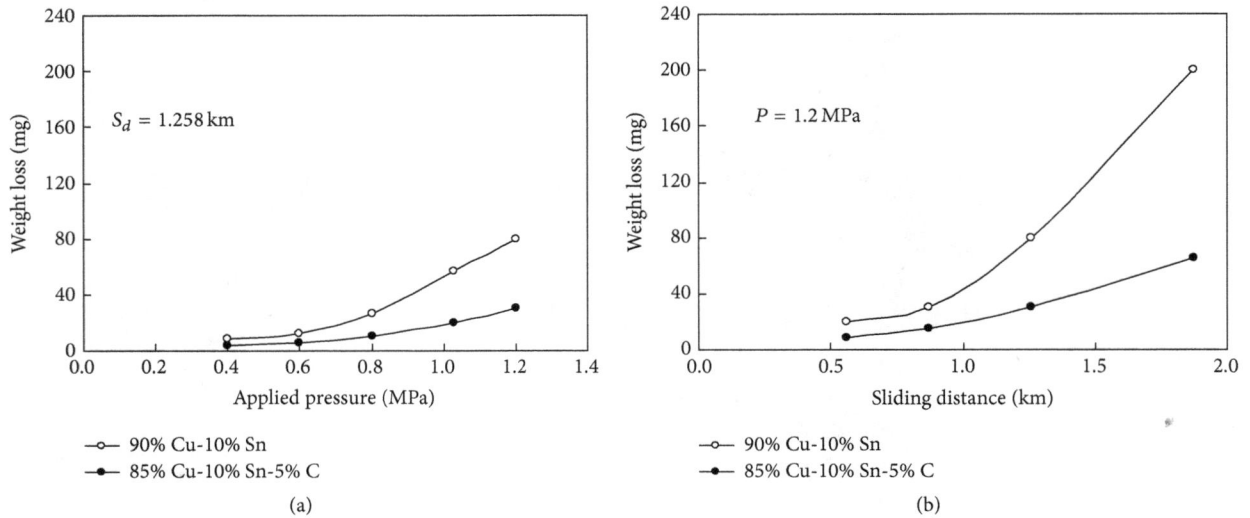

FIGURE 6: Weight losses in (mg) versus the applied pressure in MPa and the sliding distance in Km for the heat-treated (a) 90% Cu-10% Sn and (b) 85% Cu-10% Sn-5% C alloys.

TABLE 1: Vickers microhardness values in Kg/mm^2 of hot pressed Cu-Sn alloys.

Alloys	Vickers microhardness (H_v) Kg/mm^2	
	Matrix (α-Cu)	Bright (δ-phase)
90% Cu-10% Sn	112.2	246.6
85% Cu-10% Sn-5% C	83.7	162.7
80% Cu-10% Sn-10% Pb	108.9	222.2
90% Cu-10% Pb	101.4	210.3

compared to the effect of the smaller Pb particles as presented from the hardness values.

3.3. *Wear Resistance.* Wear measurements were performed by means of pin-on-disc method in dry conditions. The wear specimens were 8 mm in diameter and 12 mm in length. Surface preparation was conducted before the wear test, where each specimen was grinded with 1 μm alumina suspension. Wear tests were conducted under dry sliding conditions, applied loads of 10 N, and a constant sliding speed of 1.8 m/s. Wear loses were obtained by calculating the weight loss of the specimens before and after the testing using an electronic balance with a sensitivity of 0.1 milligram. The samples were cleaned in an acetone bath and dried using hot air before the tests to remove organic substances. The wear characteristics of 90% Cu-10% Sn and 85% Cu-10% Sn-5% C alloys were presented in Figure 6. The variations of weight loss of Cu-based alloys with the contact pressure at sliding distance of 1.258 km were presented in Figure 6(a). It should be reported that a sudden acceleration in weight loss at contact pressure 0.8 MPa for the 90% Cu-10% Sn alloy was observed. The 85% Cu-10% Sn-5% C alloy exhibited considerably high wear resistance compared to the 90% Cu-10% Sn alloy that can be seen in Figure 6(a). This may be attributed to the presence of graphite in the Cu matrix. It

was observed that for the investigated alloy, in case of 90% Cu-10% Sn alloy the aspirates of counterpart steel disc can abrade the copper alloy surface during dry sliding wear. Under a wear contact pressure of 1.2 MPa the wear rate of 90% Cu-10% Sn alloy after 1.2 km sliding distance was about 3 times higher than that of 85% Cu-10% Sn-5% C alloy. Similar results were observed for high wear contact pressure up to 1.2 MPa, as shown in Figure 6(b). This results in plastic strain localization in the subsurface region, leading to the formation of delaminating crack. The excessive delaminating of surface layers of copper alloy matrix leads to a high wear loss, which increases with increasing the contact pressure. Increasing the wear contact pressure tends to cause high plastic deformation of the matrix interface that can cause particle deformation. Thus, the wear rates of the alloys were mainly dependent on the level of the contact pressure.

The characteristics of wear resistance for both Cu-based alloys (with and without graphite particles) were presented in Figure 6(b), showing the weight loss variations with respect to sliding wear distance at constant contact pressure of 1.2 MPa. Using (1), the densification parameter (%) was calculated for the selected Cu-based alloys and presented in Table 2. It was observed the densification parameter increases with the addition of Pb significantly. For low sliding distance (0.748 and 1.2 km) the 90% Cu-10% Sn alloy gives two times higher wear rate compared to Cu-Sn-C compact. For the long sliding distance, 1.87 km, the presence of graphite leads to an increase in the wear resistance of the alloy by about three times.

The weight losses result in 90% Cu-10% Pb and Cu-Sn-Pb alloys with respect to sliding wear distance at constant contact pressure of 1.2 MPa as presented in Figure 7. The weight losses for 90% Cu-10% Pb alloy were slightly lower than that for 80% Cu-10% Sn-10% Pb alloy. Thus Sn in 90% Cu-10% Pb compacts was harmful to wear resistance.

For comparison purposes, the weight loss for Cu-based matrix alloys, at sliding distance of 1.258 km and contact

TABLE 2: The weight losses in mg at sliding distance of 1.258 km and contact pressure of 1.2 MPa and the densification parameter for all Cu-based alloys.

Alloy composition, wt.%	Weight losses, mg	Densification parameter, %
90% Cu-10% Sn	80	12.01
85% Cu-10% Sn-5% C	30	14.61
80% Cu-10% Sn-10% Pb	178	34.03
90% Cu-10% Pb	115	36.45

TABLE 3: The compressive test results for the Cu-based alloys.

Alloy	Yield strength σ_y, MPa	Ultimate strength σ_{UTS}, MPa	Elongation%
90% Cu-10% Sn	210	586	23
85% Cu-10% Sn-5% C	194	325	9
80% Cu-10% Sn-10% Pb	207	583	22.8
90% Cu-10% Pb	201	669	39

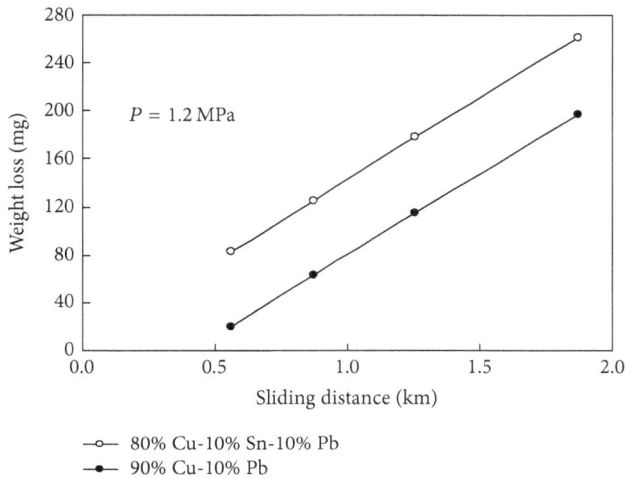

FIGURE 7: Weight losses in mg versus sliding distance in Km for the heat-treated 80% Cu-10% Sn-10% Pb and 90% Cu-10% Pb alloys.

pressure of 1.2 MPa, and the densification parameter were presented in Table 2 for all Cu-based alloys. It could be concluded from the wear resistance values for the Cu-based alloys that the addition of C elements improves the wear resistance of the alloys. On the contrary, the addition of Pb elements degrades the wear resistance of the alloys.

Low wear losses of the counter surface were found with the existence of the C large particles as affected by the particle agglomeration. The introduction of large C particles weakens the interlock inner connections between the Cu and Sn particles resulting in low hardness, but the large particles losses were reduced at these wear losses conditions. This was on the contrary when compared to the effect of introduction of the smaller Pb particles as observed in surface alloyed coated particles [30, 31].

3.4. Compression Mechanical Measurements.

Cylindrical specimens of aspect ratio of $h_o/d_o = 1.5$ (h_o and d_o were the original height and diameter of the specimen, resp.) were tested under frictionless conditions at the compression platen interface. The tests were carried out at room temperature using MTS Testing Machine (Model 610) fitted with a 160 KN load cell operating in the displacement control mode. The stress-strain responses of Cu-based alloys were measured from uniaxially compression testing performed accordingly

to ASTM standard E–9 for metals. The cross head speed was adjusted to give an average strain rate of $7.6 \times 10^{-4}\,s^{-1}$ across the specimen height. The test was terminated as the first surface crack was observed. The tests were repeated with three samples for each experiment. Figure 8 presents the experimental results of the flow curve versus the strain obtained by the compression tests for 90% Cu-10% Sn with no additions, 85% Cu-10% Sn-5% C, 80% Cu-10% Sn-10% Pb, and 90% Cu-10% Pb weight percentages alloys, respectively. The Cu-Sn alloy indicates that high strength and remarkable strain with respect to the 85% Cu-10% Sn-5% C alloy in Figure 8(a). Brittle fracture was observed for the 85% Cu-10% Sn-5% C alloy. The strength and ductility were substantially affected by the addition of Sn and Pb particles, as presented in Figure 8(b). The ductility of the 90% Cu-10% Pb alloys, as in Figure 8(a), demonstrated improvement over the 90% Cu-10% Sn alloys, as presented in Figure 8(b). In addition, the Pb addition for the Cu-Sn alloys did not show any improvements in the strength or ductility as seen in Figure 8(b). Comparison of the yield strength, ultimate strength, and the elongation percentage for the tested materials produced by the hot pressing PM technique were extracted as in Table 3.

Similarly, with the introduction to the C large particles to the Cu-Sn alloys the stress-strain response tends to decrease. As explained through the hardness the addition of C weakens the interlock inner connections between the Cu and Sn particles, whereas the addition of Pb enhances it.

3.5. Fracture Surface.

Copper particles coated with thin tin were relatively small, irregularly shaped, and tending to agglomerate. The Cu phase was dispersed with many pools or lakes present in the compression fractured samples. The high hardness may be attributed to the process of continuous crystallization during the plastic deformation. Cu-Sn alloys with the addition of C and Pb have been observed to undergo mechanically induced fine crystallization as presented in Figures 9(a) and 9(b). Fine crystal precipitation in Cu-Sn alloys was also observed within vein protrusions on the compression fracture surface and along crack propagation paths, as well as within shear bands resulting from bending [32]. The 85% Cu-10% Sn-5% C and 80% Cu-10% Sn-10% Pb alloys show an apparently classical inclined fracture surface, about 45° with the applied stress axis, which was similar to that encountered for a variety of hard metals [33] as presented in Figures 9(a) and 9(b).

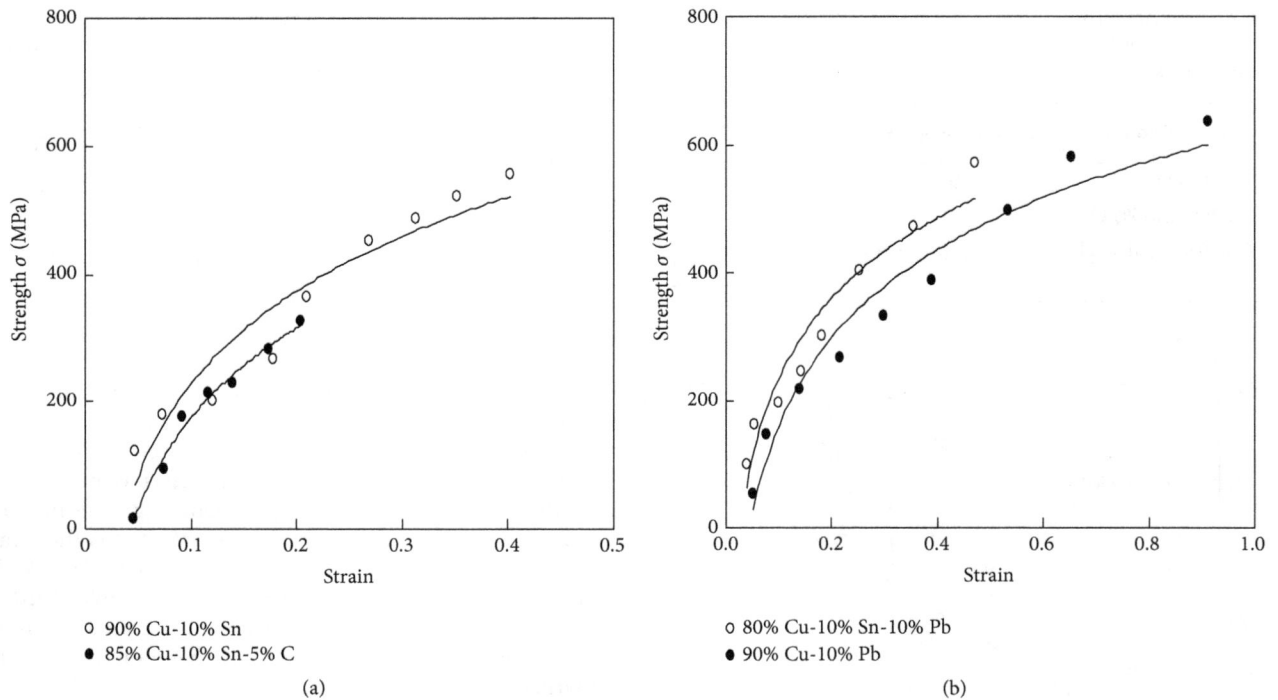

o 90% Cu-10% Sn
● 85% Cu-10% Sn-5% C

(a)

o 80% Cu-10% Sn-10% Pb
● 90% Cu-10% Pb

(b)

FIGURE 8: The stress-strain curves for the heat-treated (a) 90% Cu-10% Sn and 85% Cu-10% Sn-5% C and (b) 80% Cu-10% Sn-10% Pb and 90% Cu-10% Pb alloys.

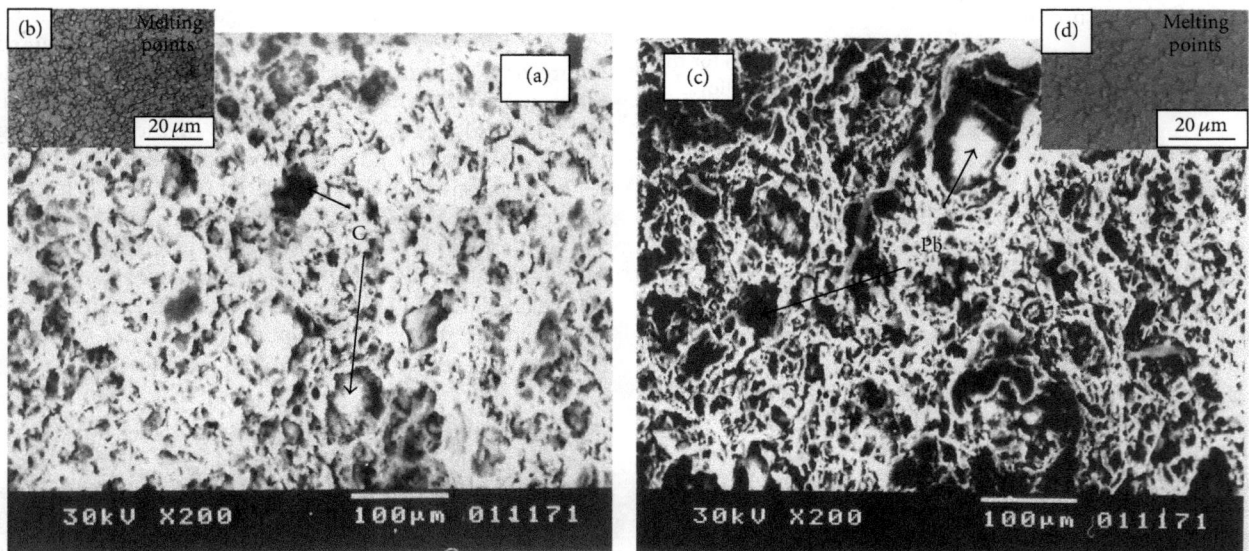

FIGURE 9: SEM image of irregularity and rough morphology of the fracture surface for (a) compression fractured 85% Cu-10% Sn-5% C alloys, (b) higher magnification showing melting droplets morphology, (c) compression fractured 80% Cu-10% Sn-10% Pb alloys, and (d) higher magnification showing melting droplets morphology.

Evidence of severe melting spots were noticed as shown in the higher magnifications of the fracture surface of 85% Cu-10% Sn-5% C and 80% Cu-10% Sn-10% Pb alloys in Figures 9(b) and 9(d). This could be due to the release of the load at the final event of fracture in this limited area. The fracture surface was linked to the formation of the very

fine particles in the alloys. This enhances the homogeneity of the deformation, leading to the formation of multiple shear planes instead of a single shear plane normally encountered in 85% Cu-10% Sn-5% C and 80% Cu-10% Sn-10% Pb alloys as presented in Figures 9(a) and 9(c). The viscosity of the Cu-Sn alloys in the region where fewer fine particles exist will be lower. This leads to a less critical shear stress and a more readily plastic deformation [34, 35].

4. Conclusions

Based on the results of the present study, the following conclusions can be summarized:

(1) The Cu-Sn powder alloys were successfully fabricated using hot pressing technique with the additions of C and Pb.

(2) The hot pressed specimens Cu-Sn alloys included intermetallic phases such as α-Cu and δ-phase, which were homogeneously distributed.

(3) As higher inner connections between Cu and Sn particles have the higher mechanical interlock of these alloys, the introduction of large C particles weakens the interlock compared to the effect of the smaller Pb particles as presented from the hardness values.

(4) The effect of adding C improves the wear resistance of the Cu-Sn alloys compared to alloys without graphite addition by three times. Compressive properties of Cu-Sn-C alloys were lower than those of Cu-Sn alloys.

(5) Significant differences in the mechanical properties such as yield strength, ultimate, and elongation percent of Cu-Sn and Cu-Sn-Pb alloys were noticed. Cu-Pb alloy had higher mechanical properties than those other compacts. Also, the wear resistance of Cu-Sn alloy was considerably higher compared to the other alloys.

(6) Evidence of severe melting spots was noticed in the higher magnifications of the compression fracture surface of 85% Cu-10% Sn-5% C and 80% Cu-10% Sn-10% Pb alloys, as it could be explained by the release of the load at the final event of fracture in this limited area.

Competing Interests

The authors declare that they have no competing interests.

References

[1] B. D. Long, R. Othman, H. Zuhailawati, and M. Umemoto, "Comparison of two powder processing techniques on the properties of Cu-NbC composites," *Advances in Materials Science and Engineering*, vol. 2014, Article ID 160580, 6 pages, 2014.

[2] H. Rudianto, G. J. Jang, S. S. Yang, Y. J. Kim, and I. Dlouhy, "Evaluation of sintering behavior of premix Al-Zn-Mg-Cu alloy powder," *Advances in Materials Science and Engineering*, vol. 2015, Article ID 987687, 8 pages, 2015.

[3] A. Es. Nassef, "Effect of Pb and C addition on mechanical behavior of hot-pressed Cu-Sn powderalloys," in *Proceedings of the International Conference on Mechanical Design and Production, MDP-8*, vol. 46, pp. 69–77, Cairo University, January 2004.

[4] M. M. De Campos and M. D. C. Ferreira, "A comparative analysis of the flow properties between two alumina-based dry powders," *Advances in Materials Science and Engineering*, vol. 2013, Article ID 519846, 7 pages, 2013.

[5] L. Robbiola, J.-M. Blengino, and C. Fiaud, "Morphology and mechanisms of formation of natural patinas on archaeological Cu-Sn alloys," *Corrosion Science*, vol. 40, no. 12, pp. 2083–2111, 1998.

[6] M. A. El-Hadek and M. Kassem, "Failure behavior of Cu-Ti-Zr-based bulk metallic glass alloys," *Journal of Materials Science*, vol. 44, no. 4, pp. 1127–1136, 2009.

[7] M. A. El-Hadek and S. H. Kaytbay, "Fracture properties of SPS tungsten copper powder composites," *Metallurgical and Materials Transactions A: Physical Metallurgy and Materials Science*, vol. 44, no. 1, pp. 544–551, 2013.

[8] M. A. El-Hadek and S. H. Kaytbay, "Al_2O_3 particle size effect on reinforced copper alloys: an experimental study," *Strain*, vol. 45, no. 6, pp. 506–515, 2009.

[9] A. F. Holleman, E. Wiberg, and N. Wiberg, "Tin," in *Lehrbuch der Anorganischen Chemie*, pp. 793–800, Walter de Gruyter, Berlin, Germany, 91th–100th edition, 1985 (German).

[10] W. Zhai, W. L. Wang, D. L. Geng, and B. Wei, "A DSC analysis of thermodynamic properties and solidification characteristics for binary Cu-Sn alloys," *Acta Materialia*, vol. 60, no. 19, pp. 6518–6527, 2012.

[11] A. A. El-Daly, A. E. Hammad, A. Fawzy, and D. A. Nasrallh, "Microstructure, mechanical properties, and deformation behavior of Sn–1.0Ag–0.5Cu solder after Ni and Sb additions," *Materials & Design*, vol. 43, pp. 40–49, 2013.

[12] H. Xie, N. Chawla, and K. Mirpuri, "Thermal and mechanical stability of Ce-containing Sn-3.9 Ag-0.7 Cu lead-free solder on Cu and electroless Ni-P metallizations," *Journal of Electronic Materials*, vol. 41, no. 12, pp. 3249–3258, 2012.

[13] B. L. Silva, N. Cheung, A. Garcia, and J. E. Spinelli, "Thermal parameters, microstructure, and mechanical properties of directionally solidified Sn-0.7 wt.%Cu solder alloys containing 0 ppm to 1000 ppm Ni," *Journal of Electronic Materials*, vol. 42, no. 1, pp. 179–191, 2013.

[14] J. H. Kim, Y. S. Jang, B. K. Park, and Y. C. Kang, "Electrochemical properties of C-coated Cu6Sn5 nanoparticles dispersed on carbon matrix prepared by spray drying process," *International Journal of Electrochemical Science*, vol. 8, no. 1, pp. 1067–1078, 2013.

[15] B. D. Polat, N. Sezgin, Ö. Keles, A. Abouimrane, and K. Amine, "Use of Cu–Sn/C multilayered thin film in lithium ion batteries," in *Proceedings of the Honolulu PRiME Meeting*, no. 10, pp. 968–968, The Electrochemical Society, Honolulu, Hawaii, USA, June 2012.

[16] J. S. Thorne, R. A. Dunlap, and M. N. Obrovac, "(Cu6Sn5)1-xCx active/inactive nanocomposite negative electrodes for Na-ion batteries," *Electrochimica Acta*, vol. 112, pp. 133–137, 2013.

[17] D. C. Walker, W. F. Caley, and M. Brochu, "Selective laser sintering of composite copper-tin powders," *Journal of Materials Research*, vol. 29, no. 17, pp. 1997–2005, 2014.

[18] J. Konstantanty, A. Bunsch, and A. Cias, "Factors affecting hardness and ductility of hot-pressed cobalt powders," *Powder Metallurgy International*, vol. 23, no. 6, pp. 354–356, 1991.

[19] M. I. Yehia and M. E. Abdel Rahman, "Production of biometallic parts from powders utilizing the powder metallurgy technique," in *Proceedings of the 3rd International Conference on Mechanical Engineering Advanced Technology for Industrial Production (MEATIP '02)*, pp. 46–55, Assiut University, Assuit, Egypt, December 2002.

[20] F. Z. Lemmadi, A. Chala, S. Ferhati, F. Chabane, and S. Benramache, "Structural and mechanical behavior during quenching of 40CrMoV5 steel," *Journal of Science and Engineering*, vol. 3, no. 1, pp. 1–6, 2013.

[21] S. F. Moustafa, S. A. El-Badry, and A. M. Sand, "The effect of graphite without and with Cu-coating on consolidation behavior and sintering of Cu-graphite Composite," in *Proceedings of the 6th International Conference on Production Engineering Design and Control (PEDAC '97)*, pp. 267–277, Alexandria, Egypt, February 1997.

[22] A. Nassef and M. El-Hadek, "Mechanics of hot pressed aluminum composites," *The International Journal of Advanced Manufacturing Technology*, vol. 76, no. 9–12, pp. 1905–1912, 2014.

[23] M. A. El-Hadek and S. Kaytbay, "Mechanical and physical characterization of copper foam," *International Journal of Mechanics and Materials in Design*, vol. 4, no. 1, pp. 63–69, 2008.

[24] A. Fathy, F. Shehata, M. Abdelhameed, and M. Elmahdy, "Compressive and wear resistance of nanometric alumina reinforced copper matrix composites," *Materials & Design*, vol. 36, pp. 100–107, 2012.

[25] F. Abd El-Salam, A. Fawzy, M. T. Mostafa, and R. H. Nada, "Resoftening and resistivity variation in thermally deformed Cu-Sn and Cu-Sn-Zn alloys," *Egyptian Journal of Solids*, vol. 23, no. 2, pp. 341–353, 2000.

[26] A. Veillere, J.-M. Heintza, N. Chandrab et al., "Influence of the interface structure on the thermo-mechanical properties of Cu–X (X = Cr or B)/carbon fiber composites," *Materials Research Bulletin*, vol. 47, no. 2, pp. 375–380, 2012.

[27] R. M. German, *Liquid Phase Sintering*, Springer Science & Business Media, Boston, Mass, USA, 2013.

[28] C. M. Sutter-Fella, J. A. Stückelberger, H. Hagendorfer et al., "Sodium assisted sintering of chalcogenides and its application to solution processed $Cu_2ZnSn(S,Se)_4$ thin film solar cells," *Chemistry of Materials*, vol. 26, no. 3, pp. 1420–1425, 2014.

[29] G. Sethi, A. Upadhyaya, and D. Agrawal, "Microwave and conventional sintering of premixed and prealloyed Cu-12Sn Bronze," *Science of Sintering*, vol. 35, no. 2, pp. 49–65, 2003.

[30] J. V. Rau, A. Latini, R. Teghil et al., "Superhard tungsten tetraboride films prepared by pulsed laser deposition method," *ACS Applied Materials and Interfaces*, vol. 3, no. 9, pp. 3738–3743, 2011.

[31] J. D. Bressan, D. P. Daros, A. Sokolowski, R. A. Mesquita, and C. A. Barbosa, "Influence of hardness on the wear resistance of 17-4 PH stainless steel evaluated by the pin-on-disc testing," *Journal of Materials Processing Technology*, vol. 205, no. 1–3, pp. 353–359, 2008.

[32] Q. K. Zhang and Z. F. Zhang, "Fracture mechanism and strength–influencing factors of Cu/Sn–4Ag solder joints aged for different times," *Journal of Alloys and Compounds*, vol. 485, no. 1-2, pp. 853–861, 2009.

[33] K. S. Kumar, L. Reinbold, A. F. Bower, and E. Chason, "Plastic deformation processes in Cu/Sn bimetallic films," *Journal of Materials Research*, vol. 23, no. 11, pp. 2916–2934, 2008.

[34] S. Kaytbay and M. El-Hadek, "Wear resistance and fracture mechanics of WC–Co composites," *International Journal of Materials Research*, vol. 105, no. 6, pp. 557–565, 2014.

[35] M. A. El-Hadek and M. A. Kassem, "Mechanics of Ti-Ni BMG-based alloys: experimental study," *Journal of Engineering Mechanics*, vol. 140, no. 1, pp. 53–60, 2014.

Permissions

All chapters in this book were first published in AMSE, by Hindawi Publishing Corporation; hereby published with permission under the Creative Commons Attribution License or equivalent. Every chapter published in this book has been scrutinized by our experts. Their significance has been extensively debated. The topics covered herein carry significant findings which will fuel the growth of the discipline. They may even be implemented as practical applications or may be referred to as a beginning point for another development.

The contributors of this book come from diverse backgrounds, making this book a truly international effort. This book will bring forth new frontiers with its revolutionizing research information and detailed analysis of the nascent developments around the world.

We would like to thank all the contributing authors for lending their expertise to make the book truly unique. They have played a crucial role in the development of this book. Without their invaluable contributions this book wouldn't have been possible. They have made vital efforts to compile up to date information on the varied aspects of this subject to make this book a valuable addition to the collection of many professionals and students.

This book was conceptualized with the vision of imparting up-to-date information and advanced data in this field. To ensure the same, a matchless editorial board was set up. Every individual on the board went through rigorous rounds of assessment to prove their worth. After which they invested a large part of their time researching and compiling the most relevant data for our readers.

The editorial board has been involved in producing this book since its inception. They have spent rigorous hours researching and exploring the diverse topics which have resulted in the successful publishing of this book. They have passed on their knowledge of decades through this book. To expedite this challenging task, the publisher supported the team at every step. A small team of assistant editors was also appointed to further simplify the editing procedure and attain best results for the readers.

Apart from the editorial board, the designing team has also invested a significant amount of their time in understanding the subject and creating the most relevant covers. They scrutinized every image to scout for the most suitable representation of the subject and create an appropriate cover for the book.

The publishing team has been an ardent support to the editorial, designing and production team. Their endless efforts to recruit the best for this project, has resulted in the accomplishment of this book. They are a veteran in the field of academics and their pool of knowledge is as vast as their experience in printing. Their expertise and guidance has proved useful at every step. Their uncompromising quality standards have made this book an exceptional effort. Their encouragement from time to time has been an inspiration for everyone.

The publisher and the editorial board hope that this book will prove to be a valuable piece of knowledge for researchers, students, practitioners and scholars across the globe.

List of Contributors

Yunlong Ding
Department of Materials Science and Engineering, Saitama Institute of Technology, Fusaiji 1690, Fukaya, Saitama 369-0293, Japan

Dongying Ju
Department of Materials Science and Engineering, Saitama Institute of Technology, Fusaiji 1690, Fukaya, Saitama 369-0293, Japan
Department of Materials Science and Engineering, University of Science and Technology Liaoning, Anshan 114051, China

M. G. Mahmoud, A. M. Samuel and F. H. Samuel
Département des Sciences Appliqúees, Université du Qúebecá Chicoutimi, Chicoutimi, QC, Canada

H. W. Doty
General Motors Materials Engineering, 823 Joslyn Ave., Pontiac, MI 48340, USA

S. Valtierra
Nemak, S.A., 66221 Garza Garcia, NL, Mexico

Gulhan Cakmak
Metallurgical and Materials Engineering, Mŭgla Sitki Kocman University, 48100 Mŭgla, Turkey

Yong-jian Fang, Xiao-song Jiang, Ting-feng Song, De-gui Zhu and Ming-hao Zhu
School of Materials Science and Engineering, Southwest Jiaotong University, Chengdu, Sichuan 610031, China

De-feng Mo
Key Laboratory of Infrared Imaging Materials and Detectors, Shanghai Institute of Technical Physics, Chinese Academy of Sciences, Shanghai 200083, China

Zhen-yi Shao
School of Materials Science and Engineering, Southwest Jiaotong University, Chengdu, Sichuan 610031, China
Department of Material Engineering, Chengdu Technological University, Chengdu, Sichuan 611730, China

Zhi-ping Luo
Department of Chemistry and Physics, Fayetteville State University, Fayetteville, NC 28301, USA

Yuta Fukuda, Masafumi Noda and Tomomi Ito
Magnesium Division, Gonda Metal Industry Co. Ltd., Sagamihara, Kanagawa 252-0212, Japan

Kazutaka Suzuki, Naobumi Saito and Yasumasa Chino
Structural Materials Research Institute of Advanced Industrial Science and Technology (AIST), Nagoya, Aichi 463-8560, Japan

Chunjiang Zhao, Zhengyi Jiang and Xiaorong Yang
School of Mechanical Engineering, Coordinative Innovation Center of Taiyuan Heavy Machinery Equipment, Taiyuan University of Science and Technology, Taiyuan 030024, China
School of Mechanical, Materials and Mechatronic Engineering, University of Wollongong, Wollongong, NSW 2522, Australia

Feitao Zhang
School of Mechanical Engineering, Coordinative Innovation Center of Taiyuan Heavy Machinery Equipment, Taiyuan University of Science and Technology, Taiyuan 030024, China

Jie Xiong
Sichuan Aerospace Special Power Research Institute, Chengdu 610100, China

Hailong Cui
School of Mechanical, Materials and Mechatronic Engineering, University of Wollongong, Wollongong, NSW 2522, Australia
Institute of Machinery Manufacturing Technology, China Academy of Engineering Physics, Mianyang 621000, China

Fawzia Hamed Basuny
Industry Service Complex, Arab Academy for Science, Technology & Maritime Transport, Abu Qir, Alexandria 21599, Egypt

Mootaz Ghazy, Abdel-Razik Y. Kandeil and Mahmoud Ahmed El-Sayed
Department of Industrial and Management Engineering, Arab Academy for Science, Technology & Maritime Transport, Abu Qir, Alexandria 21599, Egypt

Nitin D. Misal
Department of Mechanical Engineering, SVERI's College of Engineering Pandharpur, Maharashtra 413304, India

Mudigonda Sadaiah
Department of Mechanical Engineering, Dr. Babasaheb Ambedkar Technological University, Lonere, Raigad, Maharashtra 402103, India

Jinlong Chen, Hengcheng Liao and Heting Xu
School of Materials Science and Engineering, Jiangsu Key Laboratory for Advanced Metallic Materials, Southeast University, Nanjing 211189, China

Xina Huang, Lihui Lang and Gang Wang
School of Mechanical Engineering and Automation, Beihang University, Beijing 100191, China

Sergei Alexandrov
School of Mechanical Engineering and Automation, Beihang University, Beijing 100191, China
Institute for Problems in Mechanics, Russian Academy of Sciences, Moscow 119526, Russia

A. M. Al-Obaisi
Mechanical Engineering Department, College of Engineering, King Saud University, Riyadh 11421, Saudi Arabia
Mechanical Engineering Department, King Abdulaziz University, Jeddah, Saudi Arabia

E. A. El-Danaf and M. S. Soliman
Mechanical Engineering Department, College of Engineering, King Saud University, Riyadh 11421, Saudi Arabia

A. E. Ragab
Industrial Engineering Department, College of Engineering, King Saud University, Riyadh 11421, Saudi Arabia

A. N. Alhazaa
Physics & Astronomy Department, Faculty of Science, King Saud University, Riyadh 11451, Saudi Arabia
King Abdullah Institute for Nanotechnology (KAIN), King Saud University, Riyadh, Saudi Arabia

Shengfang Zhang, Wenchao Zhang, Yu Liu, Fujian Ma, Chong Su and Zhihua Sha
School of Mechanical Engineering, Dalian Jiaotong University, Dalian 116028, China

Yuri J. O. Moraes and Marcelo C. Rodrigues
Department of Mechanical Engineering, Federal University of Paraiba, SN/58.051-085 João Pessoa, PB, Brazil

Antonio A. Silva, Antonio G. B. de Lima and Paulo C. S. da Silva
Department of Mechanical Engineering, Federal University of Campina Grande, Avenue Aprígio Veloso, 882/58.429-900 Campina Grande, PB, Brazil

Rômulo P. B. dos Reis
Department of Technology and Engineering, Federal University Rural of Semi- Árido, Avenue Francisco Mota, Costa e Silva, 572/59.625-900 Mossoró, RN, Brazil

A. Luna Ramírez, Z. Mazur and V. M. Salinas-Bravo
Instituto de Investigaciones Eléctricas, Reforma 113, 62490 Cuernavaca, MOR, Mexico

J. Porcayo-Calderon
CIICAp, Universidad Autónoma del Estado de Morelos, Avenida Universidad 1001, 62209 Cuernavaca, MOR, Mexico

L. Martinez-Gomez
Instituto de Ciencias Físicas, Universidad Nacional Autónoma de México, Avenida Universidad s/n, 62210 Cuernavaca, MOR, Mexico
Corrosion y Protección (CyP), Buffon 46, 11590 Ciudad de México, DF, Mexico

Maria M. Cueto-Rodriguez, Erika O. Avila-Davila and Luis M. Palacios-Pineda
Tecnológico Nacional de México/Instituto Tecnológico de Pachuca (DEPI), Pachuca de Soto, Hgo. 42080, Mexico

Victor M. Lopez-Hirata and Maribel L. Saucedo-Muñoz
Instituto Politécnico Nacional (ESIQIE), UPALM, Ciudad de México 07300, Mexico

Luis G. Trapaga-Martinez
CIATEQ-Posgrado en Manufactura Avanzada, Querétaro, Qro. 76150, Mexico

Juan M. Alvarado-Orozco
Centro de Ingeniería y Desarrollo Industrial (CIDESI), Querétaro, Qro. 76130, Mexico

Nabeel H. Alharthi, Adel T. Abbas and Hamad F. Alharbi
Department of Mechanical Engineering, College of Engineering, King Saud University, Riyadh 11421, Saudi Arabia

Sedat Bingol
Department of Mechanical Engineering, Dicle University, Diyarbakir 21280, Turkey

Adham E. Ragab
Department of Industrial Engineering, College of Engineering, King Saud University, Riyadh 11421, Saudi Arabia

Mohamed F. Aly
Department of Mechanical Engineering, School of Sciences and Engineering, American University in Cairo, AUC Avenue, New Cairo, Egypt

L. M. C. Zarpelon, E. P. Banczek, L. G. Martinez, N. B. Lima, I. Costa and R. N. Faria
Center of Science and Materials Technology, Nuclear and Energy Research Institute (IPEN-CNEN/SP), São Paulo, SP, Brazil

Ahmed Nassef and Medhat El-Hadek
Department of Production & Mechanical Design, Faculty of Engineering, Port Said University, Port Fouad, Port Said 42523, Egypt

Index